TRENDS IN CHROMOSOME RESEARCH

Trends in Chromosome Research

EDITOR
T. SHARMA
Centre of Advanced Study in Zoology
Banaras Hindu University, Varanasi 221005, India

Springer-Verlag Berlin Heidelberg GmbH

Originally published by Springer-Verlag Berlin Heidelberg New York in 1990
Softcover reprint of the hardcover 1st edition 1990

ISBN 978-3-662-10623-5 ISBN 978-3-662-10621-1 (eBook)
DOI 10.1007/978-3-662-10621-1

Preface

These are exciting days in biology; chromosome research is no exception. Twenty years ago when Caspersson and coworkers showed that metaphase chromosomes are structurally not homogeneous through their length, a new world was in the offing. Application of various molecular techniques in chromosome research has subsequently narrowed down the gap between the levels of microscopic and molecular understanding of chromosome organization. While complexities of older questions of chromosome/chromatin organization are being understood, newer dimensions and perspectives have been gained with respect to their structure and functions. Even more, novel chromosome techniques have become an integral component of clinical and molecular genetic methodologies. The present book deals with some of the developments in this rapidly growing field, and its objective will be largely fulfilled if cytogeneticists, initiated as well as uninitiated, find the contents interesting and useful. First five chapters are devoted to the chromatin structure, its organization into metaphase chromosome, centromere and the banded regions. The next four chapters focus attention on such functional attributes of chromosomes as replication, dosage compensation and cellular response to DNA lesions. It is only recently that a molecular view of the meiotic cell division is beginning to emerge: chapter ten refers to human meiosis and the next to molecular events in meiotic prophase in the baker's yeast. Another chapter is on aneuploidy in man and the Djungarian hamster. The role of chromosome rearrangements and oncogenes in malignancy and the parallelism between the neoplastic and phylogenetic chromosomal alterations are discussed in the next two chapters. Recent introduction of potentially useful methods of chromosome isolation by flow cytometry, and mapping of structurally and functionally distinct domains on metaphase chromosomes by restriction enzyme digestion and DNase I-nick translation *in situ* have been described in chapters fifteen to seventeen. The last chapter highlights the role of repetitive DNA in evolutionary differentiation.

The present volume is a dedication to Prof. S.P. Ray-Chaudhuri who pioneered research in animal cytogenetics in India in the late forties when McKlintock's squashing technique and section cutting were the only standard methods for studying chromosomes. Trained in *Drosophila* genetics under the Nobel laureate Prof. H.J. Muller, Ray-Chaudhari took up chromosome research because facilities for setting up a lab for *Drosophila* genetics were not easy to come by. He made significant contribution to the fields of radiation and evolutionary cytogenetics. In the process he produced a number of students who have made their own mark. Many of the contributions to this volume are from his first or second generation students who consider it a privilege to be able to dedicate their contributions to him. Other authors, leading cytogeneticists from different countries, readily agreed to contribute chapters to this book, even at a very short notice. I am indeed grateful to all of them.

Prof. Ray-Chaudhuri is over 80 years, yet his interest in the subject remains unfaded. If this book succeeds in disseminating to its readers excitement of this revolutionary phase of chromosome biology we will have paid, in some measure, a tribute to this great teacher, for it was he who inspired and initiated many a generation of student in India in this science.

I take this opportunity to record my sincere thanks to Dr. Rajiva Raman without whose help, in various ways, it would have been difficult for me to bring out this volume in its present form.

Varanasi 221005 T. SHARMA

Contents

S.P. Ray-Chaudhuri
A Scholastic Builder

T.C. Hsu

Department of Cell Biology
The University of Texas
M.D. Anderson Hospital and Tumor Institute at Houston
Houston, TX 77030, USA

Some years ago, I greeted a visitor at his hotel. Since I had to take care of some property, the back of my pick-up truck was filled with small equipment items and tools. I apologized to my visitor about the mess. He took one look and commented, 'Looks like you are in the construction business.' I told him 'Spiritually I am indeed a builder.' I tried to build scientists from students at all levels (high school to graduate school), postdoctoral fellows, and visitors who came to learn and work with us.

This story probably applies to Professor S.P. Ray-Chaudhuri better than it does to me, for throughout his life he has been a scholastic builder in India. Perhaps I am not the best suited person to describe his contributions because I have not been directly associated with him professionally. However, in the few conversations I had with him, I have got a glimpse of his knowledge, enthusiasm, and personality. I have also come to know him through some of his former students, associates, and his daughter, Ratna. I gladly accepted the privilege of writing the foreword to this volume because of my respect for Professor Ray-Chaudhuri and because I admire pioneers of all kinds. This gentleman is an explorer of the first degree.

Ray-Chaudhuri was born on 15 September, 1907 in a small village of Kulcathi in East Bengal of undivided India, now Bangladesh. He received his early education in the city of Bagura in North Bengal and went to college in Calcutta when he was 16. He obtained his MSc degree in zoology from Calcutta University in 1930.

In his younger days, he once visited a fortune teller who read his palm. This fortune teller told Ray-Chaudhuri that he could become a good farmer or a good scholar. Ray-Chaudhuri farmed for three years, but a flood wiped out his crop. Perhaps to science, particularly to genetics, in India, this disaster was a blessing in disguise, for his contributions to education in India would not have otherwise been made.

In 1937, he went to the Institute of Animal Genetics, University of Edinburgh, Scotland to work toward his PhD degree under Prof. F.A.E. Crew, then the director of that institute. During that period, Prof. H. J. Muller joined the University of Edinburgh at Crew's invitation. Ray-Chaudhuri took this valuable opportunity

S.P. Ray-Chaudhuri

to work under Muller and started his research in *Drosophila* genetics. Since he had no previous experience with *Drosophila*, it was difficult for him to select a research problem. Muller suggested a project on mutation induction with radiation of extremely low intensity. His results, presented in his dissertation, 'Validity of the Bunson Roscoe Law in the Production of Mutation with Radiation of Extremely Low Intensity', surprised many geneticists but were later confirmed.

Upon his return from Edinburgh to India in 1940, Ray-Chaudhuri was appointed lecturer in the Department of Zoology, Calcutta University. He started teaching genetics to postgraduate students but could not embark on research in *Drosophila* genetics because there were no facilities to breed the flies. Being an enterprising researcher, he led his graduate students to work on the chromosome cytology of insects, particularly grasshoppers. Like many cytologists who became involved in chromosome research, he became addicted to it. Ray-Chaudhuri worked on cytogenetic problems throughout his professional life, even though he later established a laboratory for *Drosophila* gentics. Because of his love for chromosome and evolutionary research, he became friends with such authorities as Th. Dobzhansky, Julian Huxley, and M.J.D. White.

The story behind his *Drosophila* laboratory is unusual. During the forties J.B.S. Haldane, then professor at the University of London, visited India frequently. Haldane was an adjudicator of Ray-Chaudhuri's PhD thesis and was impressed with his research work. In one of Haldane's visit to Calcutta, he found that Ray-Chaudhuri was not able to work on *Drosophila*. He then wrote to Sir J. C. Ghosh, then Vice Chancellor of the Calcutta University, requesting that a *Drosophila* genetics laboratory be set up in order for him to work during his stay in Calcutta. Haldane also suggested to Sir Ghosh that Ray-Chauduri be consulted on the laboratory's construction. The vice chancellor promptly made arrangements to establish such a laboratory in the Zoology Department. Haldane returned to India a few months later and brought some mutant stocks of *Drosophila* to start Ray-Chaudhuri's research work. When the vice chancellor asked Haldane when he would start his research, he quickly replied that the laboratory was meant for Ray-Chaudhuri, not for himself. Since that time Ray-Chaudhuri has had a close working relationship with Haldane.

Research in *Drosophila* genetics and insect cytology dominated his efforts during his Calcutta days. Ray-Chaudhuri ventured into investigation on human cytology in 1948–9, however, when he led one of his students to take a look at the chromosomes of endometrium. They obtained some metaphase figures, and a picture of a metaphase was subsequently hung on the wall of the obstetric clinic. Nevertheless, no one expressed much enthusiasm until Prime Minister Nehru, who was a scientist by training, saw the photograph in a visit to the hospital. Nehru asked to meet the scientist, shook hands with Ray-Chaudhuri and spoke some words of encouragement. Ray-Chaudhuri felt so honoured that he would not shake hands with anyone else until after he went home to wash his hands, and he saved the water to scatter around his house!

Ray-Chaudhuri was elected Fellow of the Indian National Science Academy in 1956 and was appointed Professor and Head of the Department of Zoology of the Banaras Hindu University in 1960. It was at this post that Ray-Chaudhuri made his greatest contributions to education, administration, and research until his retirement in 1971. He built the department by introducing new programmes, recruiting capable faculty members to represent different disciplines of modern zoology, and improving the department to become the best of the country. In his own research, he not only established a *Drosophila* laboratory but also expanded cytological investigations from insects to reptiles, birds, mammals and humans. His relentless drive for perfection, continual desire for exploring new research fields, and unselfish personality inspired many young biologists to his department, students as well as nonstudents. His administrative and educational activities significantly reduced his own research time, but he considered such a sacrifice necessary for the good of biological science, particularly genetic science, in his country.

Following his retirement, he has lived in Calcutta, and has remained active in promoting science and offering advice and inspiration to the younger generations. During a visit to my laboratory in 1980, Ray-Chaudhuri, then 73 years old, expressed a great interest in our research projects at that time, both in concept and in results. It is fitting to celebrate his 80th birthday with this volume of papers contributed by active investigators working in the broad field of chromosome cytology and cytogenetics. Some of the contributors were Ray-Chaudhuri's former students who have become authorities in their respective areas of research. Without question, some of the topics under review here did not even exist during his teaching days; but their presence in this volume attests to the notion that science cannot and should not stand still. Prof. Ray-Chaudhuri must be proud to see the blossoms of the seeds he sowed decades ago.

1 DNA Organization in the Interphase Nucleus and Metaphase Chromosome

KENNETH J. PIENTA* AND DONALD S. COFFEY

Departments of Oncology and Urology
The Johns Hopkins University School of Medicine
600 N. Wolfe Street
Baltimore, MD 21205, USA

Introduction

The higher-order structure of DNA within the nucleus and within the metaphase chromosome still remains an unsolved problem in molecular biology. Molecular biologists still struggle to explain DNA properties and functions within the topological constraints of the nucleus. In this chapter, we will present a concept for understanding the organization of DNA in the interphase nucleus and chromosome. We will also review the nuclear matrix as the unifying structure by which DNA structure and function are organized and regulated.

The Organization of DNA

A mammalian nucleus contains 6×10^9 base pairs (b.p.) of DNA which is equivalent to approximately 2 metres of DNA. To fit this 2-metre strand of nucleotides into an average nucleus of 10 micron diameter means that the DNA must be reduced 20,000-fold in linear packaging. In recent years, four higher order levels of DNA organization have been identified, including the nucleosome, the 30 nanometer (nm) solenoid, the DNA loop-domain, and the chromosome. Pienta and Coffey (1984) have proposed a model for depicting the packing of DNA into chromosomes (Figure 1). The first order structure of DNA consists of the 2 nm DNA double helix. This double helix is wound twice around histone octamer complexes, forming nucleosomes that contain approximately 160 b.p. of DNA. These complexes were first visualized by Olins and Olins (1979) and comprise the second order structure of DNA. Nucleosome formation packs the DNA 6-fold (160 b.p. DNA $\times 0.34$ nm/b.p. $= 54.4$ nm of DNA reduced to 10 nm).

The third order structure of DNA is formed when the nucleosomes wind in a solenoid fashion with six nucleosomes per turn to form a 30 nm filament. This structure has been visualized by Finch and Klug (1976) and results in the packing of DNA another 6-fold. The exact order

*Address correspondence to.

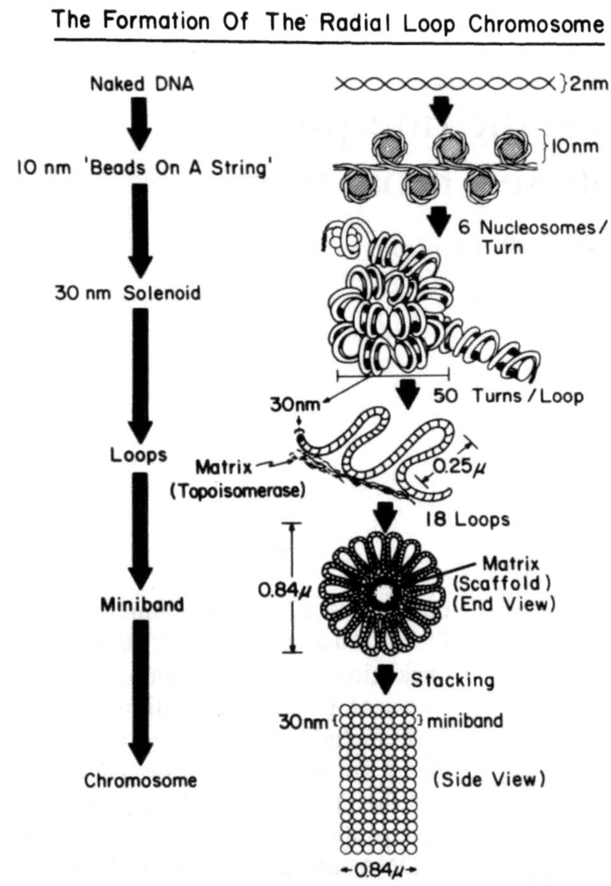

Base Pairs per Turn	Packing Ratio
10 b.p.	1
80 b.p (160 b.p. per 2 turns)	6 - 7
1,200 b.p. (per turn)	40±
60,000 b.p. (per loop)	680
1.1×10^6 b.p.± (per miniband)	1.2×10^4
18 loops/ Miniband	1.2×10^4

FIGURE 1: Schematic of the organization of DNA within a chromosome. The double helix of DNA is wrapped twice around the histone octamer to form a nucleosome. Six nucleosomes form each turn of a solenoid and the 30 nm filament. The 30 nm filament forms the loop domain of DNA attached at the base to the nuclear matrix that contains topoisomerase II. Eighteen of the loops form a radial array around the core of the chromosome to produce a miniband. The continuous formation of these radial loops produces a stack of minibands to form a metaphase chromatid. Modified from the study of Pienta and Coffey (1984).

of the 6 nucleosomes within the 30 nm filament is still a controversial topic.

The fourth order structure of DNA is the loop-domian. Cook *et al.* (1976) were the first to propose that loop structures play an important role in the superhelical organization of eukaryotic DNA. Loop-domains have been observed in both interphase nuclei (Vogelstein *et al.*, 1980) and metaphase chromosomes (Adolph, 1980; Adolph and Kreisman, 1983; Earnshaw and Laemmli, 1983). These loops are formed by the 30 nm solenoid filament as it winds its way through attachment points on the nuclear matrix. The nuclear matrix comprises the residual skeletal framework of the nucleus. Several different investigators have measured the length of these loops (Igo-Kemenes *et al.*, 1982; Georgiev *et al.*, 1978; Hancock and Boulikas, 1982; Earnshaw and Laemmli, 1983) and they range in size between 10,000 and 180,000 b.p. with an average of 63,000 b.p. (Pienta and Coffey, 1984). The functional significance of DNA loop-domains has been reported by Pardoll *et al.* (1980) and Vogelstein *et al.* (1980) when they demonstrated that loop-domains were equivalent to replicons, the basic lengths of DNA synthesized as continuous units. It is also intriguing to note that the number of loops in a diploid nucleus, 100,000 (6×10^9 b.p. per nucleus/6×10^4 b.p. per loop), is very close to the accepted number of approximately 100,000 genes within the human genome. The DNA loop-domain achieves a packing ratio of approximately 700 because 60,000 b.p. of DNA is packed into a loop of 30 nm when the loop is lying flat.

The fifth order structure of DNA is formed by the 30 nm solenoid filament as the DNA arranges to form metaphase chromosomes. Pienta and Coffey (1984), using the radial-loop model for chromosomes proposed by Marsden and Laemmli (1979) have predicted that the DNA loops arrange radially around the chromatid central axis with approximately 18 loops per turn, forming a previously undefined structure termed the miniband. The miniband achieves a packing ratio of 12,000 because it packages 1.1×10^6 b.p. of DNA into a 30 nm cross section slice of chromosome. The organization of the miniband allows DNA to fit within the nucleus with ample room to spare, utilizing only 20 per cent of the nuclear volume. The miniband is one million b.p. and is equivalent in b.p. to the centimorgan.

The sixth order structure of DNA is the metaphase chromosome. The chromosomes are formed by stacking the minibands along a central axis. Chromosomes bearing this type of arrangement have been visualized by electron micrographs by Sedat and Manuelidis (1979) and Utsumi (1981).

This ordering of DNA structure provides a simple construct to DNA packaging within the nucleus. This structural integrity is demanded by the functional requirements of replication and transcription which are placed on the DNA apparatus by the cell. It is clear that a system must exist by which the DNA is ordered and processed. In 1974, Berezney and Coffey isolated an insoluble nuclear skeleton which they termed the nuclear matrix. The nuclear matrix is defined by Nelson *et al.* (1986) as 'the dynamic structural subcomponent of the nucleus that directs the functional organization of DNA into domains and provides sites for the specific control of nucleic acids' (Figure 2). The matrix consists of only 10 per cent of the total nuclear protein mass, and has been identified as an important component in a variety of biologic processes. The nuclear matrix orders DNA replication by virtue of containing fixed sites for DNA synthesis and alpha DNA polymerase

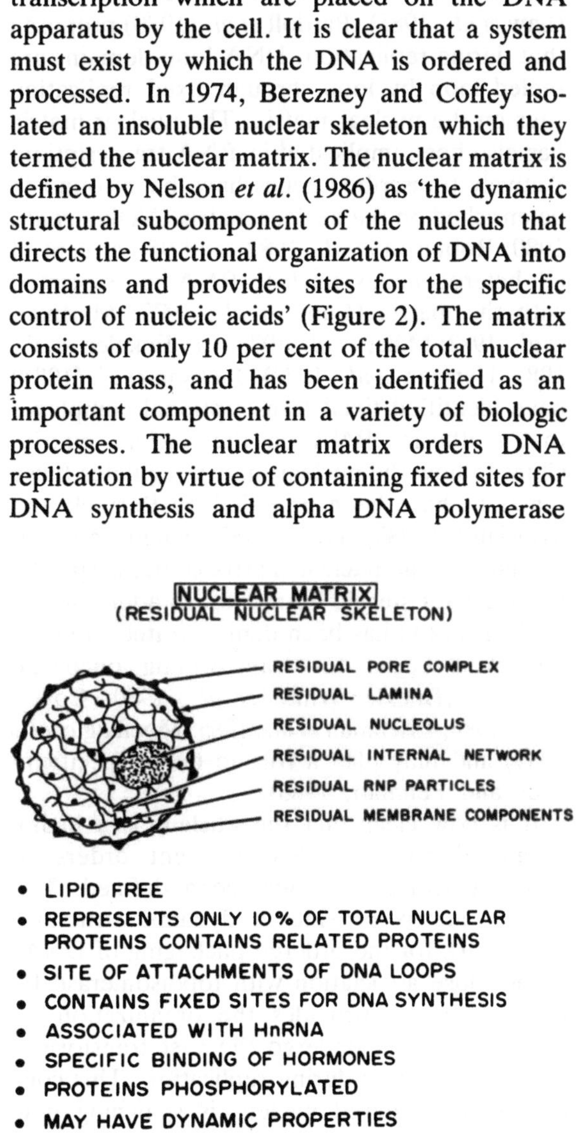

Fig. 2: A schematic of the structure and properties of the nuclear matrix (from Pienta and Coffey, 1984).

(McCready *et al.*, 1980; Smith *et al.*, 1980, 1984; Pardoll, 1980). In addition, Type II DNA topoisomerase has been associated with the matrix during interphase (Halligan *et al.*, 1984; Nelson *et al.*, 1986, 1987; Berrios *et al.*, 1984) and as part of the chromosome scaffold (Earnshaw *et al.*, 1985). Pardoll *et al.* (1980) proposed that during replication, DNA loop-domains are reeled loop by loop through fixed replication sites on the nuclear matrix. The nuclear matrix has also been implicated in RNA transcription. Actively transcribed genes have been found to be enriched on the nuclear matrix (Nelkin *et al.*, 1980; Robinson *et al.*, 1982) and 95 per cent of all heterogeneous nuclear RNA is associated with the matrix (Long *et al.*, 1979; Mariman *et al.*, 1982; Miller *et al.*, 1978; van Eekelen *et al.*, 1981; Herman *et al.*, 1978). Jackson *et al.* (1981) have identified fixed transcriptional complexes that synthesize RNA on the nuclear matrix. Clues to how the nuclear matrix regulates these functions have been provided by Barrack and associates (1980, 1983, 1983) when they reported that the nuclear matrix contains specific receptors for steroid hormones. In addition, the nuclear matrix has been demonstrated to be an important target for transforming oncogene products (Buckler-White *et al.*, 1980; Covey *et al.*, 1984; Eisenman *et al.*, 1985; Moelling *et al.*, 1984) and may play a role in tissue specificity (Fey and Penman, 1988).

It is now clear that the nucleus is a highly organized structure. Six different orders of DNA structure have now been defined. The nuclear matrix provides a dynamic structural framework for the ordered packaging of DNA. By its close association with topoisomerase II, this framework provides the organization by which DNA can undergo the vast transformations which occur during replication. This concept of an ordered nucleus is spawning new research into the interrelationships between structure and biological processes with the nucleus.

Acknowledgements

Work performed in the authors' laboratories was supported by NIH grant AM22000. The authors thank Mrs Tammy Deitrich for her assistance in the preparation of the manuscript.

References

Adolph KW (1980). Isolation and structural organization of human mitotic chromosomes. Chromosoma 76; 23–33.

Adolph KW, Kreisman LR (1983). Surface structure and isolated metaphase chromosomes. Exp. Cell Res. 147; 155–66.

Barrack ER (1983). The nuclear matrix of the prostate contains acceptor sites for androgen receptors. Endocrinology 113; 430–2.

Barrack ER, Coffey DS (1980). The specific binding of estrogen and androgens to the nuclear matrix of sex hormone responsive tissues. J. Biol. Chem. 255; 7265–75.

Barrack ER, Coffey DS (1983). Hormone receptors and the nuclear matrix. In Gene Regulation by Steroid Hormones II (eds. Roy AK and Clark JH), Springer-Verlag, New York; 239–66.

Berezney R, Coffey DS (1974). Identification of a nuclear protein matrix. Biochem. Biophys. Res. Commun. 60; 1410–17.

Berezney R, Coffey DS (1975). Nuclear protein matrix: Association with newly synthesized DNA. Science 189; 291–3.

Berezney R, Coffey DS (1977). Nuclear matrix: Isolation and characterization of a framework structure from rat liver nuclei. J. Cell Biol. 73; 616–37.

Berrios M, Osheroff N, Fisher PA (1984). Type II DNA topoisomerase is a major polypeptide component of the Drosophila nuclear matrix. J. Cell Biol. 99; 128a.

Buckler-White A, Humphrey GW, Pigiet V (1980). Association of polyoma T antigen and DNA with the nuclear matrix from lytically infected 3T6 cells. Cell 22; 37–46.

Cook PR, Brazell IA, Jost E (1976). Characterization of nuclear structures containing superhelical DNA. J Cell Sci. 22; 303–24.

Covey L, Choi Y, Prives C (1984). Association of simian virus 40 T antigen with the nuclear matrix of infected and transformed monkey cells. Mol. Cell. Biol. 4; 1384–92.

Deppert W (1978). Simian Virus 40 (SV40)-specific proteins associated with the nuclear matrix isolated from adenovirus type 2-SV40 hybrid virus-infected HeLa cells carry SV40 U-antigen determinants. J. Virol. 26; 165–78.

Earnshaw WC, Laemmli UK (1983). Architecture of metaphase chromosomes and chromosome scaffolds. J. Cell Biol. 96; 84–93.

Earnshaw WC, Heck MMS (1985). Localization of topoisomerase II in mitotic chromosomes. J. Cell. Biol. 100; 1716–25.

Earnshaw WC, Halligan B, Cooke CA, Heck MMS, Liu LF (1985). Topoisomerase II is a structural component of mitotic chromosome scaffolds. J. Cell. Biol. 100; 1706-15.

Eisenman RN, Tachibana CY, Abrams HD, Hann SR (1985). v-myc- and c-myc- encoded proteins are associated with the nuclear matrix. Mol. Cell. Biol. 5; 114–26.

Fey EG, Penman S (1988). Nuclear matrix proteins reflex cell type of origin in cultured human cells. Proc. Natl. Acad. Sci. U.S.A. 85; 121–5.

Finch JT, Klug A (1976). Solenoid model for superstructure in chromatin. Proc. Natl. Acad. Sci. U.S.A. 73; 1897–901.

Georgiev GP, Nedspasov SA, Bakayev VU (1978). Supranucleosomal levels of chromatin organization. In The Cell Nucleus, 6 (ed. Busch H), New York, Academic Press; 3–4.

Halligan BD, Small D, Vogelstein B, Hsieh TS, Liu LF (1984). Localization of type II DNA topoisomerase in nuclear matrix. J. Cell. Biol. 99; 128a.

Hancock R, Boulikas I (1982). Functional organization in the nucleus. Int. Rev. Cytol. 79; 165–214.

Herman R, Weymouth L. Penman S (1978). Heterogeneous nuclear RNA-protein fibers in chromatin-depleted nuclei. J. Cell. Biol. 78; 663–74.

Igo-Kemenes I, Horz W, Zachau MG (1982). Chromatin. Ann. Rev. Biochem. 51; 89–121.

Jackson DA, McCready SJ, Cook PR (1981). RNA is synthesized at the nuclear cage. Nature 292; 552–5.

Laemmli UK (1979). Levels of organization of the DNA in eukaryotic chromosomes. Pharmacol. Revs. 30; 469–76.

Long BH, Huang CY, Pogo AO (1979). Isolation and characterization of the nuclear matrix in Friend erythroleukemia cells: Chromatin and hnRNA interactions with the nuclear matrix. Cell 18; 1079–90.

Mariman ECM, van Eekelen CAG, Reinders RJ, Berns AJM, van Vernooij WJ (1982). Adenoviral heterogeneous nuclear RNA is associated with the host nuclear matrix during splicing. J. Mol. Biol. 154; 102–19.

Marsden MPF, Laemmli UK (1979). Metaphase chromosome structure: Evidence for a radial loop model. Cell 17; 849–58.

McCready SJ, Godwin J, Mason DW, Brazell IA, Cook PR (1980). DNA is replicated at the nuclear cage. J. Cell. Sci. 46; 365–8.

Miller SJ, Huang C, Pogo AO (1978). Rat liver nuclear skeleton and ribonucleoprotein complexes containing hnRNA. J. Cell Biol. 76; 675–91.

Moelling K, Benter T, Bunte T, Pfaff E, Deppert E, Egly JM, Miyamoto NB (1984). Properties of the myc-gene product: Nuclear association, inhibition of transcription and activation in stimulated lymphocytes. In Current Topics in Microbiology and Immunology, Vol. 113, Springer-Verlag, Berlin; 198–207.

Nelkin BD, Pardoll DM, Vogelstein B (1980). Localization of SV40 genes within supercoiled loop domains. Nucleic Acids Res. 8; 5623–33.

Nelson WG, Cho KR, Hsiang Y-H, *et al.* (1987). Growth-related elevations of DNA topoisomerase II levels are found in Dunning R3327 rat prostate adenocarcinomas. Cancer Res. 47; 3246–50.

Nelson WG, Liu LF, Coffey DS (1986). Newly replicated DNA is associated with DNA topoisomerase II in cultured prostatic adenocarcinoma cells. Nature 322; 187–9.

Nelson WG, Pienta KJ, Barrack ER, Coffey DS (1986). The role of the nuclear matrix in the organization and function of DNA. Ann. Rev. Biophys. Chem. 15; 457–75.

Olins AL, Olins DE (1974). Spherical chromatin units (nubodies) Science 183; 330–2.

Olins AL, Olins DE (1979). Stereoelectron microscopy of the 25 nm chromatin fibers in isolated nuclei. J. Cell Biol. 81; 260–5.

Pardoll DM, Vogelstein B, Coffey DS (1980). A fixed site of DNA replication in eukaryotic cells. Cell 19; 527–36.

Paulson JR, Laemmli UK (1977). The structure of histone-depleted metaphase chromosomes. Cell 12; 817–28.

Pienta KJ, Coffey DS (1984). A structural analysis of the role of the nuclear matrix and DNA loops in the organization of the nucleus and chromosome. J. Cell Sci. Suppl. 1; 123–35.

Robinson SI, Nelkin BD, Vogelstein B (1982). The ovalbumin gene is associated with the nuclear matrix of chicken oviduct cells. Cell 28; 99–106.

Sedat J, Manuelidis (1979). A direct approach to the structure of eukaryotic chromosomes. CSHSQB XII; 331–50.

Smith HC, Berezney R (1980). DNA polymerase is tightly bound to the nuclear matrix of actively replicating liver. Biochem. Biophys. Res. Commun. 97; 1541–7.

Smith HC, Puvion E, Buchholtz LA, Berezney R (1984). Spatial distribution of DNA loop attachment and replicational sites in the nuclear matrix. J. Cell Biol. 99; 1794–1802.

Utsumi KR (1981). Studies on the structure of chromosomes II. Chromosome fibers as revealed by scanning electron microscopy. Cell Structure and Function 6; 395–401.

van Eekelen CAG, van Venrooij WJ (1981). hnRNA and its attachment to a nuclear protein matrix. J. Cell Biol. 88; 554–63.

Vogelstein B, Pardoll DM, Coffey DS (1980). Supercoiled and eukaryotic DNA replication. Cell 22; 79–85.

2 On Interspersed Repetitive DNA Sequences in Animals

JÖRG T. EPPLEN AND ROLAND STUDER

Junior Cancer Research Unit
Max-Planck-Institut für Immunbiologie
D 7800 Freiburg, FRG
and also
Max-Planck-Institut für Psychiatrie
Am Klopferspitz 18A
8033 Planegg-Martinsried, FRG

Introduction

Most present-day prokaryotes are 'streamlined' organisms with small genomes that consist predominantly of unique or low-copy DNA sequences. In contrast, eukaryotic genomes are usually orders of magnitude larger. Much of the eukaryotic DNA present in the genome appears to be in excess of the minimum required—and is therefore called 'secondary DNA' (Hinegardner, 1976). Repetitive sequences constitute a major part of this secondary DNA. Since Britten and Kohne (1968) discovered that repetitive DNA makes up about one third of many eukaryotic genomes, it has been a major scientific challenge to understand the phenomenon of repeated sequences and the mechanisms responsible for generating the necessary sequence amplifications during evolution. Though enormous amounts of information have been accumulated on the subject, some members of the scientific community tend to ignore any relevance of repetitive DNA. Consequently there is a considerable range of opinion on the meaning of repeated sequences. Statements are quite common like repetitive DNA is 'still in search of a function' (Lewin, 1982), is 'selfish' and has no function (Doolittle and Sapienza, 1980; Orgel and Crick, 1980), and many a creed culminates in the 'meaninglessness of simple repeats' (Doolittle, 1987). In this situation objective treatment of the subject is required as well as balanced commentary without tendentious semantics. It is truly difficult to judge the function of repetitive DNA since these sequences are so tremendously heterogeneous. Hence it is useful to classify them according to their structure, distribution and reiteration frequency (see e.g. Schmidtke and Epplen, 1980; Jelinek and Schmid, 1982). We ask here questions on the origins of repetitive DNAs but limit our discussion to selected repetitive elements that are *interspersed* in eukaryotic genomes. First we concentrate on interspersed simple, tandemly repeated DNA and subsequently we deal with *l*ong and *s*hort *i*nterspersed *n*ucleotide *e*lements —LINEs and SINEs (Singer, 1982).

Interspersed Simple Tandemly Repeated Sequences

In an era when powerful industrial and/or governmental organizations intend to determine the entire nucleotide sequence of human individuals, it seems appropriate to question the validity and feasibility of such an approach for major parts of the heterochromatin and the highly repeated satellite DNA. 'Typical satellite DNA is composed of short, tandemly repeated sequences each usually less than 1000 nucleotide base pairs in length and present in more than 10^5 copies per haploid genome' (Macgregor and Sessions, 1986). Large blocks of these sequences are often concentrated in specific parts of the chromosome complement (Miklos and Gill, 1982). This subject was extensively covered in an excellent and eloquent review (John and Miklos, 1979) and even nearly ten years later we are not able to add substantially new or divergent evidence to the field of typical satellite DNAs. The ensuing debate, however, was biased: satellite DNA and repetitive DNA as a whole was largely regarded as 'selfish', 'parasitic' and/or 'junk' DNA (see e.g. Doolittle and Sapienza, 1980; Orgel and Crick, 1980). The last designation, junk, was coined by Ohno (1972) with an interpretation deviating from the one implied by others (Ohno, personal communication). We would like to re-emphasize here the original/actual meaning: 'junk' designates items which have become useless over the years but originally had a function. We consider certain repetitive DNAs to be evolutionary relics in agreement with the viewpoints of some of the other authors. The reasons for this short discussion of nomenclature will become obvious below where CACA and GATA/GACA repeats are treated paradigmatically for interspersed simple repeated sequences.

How might simple repetitive sequences have been generated? Different mechanisms have probably contributed simultaneously to generate the tandem sequence arrays. These could include: several consecutive unequal crossing-over events between initially repetitive or at least duplicated sequences (Smith, 1976), slipped-strand mispairing (Streisinger *et al.*, 1966; Kornberg *et al.*, 1964; Tautz and Renz, 1984; Levinson *et al.*, 1985), a 'rolling circle' mechanism (Hourcade *et al.*, 1973) and 'aberrant *in-situ* replication' (Schimke, 1982). Any of the aforementioned random mechanisms can produce two kinds of the simplest possible repeats when the unit is one base (e.g. A and C; T and G are on the complementary strand). Obviously poly $(dA)^+$ stretches are adequately explained by reinsertion into the genome of poly $(A)^+$ mRNAs after their initial reverse transcription. Poly $(dG)^+$ can initially be polymerized by terminal nucleotidyl transferase (Alt and Baltimore, 1982) as exemplified in the antigen recognition portions of immunoglobulin and T lymphocyte antigen receptor genes. Redistribution and amplification requires additional mechanisms such as the ones mentioned above. Apart from the homopolymers, how many different simple repeats are principally possible? If the basic simple repeat unit is two bases long, six different permutations are unique (AA, AC, AG, AT, CC, CG). All the other possibilities are covered by shifting the phase by one nucleotide or are represented on the complementary strand, respectively. Using three bases there are 18 *different* base patterns possible out of the 64 different base triplets. It seems obvious that the formula $4^{n-1}+2$ describes the number of different simple repeats with n nucleotides of length (for $n \geqslant 2$). For tetrameric repeats 66 $[4^{4-1}+2]$ different simple tandem units exist. Higher eukaryote genomes have so far not been screened systematically for the presence of each of the theoretically possible sequence permutations. Yet it is safe to state that many of these combinations are not present in the investigated eukaryote genomes in substantial amounts. Furthermore sequence specific amplification mechanisms for selected nucleotide combinations are not known. How then can the preponderance of certain simple repeats be explained? Since there is no obvious preferential mechanisms for generating simple sequences, the spreading of certain simple sequences is the result of chance, e.g. by being included in a retrotransposable unit.

(a) CACA simple repeats

One of the most frequently occurring simple DNA sequences is the alternating purine pyrimidine co-polymer $(CA)_n$ first identified in spacer DNA of the murine rRNA (Arnheim *et al.*, 1980). Members of the poly (CA) family of repeats are present in all eukaryotic DNAs (Miesfield *et al.*, 1981; Hamada *et al.*, 1982). Rogers (1983b) has suggested that CACA sequences may have arisen at staggered breaks in the target DNA by 3′ polymerization of $(CA)_n$ onto one strand or by insertion and ligation of pre-existing CACA-stretches. Such a process is unprecedented so far and the respective dinucleotidespecific polymerase activity is unknown. Nevertheless once invented and present in the genome, we entertain the idea that CACA blocks spread and insert by a retroposon-like mechanism. Though a 'retrogene' generation for the CACA family is not generally accepted (e.g. Sun *et al.*, 1984), few tandem copies would suffice to achieve the end result. Evidence in favour of such a mechanism is manifold (Gebhard and Zachau, 1983; for a sequence compilation see Rogers, 1983b; Greaves and Patient, 1985).

The transcriptional activity of the CACA/TGTG simple repeats is limited and due to RNA polymerase II. Both strands are equally represented in e.g. nuclear and cytoplasmic transcripts of Hela cells. As for other simple repeated sequences it has been suggested that CACA/TGTG elements are adventitiously included in longer transcription units and thus may survive RNA processing (Sun *et al.*, 1984). CACA/TGTG blocks are indeed often seen in the vicinity of many genes including amongst others those belonging to the immunoglobulin gene superfamily (Hunkapiller and Hood, 1986). We also stumbled upon several additional CACA blocks in the human and mouse T-lymphocyte receptor α- and β-chain loci (Hinkkanen *et al.*, 1988; Epplen and Becker, unpublished observations): $(TG)_{21}$ was found in an intron of a human variable (V_β) element, $(CA)_{24}$ is present in the vicinity of a novel mouse joining (J_α) segment (for a review on the T-lymphocyte receptor gene organization see Epplen *et al.*, 1987). The meaning of these sequence associations remains elusive so far, especially with respect to the regulation of transcription. It has been demonstrated, however, that $(TG)_n$ repeats have in certain expression vectors a weak enhancer-like activity (Hamada *et al.*, 1984).

Although one could imagine a variety of possibilities, isolated CACA tetramers could serve as candidate precursor elements for the GACA simple quadruplet repeats. A potential additional transition mutation in GACA generates the GATA element. Again amplification and redistribution modes are required to amplify and spread the initial simple repeats. The inclusion of CACA sequences into the unit of unrelated retrogenes would be perhaps the most straightforward explanation.

(b) GATA/GACA simple repeats

Initially GATA/GACA simple quadruplet repeats were isolated from female Colubrid snakes in the form of sex-specific satellites and called Bkm DNA (from *b*anded *k*rait *m*inor satellite; Singh *et al.*, 1980). The identification of the GATA/GACA components in the satellite was subsequently possible in cloned *Elaphe radiata* DNA (Epplen *et. al.*, 1982).

Tandem arrays of GATA/GACA repeats have been demonstrated in every eukaryote investigated so far (see e.g. Epplen and Ohno, 1987). The smallest and fundamental unit of amplification is the tetrameric repeat, but the final stretches of tandemly repeated sequences may be extremely large. The possible mechanisms for this amplification have been discussed above but are also not known precisely for GATA/GACA repeats. Independent mechanisms may contribute to the phenomenon of random distribution (see below). As a result of spreading mechanisms, the GATA/GACA elements are randomly interspersed throughout the autosomal chromosomes in several mammalian and reptile species (Epplen and Ohno, 1987). Often, the organization of GATA/GACA repeats varies between the heteromorphic sex chromosomes within a certain species. Unequal crossing-over between homologous loci on the heteromorphic sex chromosomes cannot account for the difference in GATA/GACA repeat length, since these sequences are located outside the meiotic pairing regions. The gonosomal GATA/GACA distributions in mice and snakes have suggested to some that these sequences are involved in primary sex-determination (e.g. Singh *et al.*, 1980). Today we think it more likely that rather than determining sex themselves GATA/GACA sequences are closely linked to sex-determining genes (e.g. in mouse).

Accumulation of GATA/GACA sequences on the heteromorphic sex chromosomes is not the only interesting phenomenon observed in the investigation of these elements. In the mull-vole *Ellobius lutescens* for example the sex chromosomes have not been positively identified. This species is endowed with 17 chromosomes whereby the singular chromosome number 9 does not pair during meiosis. We and others initially had entertained the suspicion that chromosome 9 is the X chromosome because GATA/GACA repeats were significantly reduced on the presumptive X chromosome as demonstrated by oligonucleotide hybridization *in situ*. Recently we were able to confirm this view by somatic cell genetic methods (Vogel *et al.*, 1988). In another rodent, *Microtus aggrestis*, the extremely large X chromosome contains an enormous block of constitutive heterochromatin. Here, in contrast to *E. lutescens*, we found a very high concentration of GATA/GACA sequences in the X-chromosomal heterochromatin by *in-situ* hybridization as well as in the DNA of somatic cell hybrids containing only the *M. aggrestis* X chromosome (Nanda, Studer, Sperling and Epplen, unpublished observations). Thus in rodent evolution there are mechanisms at work that result in extreme variations of gonosomal GATA/GACA sequences as compared to the more or less even distribution on autosomal chromosomes.

GATA/GACA are ubiquitously interspersed on the autosomes in many reptile, bird and mammalian species though there are again exceptions to the rule. Recently Traut (1987) showed the presence of significant GATA stretches in only very few blocks or restriction fragments in an insect, the moth *Ephestia kuehniella*. This result was confirmed with base-specific GATA/GACA-oligonucleotide probes (Traut and Epplen, unpublished results). Ubiquitous autosomal interspersion of GATA/GACA sequences in man and mouse, on the other hand, allows the production of individual-specific DNA fingerprints in these species (Ali *et al.*, 1986; Epplen and McLaren, 1987). Although several mechanisms for tandem sequence amplification have already been discussed, we favour an additional possibility to generate homogeneous GATA/GACA sequence distribution over the whole genome. It has been shown that GATA/GACA sequences are transcribed in many mouse tissues (Schäfer *et al.*, 1986a). Sequence analysis of several independent cDNA clones revealed flanking inverted repeats, in addition to the demonstration of GATA/GACA elements. Furthermore these cDNA clones show several sequence characteristics of transposable elements such as inverted repeats and consensus ends (Epplen *et al.*, 1983). Since the GATA/GACA repeats are transcribed and polyadenylated they are potential candidates for reverse transcription and may re-enter the vertebrate genome to generate their random interspersion pattern.

With respect to sequence-specific functions of the simple GATA/GACA elements, studies on the major histocompatibility complex genes of the mouse should be critically evaluated. Two groups arrived at the conclusion that $(GACA)_6$ sequences were spatially connected to hotspots of recombination (Steinmetz *et al.*, 1986; Kobori *et al.*, 1986). In a completely unrelated investigation on the sex-reversed mutation (sxr) of the mouse, we found an exceedingly high rate of unequal recombination of the murine Y chromosome in fertile XY^{sxr} carriers (Epplen and McLaren, 1987). The sxr region is tightly linked to long GATA/GACA containing DNA stretches. Nevertheless, we cannot imagine the significance of GATA/GACA repeats themselves as being preferred sites of recombination, simply because from their preponderance in eukaryote genomes (more than 10,000 copies per haploid genome; Schäfer *et al.*, 1986b): (1) one would predict excessive recombinatorial events (also between non-homologous chromosomes), and (2) one would not expect the observation that only one of the GATA/GACA elements is involved in the unequal recombination in XY^{sxr} carrier mice (Epplen and McLaren, 1987).

'Interspersed Nucleotide Elements' (Singer, 1982)

Among the different categories of repeated DNA sequences, another striking and interesting one is the interspersed non-tandemly organized repetitive DNA elements. These latter DNA sequences are characteristic of all investigated eukaryotic genomes, yet no simple description is possible, because again these sequences are themselves extremely heterogeneous (Jelinek and Schmid, 1982). A now widely used classification scheme was proposed by Singer (1982). It distinguishes repetitive elements simply on the basis of their physical lengths: short versus long interspersed nucleotide elements [SINEs, <500 basepairs (bp); LINEs, several kilobases (kb)]. Both SINEs and LINEs share common features like A-rich tracts at their 3′ ends, RNA polymerase II/III promoter structures and flanking short direct repeats. Furthermore, both types of elements occur very abundantly in the genome. These facts immediately suggested that an RNA intermediate exists which can be reverse-transcribed and reinserted in e.g. several mammalian genomes (for review see Singer and Skowronski, 1985; Rogers, 1985a,b; Rogers, 1986; Weiner *et al.*, 1986; Hardman, 1986). Therefore, such elements have been conceptually linked to transposable elements and termed retrotransposons (Rogers, 1983a). We discuss here specifically the organization and the amplification as well as the possible significance of the SINEs and LINEs in an evolutionary context.

(a) SINEs

Since most of the SINEs share a common Alu I restriction enzyme recognition site, these elements are members of the so-called Alu family. The Alu sequences of primates are the most extensively studied of all SINEs and can therefore serve as prototype for the whole class of these elements. Alu sequences are very abundantly interspersed throughout the primate genome and differ on average by 13 per cent from the consensus sequence (Schmid and Jelinek, 1982). Renaturation rate studies revealed e.g. about 500,000 copies per haploid human genome. Given even distribution and a haploid genome size of approximately 3.5×10^6 kb, the calculated average distance between two elements comprises less than 10kb. Individual human Alu sequences are approximately 300 bp in length (in total ~5 per cent of the cellular DNA content). These elements are usually composed of two related segments, each about 130 bp long and terminated by a short A-rich segment. These sequences are obviously duplicated in tandem but the 3′ parts contain an additional internal segment of some 30 bp each. The rodent equivalent to the human Alu sequence is a monomer of roughly 130 bp homologous to the 5′ part of the dimeric Alu element. Collectively these short repetitive rodent sequences comprise the so-called B1 family (Schmid and Jelinek, 1982; Rogers, 1985a; Weiner *et al.*, 1986). It has been suggested that both Alu and B1 sequences are processed 7SL RNA genes (Ullu and Tschudi, 1984; Brown, 1984). 7SL RNA is a constituent of the *s*ignal *r*ecognition *p*article (SRP) which comprises six different polypeptides and one molecule of 7SL RNA. The SRP mediates translocation of secretory proteins across the endoplasmic reticulum. The assumption that Alu and B1 sequences are derived from the 7SL sequence, rather than vice versa, rests primarily on the remarkable evolutionary conservation of 7SL RNA in different taxa in contrast to the lack of conservation of the respective SINEs. While a central region of 155 nucleotides is unique to the 300 bp long 7SL RNA, approximately 100 nucleotides at the 5′ end and 45 nucleotides at the 3′ end of the RNA are homologous to the human Alu dimeric consensus sequence. Therefore, one could argue that the putative ancestral 3′-Alu monomer could have been generated by differential RNA splicing which excised 155 internal nucleotides and by the acquisition of an A-rich tract at the 3′ end of the native RNA sequence. The Alu sequence could in principle also have been produced at the DNA level through a non-homologous recombination event within or between 7SL RNA genes. Because of their internal promoters at the 5′ ends, many Alu and some B1 elements indeed can function as polymerase III transcription units. As already mentioned above, most of the Alu-like family members terminate with an A-rich sequence. Both findings are consistent with the dispersal of Alu sequences throughout genome by reverse-transcribed RNA. Furthermore, Alu-like sequences are flanked by short terminal repeats of 7 to 20 bp. Such short terminal repeats are usually found at the ends of transposable elements suggesting a mechanism of staggered breaks at the site of insertion with the result of target site duplications surrounding the mobile element after insertion in the genome (Sharp, 1983). On the basis of these data, several groups (Van Arsdell *et al.*, 1981; Jagadeeswaran *et al.*, 1981) have proposed a mechanism of transposition by self-primed reverse transcription to explain the high abundance of Alu sequences in the genome. This model requires an internal RNA polymerase III promoter which is located in such a way that transcription of the element begins at the 5′ end and runs through the entire element including the 3′ A-rich tract. Transcription is proposed to terminate at variable sites in adjacent nonrepetitive DNA within a U-rich sequence which is characteristic for RNA polymerase III termination sites. If the 3′ oligo (U) sequence of the RNA transcript pairs with the most 3′ internal oligo (A) stretch to prime cDNA synthesis by reverse transcription, the flanking non-repetitive DNA would be looped out in a hairpin structure. The result would be a reverse transcript without flanking non-repetitive DNA that conceals its origin. Since the promoter would be transposed with the repeat, each new copy has the potential to be transcribed, reverse-transcribed and thus retrotransposed (Deininger and Daniels, 1986; Weiner *et al.*, 1986). A separate model to explain the self priming process for reverse transcription and subsequent integration mechanism in the genome makes use of the frequent $(NA_x)_y$ tails (N most often being C) predominantly within the 3′ on the 3′ A-rich region of SINEs (Rogers, 1985a). This model includes 3′ tailing of a nick in chromosomal DNA with a sequence, e.g. $(GT_x)_y$ which is complementary to the $(CA_x)_y$ part present in the RNA intermediate of a repetitive element. Subsequent pairing of RNA and chromosomal DNA would suffice to prime cDNA synthesis (5′ to 3′) on the 3′ A-rich part of the RNA. A second nick on the complementary DNA strand, which is staggered to the first one, is necessary to fix the mRNA sequence in chromosomal DNA. After ligation of newly synthesized cDNA into chromosomal DNA, repair enzymes could replace the RNA template with DNA. In all models the possible source of reverse transcriptase remains elusive. The only well characterized biochemical process for copying DNA from RNA is catalyzed by viral reverse transcriptase. Hence, a plausible source in mice are the so-called *I*ntracisternal *A* *p*articles (IAPs) which contain reverse transcriptase activity. The encoding genes behave like proviruses but are particularly active in early embryos. Furthermore, it was shown that these IAPs are abundantly expressed in many mouse tumour cells (Temin, 1985; Katzir *et al.*, 1985; Horowitz *et al.*, 1984). The first discovery of 7SL RNA as a component of avian and murine retroviral particles points into the same direction (Ullu and Tschudi, 1984). Alternatively, novel eukaryotic enzymes with similar properties could also be responsible for retrotransposon synthesis.

Several mammalian Alu-like families have been shown to be probably derived from tRNA genes (Daniels and Deininger, 1985; Rogers, 1985b; Deininger and Daniels, 1986). These include the C family of artiodactyls (cows and goats), the Monomer family of the bushbaby Galago, the rodent B2 or Type II family and the rat *id*entifier (*ID*) or R.dre.1 family. The assignment of their orgin is based on the homology (about 70 per cent) between the consensus sequence of these different SINE families and the corresponding tRNA genes. Theoretically, this high homology allows folding into a secondary structure similar to that of typical tRNA molecules. One could imagine that these tRNA-like secondary structures could be recognized by reverse transcriptase, especially since it is known that tRNAs serve as primers for reverse transcription in retroviruses.

(b) LINEs

A thorough understanding of LINEs has been hampered mainly by their enormous complexity in the absence of virtually any clues on their function. So far it is clear, however, that the various repetitive DNA fragments produced by certain restriction endonuclease belong to larger repetitive units—at least in primates and rodents. Among these are the Hind III fragments, the primate Kpn I fragments, the mouse BamH I fragments (or MIF-I = *m*ouse *i*nterspersed *f*ragment *I*), the EcoR I fragments and the mouse R elements. All of these elements had previously been considered to be separate repeat families. Hybridization as well as sequence data support the assertion that these fragments are parts of longer units termed LINEs (Singer and Skowronski, 1985). Voliva and Coworkers (1983) have simplified the nomenclature. Independent of the predominant restriction fragments of each species, the family is called L1 for *L*INE family *1*. A two-letter suffix identifies the genus and species e.g. L1 Md for *M*us musculus *d*omesticus. Additional letters and numbers can be used to specify individual family members.

Comparison of L1-like sequences of four mammalian orders (carnivore, lagomorph, rodent and primate) showed that all these sequences are related to each other (Fanning and Singer, 1987). Additional hybridization studies revealed that L1 sequences are present throughout marsupial and placental orders. This suggests that the L1 family has been present in early mammals and has been conserved during their evolution. The fact that L1 members are more homologous when compared within species rather than between species leads to the conclusion that the L1 family has undergone concerted evolution in the mammalian lineages and perhaps descended from an ancestral protein that was present in the progenitors of mammals (Burton *et al.*, 1986). Each mammalian species has perhaps only one highly abundant LINE family (Singer and Skowronski, 1985). Sequence analysis of primate and rodent L1 sequences has shown a homology of about 70 per cent which extends for at least 1.4 kb. Further comparison of several primate and rodent L1 sequences revealed that both families differ at their extremes but they are internally more or less homologous depending on the region (Burton *et al.*, 1986).

Like the SINEs, the L1 family members comprise 1–8 per cent genomic DNA. If one supposes a random distribution of 2×10^4 copies in the genome (total size in mammals approximately 3.5×10^6 kb), one L1 element is expected every 175 kb. But not all L1 sequences are randomly distributed in the genome. The human β-globin gene cluster is only 60 kb in length but it contains nine L1 members (Hattori *et al.*, 1985). In contrast to their shorter relatives, the SINEs, L1 family members are highly heterogeneous in size and only a few are 6–7 kb in length. Most of the members are randomly truncated at their 5′ ends. Therefore sequences of the 5′ end are much less frequent (about 10^4 copies per haploid genome) than the sequences of the extreme 3′ end (about 10^5 copies) (Singer and Skowronski, 1985; Rogers, 1985a). As already mentioned above, some L1 elements have structural characteristics associated with certain mobile DNA sequences, including 3′ A-rich stretches and target-site duplications that surround some of the family members. Analogous to the Alu and Alu-like sequences, the 3′ A-rich tail is taken as indication of an RNA intermediate; the target-site duplications suggest insertion into the genome by a mechanism involving staggered breaks in the target site. Yet not all L1 units are endowed with target-site duplications and the sizes of duplication vary from one L1 to another (Voliva *et al.*, 1984). Nevertheless, these characteristics suggest a multiplication mechanism similar to that proposed for Alu sequences: transcription of a L1 element, reverse transcription into cDNA and insertion of the cDNA into staggered breaks of a target site.

RNA transcripts complementary to L1 elements have been detected in a number of murine (Jackson *et al.*, 1985) and primate (Kole *et al.*, 1983) cell lines and tissues. In general these transcripts are synthesized by polymerase II and are heterogeneous in size. Most of the transcripts are confined to the cell nucleus as is the case for heterogeneous nuclear RNA (hnRNA) but L1 transcripts are not polyadenylated. Some transcripts contain additional sequences not derived from L1 elements. This suggests that these L1 RNAs are only a part of a whole transcript that is primed outside the L1 element (Shafit-Zagardo *et al.*, 1983; Sun *et al.*, 1984). It is not clear if priming outside the L1 template is common or if L1 segments use mainly their own promotors (Loeb *et al.*, 1986). A variety of rodent and primate cell lines and tissues have been analysed for the presence of cytoplasmic poly$(A)^+$ RNA homologous to L1 sequences. Only small amounts of RNA homologus to L1 DNA could be detected in the cytoplasm and even considerably less was polyadenylated (Shafit-Zagardo *et al.*, 1983). However, recently, Dudley (1987) described two species of RNA (8 and 18 kb) in murine lymphoid cells homologous to the repetitive element L1 Md. Both RNAs were polyadenylated and cytoplasmic. In addition Skowronski and Singer (1985) described a human teratocarcinoma cell line (NTera 2 D1) which expressed a 6.5 kb cytoplasmic poly$(A)^+$ RNA which hybridized to cloned L1 DNA probes. Interestingly, this cell line expressed the highest amount of the putative L1 mRNA when it exhibited stem cell character with embryonal carcinoma morphology. Yet it expressed no L1 mRNA when differentiation was induced by retinoic acid. So far no corresponding translation product has been detected. Skowronski and Singer (1985) hypothesized that these NTera 2D1 cells behave like cells in cleavage-stage human embryos. In analogy one could imagine that putative functional genes of the L1 family are active in cleavage-stage mammalian embryos. Interestingly, the B2 murine short retrotransposons are also abundantly transcribed in early embryos (Vasseur *et al.*, 1985). In this context it is noteworthy that those mammalian embryos, which begin to transcribe the zygotic genome at the two-cell stage contain numerous retrotransposon copies. Those animals, which do not begin transcription until later in development, have only few retrotransposons. Therefore, the embryonal cleavage-stage could be the phase in which new retrotransposons appear (Rogers, 1985a). The extraordinary abundance of L1 sequences in mammalian genomes also suggests that an active dispersal process may exist in germ line cells or in their progenitors in early embryos (Skowronski and Singer, 1985). At the DNA level *o*pen *r*eading *f*rames (ORFs, defined by the distance between two termination codons in frame) of some 100 to several 1000 bp in length have been found in many L1 elements of different primate (Potter, 1984) and rodent (Martin *et al.*, 1984) species. Loeb *et al.*, (1986) described one element termed L1 Md-A2 that comprises a total ORF of about 5 kb. This ORF is split into two frames of 1137 (bases at the 5′ end) and 3900 bases (at the 3′ end). Both parts of the ORF overlapped by 14 bases. Comparison of the L1 Md-A2 sequence (6851 bp) to a consensus sequence compiled from 20 overlapping segments (Singer and Skowronski, 1985) revealed a large sequence homology. At the protein level the *ratio* of *r*eplacement to *s*ilent aminoacid substitutions was calculated (R to S ratio): silent sites (with no alteration of the aminoacids) are more frequent than replacements (where an aminoacid exchange results). On the basis of this analysis, Loeb *et al.*, (1986) conclude that essentially the entire 3900-bp ORF region is evolving as under selection for protein function. Using probes of this 3900-bp ORF region for hybridization, Burton *et al.*, (1986) could show homology in a wide cross-section of mammalian orders. Therefore, one could speculate on a similar functional gene product throughout the mammalian orders.

Recent comparisons between the conserved ORFs of L1 elements with several *bona fide* retrotransposon sequences revealed similarities in overlapping organization and size. The retrotransposons include yeast Ty1 elements, Drosophila copia-like elements and the genes of several retroviruses, such as Rous sarcoma virus. Comparison on the protein level showed sequence homology which suggested that L1 family members are closely related to reverse transcriptase genes (Loeb *et al.*, 1986; Hattori *et al.*, 1986). A translational frame-shift producing a fusion precursor polypeptide has been shown to occur within the mobile element Ty of yeast and has been proposed also for the Rous sarcoma virus. Loeb *et al.*, (1986) suggested that such a frame-shifting event also could be responsible for regulating the relative expression of the two reading frames from the L1 Md-A2 element and perhaps this could also be true for other L1 elements. Another part of L1 segments has been shown to be homologous to transferrin (and lactotransferrin) proteins (Hattori *et al.*, 1986). Furthermore Fanning and Singer (1987) have termed successively A, B, C, D, E, F and G those regions which are homologous to reverse transcriptase protein sequences. The region homologous to transferrin is termed H. These authors identified an additional region I on the amino acid level which is highly conserved in the mammalian orders of primates, rodents, lagomorphs and carnivores. Segment I shares a striking homology with the conserved domains common to many retroviral nucleic acid binding proteins. These proteins, encoded within the gag genes of retroviruses, contain less than 100 amino acids and are able to bind to single-stranded DNA and to RNA. Retroviral and related nucleic acid binding proteins have in common a unit consisting of: Cys-X_{2-4}-Cys-X_{4-16}-His (or Cys)-X_{2-8}-Cys (or His) where X may be any aminoacid. This sequence is similar to the repeated zinc-binding domains (Cys-X_4-Cys-X_3-Phe-X_5-Leu-X_2-His-X_{3-4}-His) of the transcription factor III A which are considered to serve as DNA-binding 'fingers' (Miller *et al.*, 1985; Berg, 1986). The strong conservation of region I as part of a putative L1 protein might indicate an important structural and/or functional role of the protein as a whole. On this basis it has been suggested that putative gene products of L1 have the ability to bind nucleic acids and may function as a reverse transcriptase (Loeb *et al.*, 1986; Hattori *et al.*, 1986; Fanning and Singer, 1987). Whether the two domains reside on the same protein or are separated by post-translational proteolysis as in retroviruses, remains to be elucidated. It is also possible that L1 RNA is packaged into an intracellular ribonucleoprotein particle in which L1 RNA is reverse transcribed. Thereafter, the reverse transcripts might be reintegrated into the cellular genome (Fanning and Singer, 1987). It is interesting to note that a mouse IAP retrotransposon produces particles with reverse transcriptase activity which is particularly high in early embryos and lymphoid tumour cells. Like other transposons, IAPs have been shown to participate in gene activation (e.g. the oncogene c-mos; Horowitz *et al.*, 1984). If L1 elements are indeed transcribed through their own translation products, reverse transcripts could enter the genome and establish new elements within germ-line cells. In the absence of any specific excission and destruction mechanism this could explain why respective elements are so abundant. The typically truncated 5′ ends of L1 elements could be explained by early termination of reverse transcription rather than aberrant initiation by polymerase II. Premature termination is a well known feature associated with reverse transcriptase activity in *vitro* (Voliva *et al.*, 1984; Martin *et al.*, 1984).

The function of the L1 elements in mammalian genomes is unknown. The idea that the L1 family includes functional genes does not preclude the possibility that some of the many noncoding family members influence the expression of neighbouring genes in significant ways. One truncated mouse L1 segment was shown to enhance transcription when tested in monkey cos cells using an expression vector construction (Lueders *et al.*, 1984). Given that promoter and enhancer structures of L1 sequences in vicinity of functional genes could influence gene expression *in vivo*, one could assume that such positions of L1 elements might be conserved in mammalian genomes. In this regard, the presence of L1 units in similar positions downstream of the β-globin genes in mice and man is potentially revealing (Adams *et al.*, 1980; Voliva *et al.*, 1984). L1 elements were also found on either side of and between the two fetal globin genes C_α and A_α (Shen *et al.*, 1981) suggesting that some L1 sequences are responsible for the duplication process of these genes since L1 units are often rearranged by inversions, internal deletions and permutations of the typical linear sequence (Gebhard and Zachau, 1983; Lerman *et al.*, 1983; Di Giovanni *et al.*, 1983; Potter, 1984). Hence Singer and Skowronski (1985) suggested that some L1 segments may foster rearrangements both within L1 units and within neighbouring genomic regions.

Summary

The generation and persistence of repetitive DNA sequences in eukaryotic genomes is far from understood. The inexhaustible variability of animal repetitive sequences can be categorized according to the degree of repetition, length and additional structural criteria. In contrast to the long uninterrupted stretches of repetitive sequences in classical satellite DNAs, the *interspersed* repeats of different sorts are intriguing with respect to their relation to structural genes, recombination, transcription and retrotransposition. We have discussed three kinds of repetitive DNA elements present in animal genomes namely simple sequences, SINEs and LINEs. The CACA and GATA/GACA sequence families are well-studied examples of interspersed simple repetitive elements which show evolutionary conservation as well as sequence features also found in retrotransposons. The most abundant interspersed repetitive DNA elements of higher vertebrates are divided into SINEs and LINEs. Both kinds of elements are intensively interspersed throughout the genomes but they differ in their lengths and sequence characteristics. In conclusion, it is not clear whether repetitive elements exist because of their function or merely because of their evolutionary potential as efficient self-propagating elements.

Acknowledgements

We thank Gabi Eichhorn for patiently typing and Viktor Steimle and Peter J. Nielsen for discussing this manuscript as well as Andrea Becker for skilled technical assistance.

References

Adams JW, Kaufman RE, Kretschmer PJ, Harrison M, Nienhuis AW (1980). A family of long reiterated DNA sequences, one copy of which is next to the human betaglobin gene. Nucl. Acids Res. 8; 6113–28.

Ali S, Müller CR, Epplen JT (1986). DNA fingerprinting by oligonucleotide probes specific for simple repeats. Hum Genet 74; 239–43.

Alt FW, Baltimore D (1982). Joining of immunoglobulin heavy chain gene segments: implications from a chromosome with evidence of three D-J_H fusions. Proc. Natl. Acad. Sci. USA 79; 4118–22.

Arnheim N, Seperack P, Banerji I, Lang RB, Miesfeld R, Marcu KB (1980). Mouse rDNA nontranscribed spacer sequences are found flanking immunoglobulin C_H genes and elsewhere throughout the genome. Cell 22; 179–85.

Berg JM (1986). Potential metal-binding domains in nucleic acid binding proteins. Science 232; 485–7.

Britten RJ, Kohne DE (1968). Repeated sequences in DNA. Science 161; 529–40.

Brown AL (1984). On the origin of the Alu family of repeated sequences. Nature 312; 106.

Burton FH, Loeb DD, Voliva ChF, Martin SL, Edgell MH, Hutchison CA (1986). Conservation throughout mammalian and extensive protein-encoding capacity of the highly repeated DNA long interspersed sequence one. J. Mol. Biol. 187; 291–304.

Daniels GR, Deininger PL (1985). Repeat sequence families derived from mammalian t-RNA genes. Nature 317; 819–22.

Deininger PL, Daniels GR (1986). The recent evolution of mammalian repetitive DNA elements. Trends Genet 2; 76–80.

Di Giovanni L, Haynes SR, Misra R, Jelinek WR (1983). KpnI family of long-dispersed repeated DNA sequences of man: evidence for entry into genomic DNA of DNA copies of poly (A)-terminated KpnI RNAs. Proc. Natl. Acad. Sci. USA 80; 6533–7.

Doolittle WF (1987). Genome evolution in review. Trends Genet 3; 82–3.

Doolittle WF, Sapienza C (1980). Selfish genes, the phenotype paradigma and genome evolution. Nature 284; 601–3.

Dudley JP (1987). Discrete high molecular weight RNA transcribed from the long interspersed repetitive element LI Md. Nucl. Acids Res. 15; 2581–92.

Epplen JT, Cellini A, Romero S, Ohno S (1983). An attempt to approach the molecular mechanisms of primary sex determination: W- and Y-chromosomal conserved simple repetitive DNA sequence and their differential expression in mRNA. J. exp. Zool 228; 305–12.

Epplen JT, McCarrey JR, Sutou S, Ohno S (1982). Base sequence of a cloned W-chromosome DNA fragment and identificaction of a male-specific putative mRNA in the mouse. Proc. Natl. Acad. Sci. USA 79; 3798–802.

Epplen JT, Chluba J, Hardt C, Hinkkanen A, Steimle V, Stockinger H (1987). Mammalian T-lymphocyte antigen receptor genes: genetic and nongenetic potential to generate variability. Hum Genet 75; 300–10.

Epplen JT, McLaren A (1987). $GA^{T}{}_{C}A$ simple repeated DNA as a means to dissect the sex-reversed (sxr) mutation of the mouse. In Evolutionary mechanisms in sex determination (ed. Wachtel SS). CRC Press (in press).

Epplen JT, Ohno S (1987). On DNA, RNA and sex determination. In Selected topics in molecular endocrinology (ed. Lau YCF). Oxford University Press, New York (in press).

Fanning T, Singer M (1987). The LINE-1 DNA sequences in four mammalian orders predict proteins that conserve homologies to retrovirus proteins. Nucl. Acids Res. 15; 2251–60.

Gebhard W, Zachau HG (1983). Organization of the R family and other interspersed repetitive DNA sequences in the mouse genome. J. Mol. Biol. 170; 255–70.

Greaves DR, Patient RK (1985). $(AT)_n$ is an interspersed repeat in the Xenopus genome. EMBO J. 4; 2617–26.

Hamada H, Petrino MG, Kakunaga T (1982). A novel repeated element with Z-DNA forming potential is widely found in evolutionarily diverse eukaryotic genomes. Proc. Natl. Acad. Sci. USA 79; 6465–9.

Hamada H, Seidman M, Howard BH, Gorman CM (1984). Enhanced gene expression by the poly (dT-dG)· poly (dC-dA) sequence. Mol. Cell Biol. 4; 2622–30.

Hardman N (1986). Structure and function of repetitive DNA in eukaryotes. Biochem J. 234; 1–11.

Hattori M, Hidaka S, Sakaki I (1985). Sequence analysis of a KpnI family member near the 3′ end of human β-globin gene. Nucl. Acids Res. 13; 7813–27.

Hattori M, Kuhara S, Takenaka O, Sakaki I (1986). L1 family of repetitive DNA sequences in primates may be derived from a sequence encoding a reverse transcriptase-related protein. Nature 321; 625–8.

Hinegardner R (1976). Evolution of genome size. In Molecular evolution (ed. Ayala FJ) Sinauer Associates, Sunderland, Ma; 179–99.

Hinkkanen A, Kempkes B, Steimle V, Stockinger H, Weltzien HU, Epplen JT (1988). Are there restrictions in T cell receptor diversity in responses to haptens and autoantigens? In The T cell receptor (eds Davis MM, Kappler J). CRC Press, New York, NY; 253–62.

Horowitz M, Luria S, Rechavi G, Givol D (1984). Mechanism of activation of the mouse c-mos oncogene by the LTR of an intracisternal A-particle gene. EMBO J. 3; 2937–41.

Hourcade D, Dressler D, Wolfson J (1973). The amplification of ribosomal RNA genes involving a rolling circle intermediate. Proc. Natl. Acad. Sci. USA 70; 2926–30.

Hunkapiller T, Hood L (1986). The growing immunoglobulin gene superfamily. Nature 323; 15–16.

Jackson M, Heller D, Leinwand L (1985). Transcriptional measurements of mouse repeated DNA sequences. Nucl. Acids Res. 13; 3389–403.

Jagadeeswaran P, Forget BG, Weissman SM (1981). Short interspersed repetitive DNA elements in eucaryotes: Transposable DNA elements generated by reverse transcription of RNA pol III transcripts? Cell 26; 141–2.

Jelinek WR, Schmid CW (1982). Repetitive sequences in eukaryotic DNA and their expression. Ann. Rev. Biochem 51; 813–44.

John B, Miklos GLG (1979). Functional aspects of satellite DNA and heterochromatin. Int. Rev. Cytol 58; 1–114.

Katzir N, Rechavi G, Cohen IB, Unger T, Simoni F, Segal S, Cohen D, Givol D (1985). 'Retroposon' insertion into the cellular oncogene c-myc in canine transmissible venereal tumour. Proc. Natl. Acad. Sci. USA 82; 1054–8.

Kobori JA, Strauss E, Minard K, Hood L (1986). Molecular analysis of the hotspot of recombination in the murine major histocompatibility complex. Science 234; 173–9.

Kole LB, Haynes SR, Jelinek WR (1983). Discrete and heterogeneous high molecular weight RNA's complementary to a long dispersed repeat family (a possible transposon) of human DNA. J. Mol. Biol. 165; 257–86.

Kornberg A, Bertsch LL, Jackson JF, Khorana HG (1964). Enzymatic synthesis of deoxyribonucleic acid. XVI. Oligonucleotides as templates and the mechanism of their replication. Proc. Natl. Acad. Sci. USA 51; 315–23.

Lerman MI, Thayer RE, Singer MF (1983). KpnI family of long interspersed repeated DNA sequences in primates: polymorphism of family members and evidence for transcription. Proc. Natl. Acad. Sci. USA 80; 3966–70.

Levinson G, Marsh JL, Epplen JT, Gutman GA (1985). Cross-hybridizing snake satellite, *Drosophila*, and mouse DNA sequences may have arisen independently Mol. Biol. Evol. 2; 494–504.

Lewin R (1982). Repeated DNA still in search of a function. Science 217; 621–3.

Loeb DD, Padgett RW, Hardies SC, Shehee RW, Comer MB, Edgell MH, Hutchison CA (1986). The sequence of a large L1 Md element reveals a tandemly repeated 5′ end and several features found in retroposons. Mol. Cell Biol. 6; 168–82.

Lueders KK, Fewell JW, Kuff EL, Koch T (1984). The long terminal repeat of an endogenous intracisternal A-particle gene functions as a promoter when introduced into eukaryotic cells by transfection. Mol. Cell Biol. 4; 2128–35.

Macgregor HC, Sessions SK (1986). The biological significance of variation in satellite DNA and heterochromatin in newts of the genus *Triturus:* an evolutionary perspective. Phil. Trans. R. Soc. 312; 243–59.

Martin SL, Voliva ChF, Burton FH, Egdell MH, Hutchison CA (1984). A large interspersed repeat found in mouse DNA contains a long open reading frame that evolves as if it encodes a protein. Proc. Natl. Acad. Sci. USA 81; 2308–12.

Miesfeld R, Krystal M, Arnheim N (1981). A member of a new repeated sequence which is conserved throughout eukaryotic evolution is found between the human delta and beta globin genes. Nucl. Acids Res. 9; 5931–47.

Miklos GLG, Gill AC (1982). Nucleotide sequences of highly repeated DNAs; compilation and comments. Genet Res. 39; 1-30.

Miller J, McLachlan AD, Klug A (1985). Repetitive zinc-binding domains in the protein transcription factor III A from Xenopus oocytes. EMBO J. 4; 1609-14.

Ohno S (1972). So much 'junk' DNA in our genome. In Evolution of genetic systems (ed. Smith HH). Gordon and Breach, New York.

Orgel LE, Crick FHC (1980). Selfish DNA: the ultimate parasite. Nature 284; 604-7.

Potter SS (1984). Rearranged sequences of a human KpnI element. Proc. Natl. Acad. Sci. USA 81; 1012-16.

Rogers J (1983a). Retroposons defined. Nature 301; 460.

Rogers J (1983b). CACA sequences—the ends and the means? Nature 305; 101-2.

Rogers J (1985a). The origin and evolution of retroposons. Int. Rev. Cytol 93; 187-279.

Rogers J (1985b). Origins of repeated DNA. Nature 317; 765-6.

Rogers J (1986). The origin of retroposons. Nature 319; 725.

Schäfer R, Böltz E, Becker A, Bartels F, Epplen JT (1986a). The expression of the evolutionarily conserved GATA/GACA repeats in mouse tissues. Chromosoma 93; 496-501.

Schäfer R, Ali S, Epplen JT (1986b). The organization of the evolutionarily conserved GATA/GACA sequences in the mouse genome. Chromosoma 93; 502-10.

Schimke R (1982). In Gene amplification (ed. Schimke R). Cold Spring Harbor Press, New York, 317-33.

Schmid CW, Jelinek WR (1982). The Alu family of dispersed repetitive sequences. Science 216; 1065-70.

Schmidtke J, Epplen JT (1980). Sequence organization of animal nuclear DNA. Hum Genet 55; 1-18.

Shafit-Zagardo B, Brown FL, Zavodny PJ, Maio JJ (1983). Transcription of the KpnI families of long interspersed DNA's in human cells. Nature 304; 277-80.

Sharp PA (1983). Conversion of RNA to DNA in mammals: Alu-like elements and pseudogenes. Nature 301; 471-2.

Shen S, Slighton JL, Smithies O (1981). A history of the human fetal globin gene duplication. Cell 26; 191-203.

Singer MF (1982). SINES and LINES: Highly repeated short and long interspersed sequences in mammalian genomes. Cell 28; 433-4.

Singer MF, Skowronski J (1985). Making sense out of LINES: long interspersed repeat sequences in mammalian genomes. Trends Biochem Sci 10; 119-22.

Singh L, Purdom IF, Jones KW (1980). Conserved sex-chromosome-associated nucleotide sequence in eukaryotes. Cold Spring Harbor Symp Quant Biol. 45; 805-14

Skowronski J, Singer MF (1985). Expression of a cytoplasmic LINE-1 transcript is regulated in a human teratocarcinoma cell line. Proc. Natl. Acad. Sci. USA 82; 6050-4.

Smith GP (1976). Evolution of repeated DNA sequences by unequal crossover. Science 191; 528-35.

Steinmetz M, Stephan D, Fischer Lindahl K (1986). Gene organization and recombinational hotspots in the murine major histocompatibility complex. Cell 44; 895-904.

Streisinger G, Okada Y, Emrich J, Newton J, Tsugita A, Terzaghi E, Inouye M (1966). Cold Spring Harbor Symp Quant Biol. 31; 77-84.

Sun L, Paulson KE, Schmid CW, Kadyk L, Leinwand L (1984). Non-Alu family interspersed repeats in human DNA and their transcriptional activity. Nucl. Acids Res. 12; 2669-90.

Tautz D, Renz M (1984). Simple sequences are ubiquitous repetitive components of eukaryotic genomes. Nucl. Acids Res. 12; 4127-38.

Temin HM (1985). Reverse transcription in the eukaryotic genome: retroviruses, pararetroviruses, retrotransposons and retrotranscripts. Mol. Biol. Evol. 2; 455-68.

Traut W (1987). Hypervariable Bkm DNA loci in a moth, *Ephestia kuehniella*: Does transposition cause restriction fragment length polymorphism (RFLP)? Genetics 115; 493-8.

Ullu E, Tschudi C (1984). Aly sequences are proposed 7SL RNA genes. Nature 312; 171-2.

Van Arsdell SW, Denison RA, Bernstein LB, Weiner AM, Manser T, Gesteland RF (1981). Direct repeats flank three small nuclear RNA pseudogenes in the human genome. Cell 26; 11-17.

Vasseur M, Condamine M, Duprey P (1985). RNAs containing B2 repeated sequences are transcribed in the early stages of mouse embryogenesis. EMBO J. 4; 1749-53.

Vogel W, Steinbach P, Djalali, Mehnert K, Ali, S, Epplen JT (1988). Chromosome No. 9 of *Ellobius lutescens* is the X-chromosome. Chromosoma 96; 112-8.

Voliva ChF, Jahn CL, Comer MB, Hutchison CA, Edgell MH (1983). The LIMd long interspersed repeat family in the mouse: almost all examples are truncated at one end. Nucl. Acids Res. 11; 8847-59.

Voliva ChF, Martin SL, Hutchison CA, Edgell MH (1984). Dispersal process associated with the L1 family of interspersed repetitive DNA sequences. J. Mol. Biol. 178; 795-813.

Weiner AM, Deininger PL, Efstratiadis A (1986). Nonviral retroposons: Genes, pseudogenes, and transposable elements generated by the reverse flow of genetic information. Ann. Rev. Biochem 55; 631-61.

3 Chromatin Structure and Gene Expression in Eukaryotes

SUBRATA SEN AND M. TIEN KUO

Department of Molecular Pathology
The University of Texas System Cancer Center
M.D. Anderson Hospital and Tumor Institute
Houston, Texas 77030, USA

Introduction

It has been generally thought that the mitotic chromosomes are uninemic, i.e. a single duplexed DNA molecule runs from one end of a chromatid to the other (Prescott, 1970). The first order of packaging of this DNA into chromatin is by the formation of nucleosomes, which have been observed both in interphase nuclei and in metaphase chromosomes. The basic structural organization of eukaryotic chromatin is now known to consist of these repeating subunit particles which are also referred to as nu bodies. Each subunit consists of about 160–220 base pairs (bp) of approximately two left-handed superhelical turns of DNA and an octamer core of histone proteins. The octamer core consists of two each of the histones H2A, H2B, H3 and H4. One molecule of histone H1 is bound to DNA sequences entering and leaving the nucleosome core. Limited digestion with micrococcal nuclease that cleaves between the particles gives rise to a mixture of oligonucleosomes in which the DNA length is an integer multiple of the DNA length of a nucleosome and the linker DNA. Further digestion with the nuclease gives rise to nucleosome core particles that lack histone H1 and contain about 146 to 148 bp of DNA. These core particles have been crystallized and investigated by several structural methods. Electron microscopy and x-ray diffraction techniques suggested that the particle has the shape of a disk, 57 Å thick and 110 Å in diameter (Richmond *et al.*, 1984). The DNA appears to be wound around the histone core in 1.8 turns of a flat superhelix with a pitch of 28 A. Although the x-ray diffraction resolution is now limited to 7 Å, the images combined with our knowledge of histone-histone and histone-DNA interactions provide a close approximation of the real structure of this particle.

A second level of organization concerns the folding of the linear nucleosomal array into a 300 Å solenoid fibre with 6 particles per turn in a superbead configuration. This higher order structure is maintained by the presence of histone H1 (Thoma *et al.*, 1979).

Several comprehensive reviews on chromatin have been written in recent years (Mathis *et al.*, 1980; McGhee and Felsenfeld, 1980; Cartwright *et al.*, 1982; Igo-Kemenes and Zachau, 1982; Eissenberg *et al.*, 1985; Pederson *et al.*, 1986). In order to avoid repetition, it is our intent in this chapter to discuss only the recent developments in chromatin structure. We focus on the chromatin structure of active genes with specific emphasis on the regulatory DNA elements and on high-order chromatin structure and gene expression.

Because of page limitation, only selected topics of chromatin research in these areas are discussed. The identification of a stretch of DNA that is responsible for gene activity was coincidental with the discovery of DNase hypersensitive cleavage in chromatin. It is now known that a number of nuclease hypersensitive sites (HS) contain DNA sequences responsible for gene expression. In a number of cases, we now know the precise sequences at the nucleotide level in the HS that are essential for gene expression; and the protein factors that interact with these sequences have been purified. We will review the present status of nuclease-hypersensitive cleavage in chromatin. We will then review some recent developments in portein-DNA interactions that are relevant to the HS.

Research on chromatin structure over the past decade has provided a great many of new insights into nucleosome structure. The next levels of chromatin structure which deal with how nucleosomal arrays are packaged into higher order structures are important and challenging problems in chromatin research. It has been known for some time that the chromatin fibres are folded into loops or domains within the nucleus. How is the folded chromatin formed? Is it under torsional constraint? What role do the topoisomerases play? In the second part of this chapter, we will discuss the recent developments in this area of research, particularly in reference to gene expression.

Chromatin Structure of Active Genes

Digestion of nuclei with the enzyme DNase I reveals differences in chromatin organization between the transcriptionally active and inactive genes. Active genes are always more sensitive to nuclease digestion than the inactive ones (Weintraub and Groudine, 1976; Garel and Axel, 1976). A difference in structural organization that depends on functional state of a gene is also shown in preferential solubilization of active genes by brief treatment with micrococcal nuclease (Bellard *et al.*, 1977; Bloom and Anderson, 1979) and by increased accessibility to the DNA-damaging agent (Kuo, 1981). These differences may be due to the looser packaging of chromatin containing active genes than that containing inactive genes. It has been suggested from the work of Weintraub (1984) that histone H1 cross-links adjacent nucleosomes in the inactive regions into supernucleosomal structures and that H1 is bound differently in expressed regions. It is interesting to note that nuclease sensitivity analysis can even be extended to highly condensed metaphase chromosomes where nuclease-sensitive regions can be detected by nick translation of chromosomes in solution (Kuo and Plunkett, 1985) and in spreads on slides after fixation (Gazit *et al.*, 1982). The same experimental strategy also distinguished an entire active X chromosome from the inactive X chromosomes in female mammalian cells (Kerem *et al.*, 1983). Several earlier reports had also suggested that active genes are found in modified chromatin containing highly acetylated histone, non-histone proteins like HMG 14 and 17, ubiquitinated H2A, etc. (see review by Weisbrod, 1982). It was suggested that these non-histone proteins modify the nucleosome structure in active genes, producing loosely packaged chromatin compartments.

DNase hypersensitive cleavage in chromatin

Besides the overall increased sensitivity of the active chromatin domains, the chromatin containing active genes is also characterized by the presence of specific hypersensitive sites (HS) to DNase I and other nuclease digestions. When DNA prepared from nuclei which have been digested with low level of nuclease followed by complete digestion with a restriction enzyme, is Southern blot-hybridized with a specific probe, sub-bands are often revealed on the autoradiograph. These sub-bands represent specific DNA segments produced by nuclease hypersensitive cleavage of chromatin (see review by Elgin, 1981). The locations of these HS have been mostly determined by the indirect-end-labelling method first applied by Nedospasov and Georgiev (1980) using micrococcal nuclease and by Wu (1980) using DNase I. These methods usually permit the assignments of HS with a precision no better than ± 10 bp. Other methods have been later devised to map these sites with greater precision.

High-resolution mapping of DNase hypersensitive sites

Wu (1984a, b) used exonuclease III to map the boundaries of protein complexed to hypersensitive domain in the 5′ flanking region of *Drosophila* heat shock genes. The enzyme, a 3′ to 5′ exonuclease, degrades a double-stranded DNA fragment progressively from 3′ termini until it reaches the location of specific bound protein. Two sites within the 5′ flanking region involved in the regulation of the Drosophila heat shock genes were detected. Site I, covering 22 or 28 bp, surrounds the TATA box sequence which is indispensible for the start of transcription. Site II includes 36 or 68 bp for the two genes analysed and covers the upstream heat shock consensus sequence.

Another high-resolution HS mapping method has been reported by Jackson and Felsenfeld (1985) and Weintraub (1985). It is based on the S1 nuclease mapping of the RNA initiation site developed by Berk and Sharp (1977) and makes use of a long single-stranded, terminally labelled DNA probe. The probe is annealed to DNA fragments isolated from nuclease-digested nuclei. The resulting duplexes are digested with single-strand-specific nuclease, and analysed by gel electrophoresis. These authors claimed that the precision of mapping is approximately ± 2 nucleotides in the most highly resolved portion of the gel. With this procedure, Jackson and Felsenfeld (1985) mapped the DNase I digestion pattern of the β-globin locus in chicken erythrocyte nuclei. A strong protection region bounded by nucleotides −136 and −163, and another weak protection region bounded by nucleotides −175 to −200, were detected. The latter region contains the poly (dc) string. Weintraub (1985) applied a similar procedure and mapped a major S1 cleavage site near the TATA box at about −30 bp from the adenovirus major late promoter.

Sun *et al.* (1987) developed a new protocol for high-resolution mapping of HS at the promoter region of the chicken β2 tubulin gene transfected into monkey cos-7 celss. The protocol is schematically shown in Figure 1. A DNA segment of about 400 bp containing the promoter region of the β2 gene was subcloned into M13 phage. A single-stranded DNA probe with a defined restriction enzyme cleavage site at one end was prepared from the M13 phage DNA template. The single-stranded probe and the restriction enzyme digested DNA from nuclease digested nuclei from transfected cells were sequentially annealed back to the template. After ligation between the probe and the transfected DNA, the HS was determined by DNA sequencing gel electrophoresis. Figure 2 shows the results of the high-resolution mapping of HS, which determined the 3′ boundary of the HS with single-nucleotide precision. A major

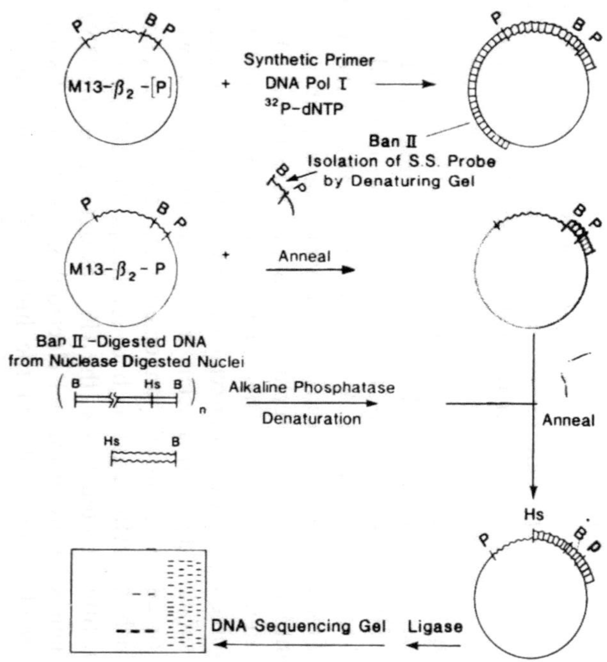

FIGURE 1: Schematic diagram of high-resolution mapping of DNase I hypersensitive cleavage sites. The promoter region of the chicken β2 tubulin gene (shown by wavy line) which contains a HS was subcloned into M13 phage DNA. Single-stranded DNA prepared from the construct was used as template for DNA synthesis and a labelled DNA primer (probe) was prepared. This primer was re-annealed with the template and the template-primer was in turn used for annealing with the restriction enzyme-digested DNA prepared from nuclease digested nuclei. After ligation, the HS can be mapped by DNA sequencing gel electrophoresis. P, PstI; B, BanII restriction enzyme cleavage sites; HS, hypersensitive site. (Taken from Sun *et al.*, (1987) with permission.)

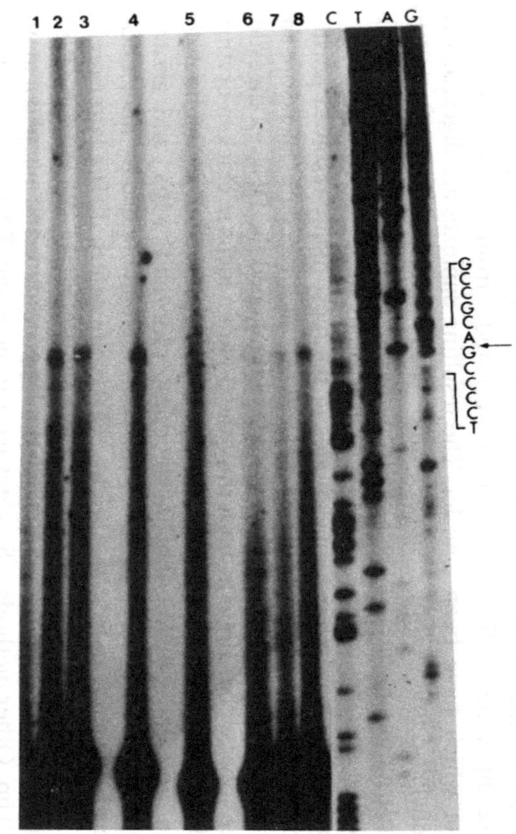

FIGURE 2: Autoradiograph showing high-resolution mapping of the DNase I hypersensitive cleavage site in the transfected DNA using the mapping strategy shown in Fig. 1. Two independent experiments are shown. Lanes 1 and 5 are DNA from nuclease untreated nuclei. Lanes 2–4 and 6–8 are from nuclei that were with various concentrations of DNase I. Arrowhead shows the unligated probe and arrows show the ligated probe (left) and the HS (right), respectively. (Taken from Sun *et al.* (1987) with permission.)

HS at − 107 nucleotide position of the β2 tubulin gene was mapped where a CACCC-like sequence is present (Myers *et al.*, 1986).

Church and Gilbert (1984) developed a genomic DNA sequencing method. This method, in combination with the high-resolution mapping methods described here, should provide a fine structural analysis of HS in chromatin.

DNase hypersensitive sites and gene activity

A number of active genes that have been investigated contain HS (see reviews by Eissenberg *et al.*, 1985; Pederson *et al.*, 1986). As examples, we discuss only three classes of genes here—constitutively expressed, developmentally regulated, and inducibly expressed genes.

Constitutively expressed genes such as the glyceraldehyde-3-phosphate dehydrogenase contain HS located at the 5′ flanking sequences that are essential for the expression of these genes (Kuo *et al.*, 1982, Sen *et al.*, 1985). Nevertheless, these HS do not persist in all cell types in which the genes are expressed. Sun *et al.* (1987) reported that the constitutively expressed β2 tubulin gene in a transformed liver cell line contains a major HS at − 107 nucleotide, whereas the same expressed gene in a lymphocyte cell line does not exhibit such a HS. Therefore, it appears that HS at the 5′ side of the constitutively expressed genes is not strictly associated with expression of the genes.

In the case of inducible genes such as the heat-shock gene of *Drosophila*, the appearance of HS seems to be correlated with heat-shock transcriptional activation of the genes (Wu, 1980). We will discuss this further later. In the developmentally regulated genes, one finds tissue-specific patterns in which the occurrence of HS correlates with gene expression. Stadler *et al.* (1980). found that upstream HS in globin genes were present in red cell nuclei of chickens but were absent from cells that do not express globin. For rat preproinsulin II gene, a similar tissue-specific expression of HS was seen by Wu and Gilbert (1981). They found DNase I sensitivity at the 5′ terminus of the gene only in the nuclei from a pancreatic B cell tumour but not in liver, spleen, kidney and brain.

In addition to sites detected immediately upstream of the transcribed region, certain genes are preceded by discrete HS sites extending several kilobases upstream of the cap site. Examples are genes for lysozymes (Fritton *et al.*, 1984), ovalbumin (Kaye *et al.*, 1984), and glyceraldehyde-3-phosphate-dehydrogenase (Kuo *et al.*, 1982; Sen *et al.*, 1985). Although a majority of the HS have been detected on the 5′ side of the transcription start site, some are on the 3′ flanking regions, e.g. conalbumin (Kuo *et al.*, 1979), α globin (Stadler *et al.*, 1980).These observations led to the idea that the presence of these sites may reflect a committed state of the active chromatin and also the presence of some signal sequences required for interaction of regulatory proteins for gene expression.

HS appears prior to or concomitant with induction of gene expression and seems to be retained in certain instances by way of a 'memory effect' even after withdrawal of the inducing stimulus and cessation of transcriptional activity of the gene. For example, when the major chicken vitellogenin gene is transcriptionally activated in the liver by treatment with 17 β estradiol, a set of three HS is induced at the 5′ end of the gene. Two of these sites are stable and are propagated even after withdrawal of the hormone. The third site is observed only during periods of hormone treatment (Burch and Weintraub, 1983). Similar propagation of HS around globin genes induced in transformed fibroblasts is also seen following removal of induction (Groudine and Weintraub, 1982). The HS are thus thought to be necessary but not sufficient for transcription.

Possible mechanism for nuclease hypersensitive cleavage in active chromatin: Is a specific nucleotide sequence or nucleosome-free organization involved?

Several published results suggest that HS in chromatin are associated with specific nucleotide sequences. Jongstra *et al.* (1984) reported that the SV40 transcriptional promoter elements, which consist of the 72-bp and 21-bp repeats, are capable of generating HS on these sequences when they were inserted into another position in the SV40 genome.

McGinnis *et al.* (1983) analysed the association of 5′ HS and the expression of glue protein gene Sgs-4 in wild-type and naturally occurring mutants of *Drosophila*. Four major HS lie within 500 bp upstream of the glue protein gene. A single base change within the HS located around −350 bp is correlated with a 50 per cent reduction in the amount of Sgs-4 RNA produced in two of the mutants. In another mutant with multiple sequence substitutions near these HS, drastic reduction in the Sgs-4 gene expression was observed, suggesting that an interplay between these HS is responsible for the expression of this gene in wild-type animals.

The role of specific DNA sequences and perhaps the secondary structure of DNA in determining the occurrence of HS has also been suggested in DNA transfer experiments both in cultured cells and in transgenic animals. Weintraub (1983) reported that when supercoiled pBR322 DNA is transfected into mouse L cells, for majority of transfected clones, supercoil induced S1-sensitive sites in transfected DNA are also sensitive in L cell nuclei., In a study utilizing a rather long stretch of a functional housekeeping gene, Sen *et al.* (1985) found that the 5′ HS for the glyceraldehyde-3-phosphate-dehydrogenase gene is dominantly present when the gene was transfected into mouse L cells.

Introduction of a hybrid mouse/human β globin gene into transgenic mice further demonstrated that tissue specificity of DNase I hypersensitivity can be maintained *in vivo* of an exogenous gene by specific DNA sequences. In erythroid cells of the transgenic animal, the exogenous hybrid gene was expressed and showed HS at the same three positions as correspondingly located in the endogenous mouse and human β globin genes. Two of the HS in the human gene, not observed in the endogenous mouse gene were found in the human portion of the hybrid globin gene in transgenic mouse erythroid cells, suggesting that differences in mouse and human DNA sequences rather than cellular factors were responsible for the additional HS in the human gene (Radice and Constantine, 1986).

What, then, is the nature of nucleotide sequences in the HS? Kuo and Plunkett (1985) used nick-translation to label HS DNA in Chinese hamster chromosomes. The labelled DNA were located where transcribed DNA sequences were present as revealed by *in situ* hybridization using cDNA from total mRNA as probe. The labelled DNA were enriched in repetitive DNA sequences. Base composition analysis showed that the labelled DNA are highly enriched in guanine and adenine residues. These results suggest that a specific subset of genomic DNA in active chromatin is associated with DNase HS.

DNA structural analyses of HS of known genes detected by the indirect-end-labelling method suggest that non-B form conformation may be present in these regions. Z DNA binding proteins have been isolated from SV40 minichromosomes and shown to bind to the HS to SV40 chromosomes (Azorin and Rich 1985). Recently, Muller *et al.* (1987) observed binding of a specific molecular probe, A-Tris- (4, 7-diphenyl-1, 10-phenanthroline) cobalt (III) which cannot bind to regular B form DNA, to the HS of SV40 DNA.

A number of HS mapped to the 5′ flanking regions of transcribing genes in chromatin have been shown to exhibit hypersesitivity to S1 nuclease when the sequence is place in supercoiled DNA (see review by Eissenberg *et al.*, 1985). S1 nuclease has been thought to be a single-strand-specific nuclease, although whether it only digests single-stranded DNA is still in doubt (Evans and Efstratiadis, 1986; Muller and Wilson, 1987). However, its ability to recognize non-B form conformation suggests that an altered DNA structure is present in HS. As we now know, certain non-histone proteins bind to DNA in HS, resulting in DNA conformational changes. We will discuss this in the following section.

There are also reports suggesting that the DNase HS located at the 5′ side of a gene merely compose a nucleosome-free region, set up for the entry of RNA polymerase and transcription factors to interact with the promoters. The exposed DNA therefore are naturally very sensitive to the DNA damaging agents. The presence of nucleosome-free regions at these HS have been noted by several experimental approaches: e.g. restriction enzyme digestion (McGhee *et al.*, 1981), electron microscopic observations (Jongstra *et al.*, 1984), and 'histone image' analyses (Karpov *et al.*, 1984). That a nucleosome-free promoter is required for initiation of transcription by RNA polymerase II has been demonstrated recently in an *in vitro* reconstituted system (Lorch *et al.*, 1987). The best evidence to support this view is in the work published by Costlow *et al.* (1985). These authors reported that the 5′ DNase I HS is always present at the same location, regardless of the changes of the sequences in it.

Nuclear protein-DNA interactions that regulate gene expression

Recent studies on the structure and function of eukaryotic genes have identified several DNA sequences, located at either side of the transcribed regions or within the transcribed sequences, that are important for regulation of gene expression. The control region that directs correct initiation (promoter) for transcription of the 5S ribosomal RNA gene (by RNA polymerse III) is located within the gene (Bogenhagen *et al.*, 1980). The promoter regions for rDNA which is transcribed by polymerase I, have been mapped within the 'nontranscribed spacer' region within−200 base pair of the 47 S rRNA transcription unit (for review see Sollner-Webb and Tower, 1986), where a transcription terminator is also located (Grummet *et al.*, 1986; Henderson and Sollner-Webb, 1986; McStay and Reeder, 1986). The promoter regions for the RNA polymerase II-encoded genes have been located generally within 100 bp upstream of the start site of transcription (for review, see Dynam and Tjian, 1985). There are also sequences located either in the proximity, or several kilobase pairs away, that serve as binding sites for transcription-regulating proteins. We limit our review to some of these examples that have been better characterized so far.

The animal viral genes. Because of their small genome sizes and relative ease of preparation, the DNA viruses have offered a model system for studies of gene expression in animal cells (for review, see Tooze, 1982). Late in infection, SV40 DNA is organized in minichromosome in the nucleus. These minichromosomes can be isolated and used for studies as isolated genes in chromatin.

Electron microscopic observations of isolated minichromosomes revealed that about 25 per cent of them contain a nucleosome-free region. This nucleosome gap is located at the origin of DNA replication. The *ori* region also contains the DNA sequences important for initiation of transcription of early and late SV40 genes (Jongstra *et al.*, 1984). Functional analyses using deletion/site-specific mutagenesis of DNA sequences at this area have identified several cis-acting DNA sequences, notably the two 72-bp and three 21-bp repeats, that are important for the transcription of these genes (Vigneron *et al.*, 1984 and for review, see Kadonaga *et al.*, 1986). DNase I footprinting analyses using nuclear extract prepared from HeLa cells indicated specific protein binding to these DNA sequences (Gidoni *et al.*, 1985; Wildeman *et al.*, 1986).

Spl, the protein factor that binds to the GGGCGG hexanucleotide (GC-box) in the 21-bp repeat elements has been purified to near homogeneity (Kadonaga *et al.*, 1986). There are six GC-boxes in the three 21-bp repeats, each exhibiting different binding affinity to the Spl. Early RNA synthesis is mediated by interaction of this protein factor with Box I, II and III, whereas transcription in the late direction is mediated by binding to Box III, V, and VI. Therefore, this factor is responsible for bidirectional SV40 transcription (Gidoni *et al.*, 1985). The GC-box has also been found on the 5′ sides of a number of cellular genes that are constitutively expressed. The expression of these genes seems to involve the interactions of Spl and the GC-box.

Factor binding to the 72-bp repeats, designated Apl, has been detected by footprint analyses (Lee *et al.*, 1987). This factor has been shown to interact with Spl in stimulating transcription activity in SV40 and other cellular genes (Angel *et al.*, 1987). The sterospecific interactions between 21-bp and 72-bp specific protein factors are important for the early and late SV40 transcription. Takahashi *et al.* (1986) found that inserting half-turns of DNA helix in between these repeats reduced transcription more than did insertions of a full helix turn. Similar effects on the early gene expression were also observed when the insertions were placed between the 21-bp repeats and the TATA box.

The heat shock responsive genes. The observation that stretches of DNA located in the vicinity of active chromatin are hypersensitive to nuclease digestion was first made in *Drosophila* heat-shock genes. These HS were subsequently mapped in the neighbourhood of heat-shock elements (HSE) upstream from these genes (Wu, 1984a,b).

The proteins that bind to the heat-shock consensus element have been purified to near homogeneity from yeast and *Drosophila* (Wiederrecht *et al.*, 1987). The proteins, also called heat-shock transcription factors (HSTF), from these two different sources have the same molecular mass (70 kDa). Purified HSTF binds co-operatively to the two contiguous HSEs. *Drosophila* and yeast HSTF occupy the same HSE in *Drosophila* and yeast hsp 70 genes, respectively, suggesting strong evolutionary conservation in HSTF structure. The *Drosophila* HSTF exhibiting similar footprints upon binding to the same HSE as described above has been purified from another laboratory. The molecular mass of this HSTF is about 110 kDa (Wu *et al.*, 1987). Whether these HSTF represent two distinct proteins, or the 70-kDa protein is a degradation product of the 110-kDa protein, is not known.

The binding of HSTF to HSE can generate conformational changes in DNA structure. Using simple gel electrophoretic mobility analyses, Shuey and Parker (1986) demonstrated that HSTF induces DNA bending upon binding to the hsp70 promoter. This is a good example of how conformational changes in the DNase HS can be mediated by the binding of a specific protein factor recognizing a specific DNA sequence.

The hormone responsive genes. Our understanding of the mechanisms of expression of hormone-regulated genes has been greatly improved by the recent success of molecular cloning of hormonal receptor genes (see reviews by Yamamoto, 1985; Green and Chambon, 1986). Sequence comparisons from the cDNA-deduced amino acid sequences revealed domains with conserved sequences in the receptor molecules: The amino acid residues located at the C terminal half of the molecules seem to be involved in hormone binding; and those located about 60 amino acids away from the hormonal binding domain towards the amino terminus are responsible for DNA binding. These two domains, together with the sequences located in between that are hydrophilic in nature, share strong sequence homology with the product of the v-erb-A, suggesting that the receptor genes are in a super-gene family.

Binding of glucocorticoid receptors to the hormonal regulatory elements has been demonstrated in a number of genes. In most cases, binding sites are also indicated by the appearance of DNase I hypersensitive cleavages (Becker *et al.*, 1984; Zaret and Yamamoto, 1984). It is still controversial, however, whether receptor-DNA binding requires the presence of the hormone.

Using a genomic footprint technique, Becker and Gloss (1986) showed that changes in the *in vivo* protein-DNA interactions within the hormonal regulatory element of the tyrosine aminotransferase gene can be detected only after hormone treatment in hepatoma cells. However, Willman and Beato (1986) demonstrated that the DNase I footprints in the vicinity of promoters as generated by the binding of receptors complexed with or without the hormone are identical, suggesting that the specific binding of receptors does not require the ligand. Similar results have been presented by Bailey *et al.* (1986) in the *in vitro* binding assays of progesterone receptor and its specific DNA binding site.

The above described, seemingly conflicting results may suggest complex protein-protein and protein-DNA interactions in the promoter regions of hormonally regulated genes. It is conceivable that transcriptional activation of gene expression by hormones also requires factors for general transcription machinery or so-called basal transcription factors. Recently, two factors that can stimulate the transcription of the chicken ovalbumin gene *in vitro* have been identified in HeLa cells. One factor binds to the chicken ovalbumin gene upstream promoter (COUP) region as revealed by DNase I footprint methods; whereas the other does not have DNA binding activity (Sagami *et al.*, 1986). The second factor was proposed to stimulate transcriptional activity through interactions with other protein factors.

Early studies using DNase I indicated that the nuclease-sensitive domain in the ovalbumin gene in chicken oviduct extends both 5′ and 3′ directions beyond the transcribed region. Both 5′ and 3′ DNase I HS have been located (Kuo *et al.*, 1979; Bellard *et al.*, 1982, 1986; Lawson *et al.*, 1982). The functional implications for most of these HS, especially those located at the 3′ side of the gene are not known. Whether these HS are generated primarily through hormone action is not known. The fact that genomic footprinting patterns showing both receptor binding sites and their neighbouring sequences in the tyrosine aminotransferase gene are affected by hormone actions, suggests a long-range effect on the chromatin structure in hormonally controlled gene expression (Becker and Gloss, 1986). Through these longrange protein-protein interactions as elegantly described in the recent review by Ptashne (1986), the distance of this propagative effect could be quite long.

The developmentally regulated genes. The developmentally regulated globin genes provide a very attractive model system for the study of chromatin structure of active genes. About 20–23 h after fertilization, chicken embryonic genes are turned on (Groudine and Weintraub, 1981). At 12 days, the adult globin genes are transcribed. Moderately DNase I sensitive domains have been detected in the active globin chromatin and HS have been located in the 5′ and 3′ flanking regions of the active chromatin (Weintraub and Groudine, 1976; Groudine and Weintraub, 1981; McGhee *et al.*, 1981).

Emerson and Felsenfeld (1984) reconstituted the chicken adult β-globin gene with histones and 0.15 M NaCl nuclear extract from 9-day-old chicken erythrocytes. The resultant complex displays nuclease hypersensitive behaviour in the same region as it does in the chromatin. When the nuclear extracts from brain or oviduct where the globin gene is not active were used, no such hypersensitive behavour was observed. These observations suggest that non-histone proteins participate in the observed nuclease hypersensitivty. Using nitrocellulose DNA binding assay, these authors calculated the binding constant of specific factor to DNA is about 1.5×10^{12} litres/mole (Emerson *et al.*, 1985). DNase I footprints showed two protection regions in the 5′ hypersensitive region: one located between −176 and −200 contained a GC homopolymer, and the other, bound by nucleotides between −135 and −163, included a long palindromic sequence. Interactions of the binding proteins with these specific DNA sequences can be determined once these protein factors are purified.

The functional implications of the 3′ HS in the chicken adult β-globin gene have been suggested in recent experiments in which deletion mutants containing various lengths of 3′ flanking sequences were transfected into an erythrocyte cell line or primary chicken erythrocytes. Choi and Engel (1986) and Hesse *et al.* (1986) independently observed that an enhancer-like sequence lying within 400 nucleotides 3′ to the polyadenylation signal is involved in determining the erythroid cell specificity of globin gene expression.

Other inducible genes. RNA complexity analyses indicated that a battery of genes can be induced to express in quiescent cells after receiving growth stimulants, such as growth factors and tumour promoter, 12-o-tetradecanoyl phorbol-13-acetate (TPA). One of the earliest genes to be activated by growth factors is the c-fos proto-oncogene, which is the cellular homolog of the transforming gene of FBJ-osteosarcoma virus (Curran *et al.*, 1983). Nuclease digestion of chromatin showed a HS at the regulatory upstream sequences in the serum-starved and serum stimulated mouse NIH 3T3 cells (Renz *et al.*, 1985). Using mobility shift in gel electrophoresis and nuclease protection/methylation interference assays, several laboratories have reported specific factors binding to a 22-bp dyad assigned to −293 bp of the c-fos gene (Gilman *et al.*, 1986; Prywes and Roeder, 1986; Treisman, 1986; Greenberg *et al.*, 1987). The mechanism by which this protein regulates c-fos gene expression can be studied once the protein is purified.

Another interesting inducible gene is the metallothionein (MT) gene. MT is a multigene family consisting of several members in vertebrates (for reviews, see Karin, 1985 and Hamer, 1986). The MT genes are considered 'housekeeping' genes, because of their expression in almost all cells. The exact functional role of these proteins has not been elucidated, although it has been speculated to serve as a detoxifying protein, stress protein, and radical scavenger. The expression of MT genes can be induced in cultured cells and in animals by a variety of treatments, including heavy metals, acute stresses, interferon, and glucocorticoids. The upstream MT gene regulatory sequences involved in these inductions have been determined in transfection expression assays using deletion in the cloned MT gene. The elements that are responsive to heavy metals, glucorticoid, and interferon inductions are non-overlapped (see review by Hamer, 1986); and the induction events seem to be independent to each other (Morris and Huang, 1987). That trans-acting factors are involved in the transcriptional activation of MT gene expression, probably through interactions with cis-acting control elements, has been suggested by the findings that DNase I HS are located at these regions of the active MT genes (Senear and Palmiter, 1982; MacArthur and Liebeman, 1987). A protein factor binding to the metal-inducible element of the MT gene has been suggested by Exo III mapping and gel electrophoretic mobility shift methods. The binding of this putative factor to DNA requires metallions (Hamer, personal communication). The involvements of Spl and Apl factors in the basal level of expression of human MT gene have been reported by Lee *et al.* (1987). Interestingly, phorbol ester induction of expression of the MT^{IIA} and a number of other eukaryotic genes, also involves the binding of Apl to the promoter of this gene (Angel *et al.*, 1987).

Perhaps the best characterized eukaryotic transcription activation systems so far are the yeast GCN4 and GAL4 genes. GCN4 protein is required for transcription induction of a number of amino acid biosynthetic enzymes in response to amino acid starvation; while GAL4 product is responsible for galactose catabolism (see Struhl, 1987 for review). Genes encoding these proteins have been cloned and the functional domains in these proteins have been mapped using genetically engineered mutant proteins. These proteins contain their own DNA-binding domains that recognize the specific upstream activating sequences of the yeast genes. However, DNA-binding and transcriptional activation are separable functions. GCN4 contains a specific DNA-binding domain residing in the 60 C-terminal amino acids and a transcription activation domain of 19 amino acids that is centrally located (Hope and Struhl, 1986). The DNA-binding domain in GAL4 activator is located within the 147 amino acid residues from the N-terminus; whereas two short amino acid sequences, separated by 592 amino acids, are involved in transcriptional activation (Ma and Ptashne, 1987). These functional dissections reveal an intriguing model of DNA-protein and protein-protein interactions in transcription activation in yeast genes (Struhl, 1987).

Other transcription activation protein factors. In addition to the TATA consensus sequences and the GC-box that usually are located at the 5′ proximity of transcription start sites, another sequence motif, the CCAAT box, that is required for promoter function has been identified in a number of cellular genes, including the globin gene, the heat-shock gene, the thymidine kinase gene, and the chicken β2 tubulin gene. A transcription factor CTF that interacts this CCAAT box has been purified by oligonucleotide-affinity chromatography to near homogeneity. CTF purified from HeLa cells consists of a family of proteins with molecular mass ranging from 52 kDa to 60 kDa. By a number of biochemical criteria, CTF has been shown to be identical to the nuclear factor (NF-1) (Jones *et al.*, 1987). NF-1 has been shown to bind with high affinity to a specific sequence element at the adenovirus origin of replication (Rosenfeld and Kelly, 1986) and upstream of several cellular genes such as chicken lysozyme, human IgM, and human c-myc gene. The binding sites of the cellular genes coincide in most cases with DNase I hypersensitive regions. The binding of NF-1 is absolutely required for efficient initiation of adenoviral DNA replication (Gronostajski *et al.*, 1985; Bernstein *et al.*, 1986). These observations raise the possibility that a common protein factor may participate in multiple cellular processes. Incidentally, a NF-III with stimulating adenovirus DNA replication activity has recently been identified to bind at adenovirus origin very close to the NF-1 binding site. Whether NF-III has multiple cellular functions is not known (Pruijn *et al.*, 1986).

Other important cis-acting DNA sequences that can both positively and negatively influence the level of transcription are the enhancer elements. Enhancers were originally identified in viral genome, and subsequently in cellular genes (see reviews by Serfling *et al.*, 1985); and Sassone-Cori and Borrelli, 1986). The locations of these enhancers in most cases are coincident to those of nuclease HS. There is good evidence that protein factors bind to these sequences.

Church *et al.* (1985), using the genomic footprint method, detected protein protection in the immunoglobulin heavy chain enhancer. Sen and Baltimore (1986a), using gel electrophoretic mobility shift assay, identified five protein factors that interact with heavy and *k*-light chain enhancers, factor NF-A which recognizes octonucleotide 5′-ATTTGCAT-3′, and three other factors that recognize specific nucleotide sequences and are ubiquitous. Factor NF-*k*B, which recognizes an 11-bp binding site, was thought to be confined to the k-producing B cells, but was later found to be inducible in phorbol-ester treated nonlymphoid (HeLa) cells (Sen and Baltimore, 1986b). Recently, NF-*k*B has been suggested to be involved in activation of expression of human immunodeficiency virus (Nabel and Baltimore, 1987).

Protein factors bound to viral enhancer elements have been partially purified from HeLa cells (Wildeman *et al.*, 1986) and from rat liver (Johnson *et al.*, 1987). No report for purification of Ig-enhancer binding proteins has been published.

High Order Chromatin Structure of Active Genes

Topoisomerases and the chromatin structure

The topoisomerases are a group of enzymes that catalyse the topological interconversions of DNA. Essentially two types of topoisomerases have been described: the type I topoisomerase introduces a transient single-stranded DNA break followed by passing another strand through the break; the type II enzymes catalyse transient double-stranded breakage and passage of the double-stranded DNA segment through the break. These two types of isomerases have been discovered in both prokaryotic and eukaryotic cells. By these transient DNA breakage-rejoining reactions, supercoiled DNA can be either relaxed or generated (see reviews by Liu 1983; Wang, 1985).

DNA topoisomerases play an important role in gene expression in prokaryotic cells. However, in eurkaryotic cells, the association of topoisomerases with active chromatin is only suggested by indirect evidence. By using immunofluorescent staining of Drosophila polytene chromosomes with rabbit antibodies specific to Drosophila DNA topoisomerase I, Fleischmann *et al.*, demonstrated that this enzyme is preferentially associated with active loci *in vivo.* Using antibody against chicken topoisomerase I, Muller *et al.* (1985) showed that the enzyme is enriched in the nucleus of chicken cells. More direct evidence that topoisomerase is associated with active chromatin was recently published by Gilmour *et al.* (1986). These authors used ultraviolet light to cross-link protein and DNA in living cells, the topoisomerase I-DNA adduct was then immunoprecipitated by antibody against Drosophila topoisomerase I. The precipitated DNA was analysed by blot-hybridization with appropriate DNA probes. It was found that the transcribed sequences (hsp 70 upon heat induction), but not the non-transcribed flanking regions, were enriched in this DNA.

The major chromosome scaffold component, SC1 (170 kDa), has been identified to be topoisomerase II. Using antibody against this protein, as well as that raised against purified topoisomerase II, it was shown that these antibodies react with chromosome scaffolding proteins of the de-histonized metaphase chromosomes (Earnshaw *et al.*, 1985).

The attempt to localize topoisomerase II at gene level was taken mainly by mapping the topoisomerase-II mediated cleavage sites on chromosomal DNA. The mapping strategy was based on the observations that detergent (e.g. sodium dodecyl sulphate) and anti-tumour drugs (e.g. VM-26) can interfere with the strand breakage-rejoining catalytic cycle of the enzyme. Additions of these agents result in cleavage of the DNA and covalent attachment of the enzyme to the cut DNA ends. By this approach, the topoisomerase II cleavage sites have been mapped to non-transcribed spacer segments and close to the 5′ and 3′ boundaries of *Drosophila* heat-shock gene (Udvardy *et al.*, 1985) and Dictyostellium ribosomal RNA genes (Ness *et al.*, 1986). Yang *et al.*, (1985) used a similar approach to *in vivo* map DNA topoisomerase II specific cleavage sites around the DNase I HS on the SV40 chromatin isolated from virus-infected monkey cells. Interestingly, the frequency of topoisomerase II cuttings was significantly increased late in infection, suggesting the association of topoisomerase II with DNA replication.

Topological tension of chromatin and gene activity

The results described above raise the possibility that the transcribing genes might be physically separated by the anchoring of the flanking sequences to a skeletal structure. This could explain how a gene might be activated by relaxation of its chromatin constraint without necessarily influencing its neighbouring genes.

Topological tension of DNA can be generated by DNA supercoiling. DNA gyrase, a type II topoisomerase, is the enzyme responsible for producing negative DNA supercoiling (by reducing the linking number of double-stranded circular DNA) in prokaryotic DNA in an ATP-dependent fashion. The antibiotic novobiocin is known to inhibit gyrase activity by inhibiting its ATPase activity. In eukaryotic cells, however, DNA gyrase-like activity has not been detected.

The indications that topoisomerases affect transcription through their effects on DNA superhelicity have been discussed in a number of reviews (Gellert, 1981; Liu, 1983; Fisher, 1984; 1983; Wang, 1985). It is generally accepted that topological tension of DNA template plays an important role in the control of gene expression in prokaryotes. In eukaryotes, however, the relationship between topological tension of chromatin and transcriptional complex is still not well settled.

The possibility that the supercoiled state of chromatin template influences gene expression in eukaryotic cells has been suggested from the following findings. (1) Several 'single-strand-specific' nuclease Sl cleavage sites have been mapped at the active chromatin region. These presumptive single-stranded regions are thought to be generated due to torsional constraint of chromatin. (2) Circular DNA carrying the thymidine kinase gene injected into *Xenopus* oocytes is transcribed, whereas linearizing the injected DNA by subsequent injection of restriction enzymes abolishes the transcription (Harland *et al.*, 1983); a similar result has been demonstrated in transfection of supercoiled DNA into mammalian cells (Weintraub *et al.*, 1986). (3) A subpopulation (2–5 per cent) of the SV40 minichromosomes that contains transcription complexes appears to be under torsional stress (Luchnik *et al.*, 1982).

In a series of experiments, Worcel and his collaborators showed that *Xenopus* 5S ribosomal RNA gene injected into nuclei of *Xenopus* oocytes is assembled into two distinct types of chromatin: one, called static chromatin, is presumably transcriptionally inactive; and the other, called 'dynamic' chromatin, is transcriptionally permissive. The static chromatin contains conventional nucleosomes that give regular DNA repeat length while the dynamic chromatin contains open nucleosomes that can be split into half-nucleosome DNA repeat lengths upon nuclease digestion (Ryoji and Worcel, 1985). The positive transcription factor TFIII induces 5S DNA-specific supercoiling in a reaction mixture containing oocyte extract supplemented with ATP and Mg^{++} (Kmiec and Worcel, 1985; Kmiec *et al.*, 1986). The supercoiling reaction can be inhibited by a relatively high concentration (0.25 mg/ml) of novobiocin. These studies led to a postulation that dynamic (active) chromatin is assembled under topological tension.

Mutations in the genes for DNA topoisomerases I and II have been identified in yeasts. These mutants have provided good insight into the physiological roles of topoisomerases in lower eukaryotes. Top I mutants exhibit no obvious effects on cell growth, suggesting that topoisomerase I is not essential for viability. Top II mutants, however, show genetic defects in completing DNA replication and the segregation of daughter chromosomes (see review by Wang, 1985). Recent result show that top I top II double mutants display drastic defects on DNA replication and rRNA synthesis; however the effects on poly (A) + RNA and transfer RNA synthesis in these double mutants are not as significant as that for the rRNA synthesis (Uemura and Yanagida, 1986; Brill *et al.*, 1987). Defects in either one of the two enzymes produces no such effect, suggesting that DNA replication and at least ribosomal RNA synthesis only require an active topoisomerase which presumably acts as a swivel to relieve torsional stress.

The results discussed above suggest an involvement of topoiosomerase in active chromatin. However, recent results published by Petryniak and Lutter (1987) suggested that the SV40 transcription complex contains no detectable torsional tension. These authors showed that when the transcription complex prepared from SV40-infected cells was treated with topoisomerase I or DNase I, no change in mobility of the complex was detected in sucrose gradient and gel electrophoresis. If the complex has been under torsional stress, these treatments would have relaxed the topology of the template and caused the mobility shift.

The association of negative supercoiling in transcriptionally active 5S RNA chromatin assembled by *Xenopus* oocyte extract described by Worcel and associates and mentioned above has been questioned recently (Wolffe *et al.*, 1987).

Chromatin domain structure and gene expression

How the chromatin fibres inside the nucleus are ordered so that proper nuclear physiology such as DNA replication, gene transcription, and RNA processing can be regulated, is an important subject in the study of chromatin physiology. Cytogenetic observations under light microscopy often reveal the 'aggregation' of constitutive heterochromatin assigned to the centromeric areas of chromosomes. As for euchromatin, especially the active chromatin, the same idea of compartmentalized chromatin domains cannot be easily refuted.

Efforts to study ordered chromatin structure have resulted in the observations that the chromatin fibre of interphase nuclei is constrained in closed loops. The size of these looped chromatin fibres range from a few kb to 220 kb. The bases of these loops are thought to attach to a framework inside the nucleus, variously called the nuclear matrix, nucleoid, nuclear cage, or nuclear scaffold. Over the years, nuclear matrix has been proposed to play important functions in nucleus, such as DNA replication and transcription (for review, see Eissenberg *et al.*, 1985).

While the idea that nuclear matrix may serve as a framework for the internal structure of the interphase nucleus is reasonable, its exact organization inside the nucleus is still unknown. A high molar salt-, nuclease-, and detergent-resistant subnuclear material has most often been referred to as nuclear matrix. Whether the enrichment of RNA transcripts in this nuclear matrix fraction is due to contamination of transcription complex is not known because the complex is resistant to treatments with high molar salts and detergents (Kuo, 1982).

Perhaps the best characterization of matrix

(scaffold) from interphase nuclei and metaphase chromosomes has been done by Laemmli and his co-workers (see review, Gasser and Laemmli, 1987). According to these authors, the scaffold is made up of some 20 30 non-histone protein to which a halo of looped DNA ranging from 30 to 90 kb is attached. The scaffolding structures are dissociable by metal chelators and reducing agents, suggesting the involvement of metal ions and disulphide linkages in the scaffolding structures. Two high-molecular mass proteins of 170 kDa and 135 kDa were proposed to be primarily involved in the metalloprotein interactions. The 170 kDa scaffolding protein has been identified to be topoisomerase II (Earnshaw *et al.*, 1985; Gasser *et al.*, 1986). Analysis of DNA-scaffolding protein interactions resulted in identification specific DNA sequences attached to the nuclear scaffold. Within 320 kb of a genetically well-defined region of the *Drosophila* chromosome, Mirkovitch *et al.* (1986) showed that four putative attachment sites are present; these attachment sites define looped DNA containing at least one and up to several genetic units each. The loop structure is not correlated with the distribution of chromomeres in salivary gland chromosomes. In the studies with some well-defined genes, scaffold attachment DNA tend to be found close to the promoter elements, and the upstream/enhancer like elements in *Drosophila* developmentally regulated genes (*Adh*, *Sqs-4*, and *ftz*). The locations of the putative attachment DNA to the neighbourhood of enhancer DNA containing topoisomerase II consensus sequences have been also found in the mouse immunoglobulin heavy chain and *k*-light chain loci (Cockerill and Garrard, 1986; Cockerill *et al.*, 1987). However, not all the putative attachment DNA have sequences exhibiting a strong homology to the topoisomerase II consensus cleavage sequences (Cockerill *et al.*, 1987).

Our knowledge about the higher-order chromatin domain beyond the looped structure is still very scanty. The evidence that each interphase chromosome occupies a specific territory inside the nucleus has been provided by information from reconstruction of optical-sectioning fluorescent microscopy (Agar and Sedat, 1983). The territorial occupancy of interphase chromatin has been suggested to be tissue-specific (Manuelidis, 1984). Thus, one can imagine that the active chromatin in one cell type may be in different compartments from the other. Using nick-translation of interphase nuclei, Hutchinson and Weintraub (1985) labelled active chromatin in the nuclear periphery of mouse L cells whereas the labelled chromatin is at the borders of condensed chromatin masses along interchromatin channels. How the chromosomes 'talk' to each other and become reoriented in different tissues remains to be determined.

Conclusion

We describe in this chapter some selected topics on chromatin structure and gene expression. Over the past decade, recombinant DNA technologies have revolutionized the concept of eukaryotic genes. We now know in great detail the structure and function of structural genes in eukaryotic cells. The *cis*-acting elements that are responsible for expression of these genes have been identified, and a number of *trans*-acting factors that interact with these elements have been purified. We can envision that in the near future genes encoding these factors will be cloned and circuitry of gene regulations will be better defined.

In order to fully understand how a gene is functioning in eukaryotic cells, one can expect that the studies will have to be eventually done at the chromatin level, either in a reconstituted in vitro system or in a relatively simpler in vivo system (e.g. animal viral or yeast systems). From these studies, one would expect to gain insights into how other chromosomal proteins (nonhistone proteins, RNA polymerases and histones) come into play in regulating gene expression in tissue-specific and temporal manners.

The research leading to understanding higher-order chromatin structure of active genes is a challenge and still has a long way to go. In a mammalian nucleus that contains such an enormous amount of DNA and proteins in weight, it is always a big concern that experimental artifacts could be produced when one tries to 'open it up' for studying its interior structure. Perhaps this area of research needs a major advance, much like the discovery of nucleosome in the 1970s. The studies on chromatin looped domains and its scaffolding entities have offered some impressive new insights on this subject. The story beyond this remains to be seen.

Acknowledgements

We are indebted to our colleagues who supplied preprints or unpublished information of their works to Walter Pagel and Larry Teeter. We are also indebted to Walter Pagel and Larry Teeter for reading the manuscript and to Linda Jackson and Tammy Trlicek for their secretarial help. Our research and time spent on this review has been supported by the Robert A. Welch Foundation (G 831) and the National Institute of General Medical Sciences (GM28573).

References

Agard DA, Sedat JW (1983). Three dimensional architecture of a polytene nucleus. Nature 302; 676–81.

Angel P, Imagawa M, Chiu R, Stein B, Imbra RJ, Rahmsdorf HJ, Jonat C, Herrlich P, Karin M (1987). Phorbol ester inducible genes contain a common cis element recognized by a TPA-modulated trans-acting factor. Cell 49; 729–39.

Angel P, Allegretto EA, Okino ST, Hattori K, Boyle WT, Hunter T, Karin M (1988). Oncogene jun encodes a sequence-specific transactivator similar to AP-1. Nature 232; 166–70.

Azorin F, Rich A (1985). Isolation of Z-DNA binding proteins from SV40 minichromosomes: Evidence for binding to the viral control region. Cell 41; 365–74.

Bailey A, LePage C, Rauch M, Milgrom E (1986). Sequence specific DNA binding of the progesteron receptor to the uteroglobin gene effects of hormone antihormone and receptor phosphorylation. EMBO J. 5; 3235–41.

Becker PB, Gloss B (1986). In vivo protein-DNA interactions in a glucocorticoid responsive element require the presence of the hormone. Nature 324; 686–8.

Becker PB, Renkawitz R, Schultz G (1984). Tissue-specific DNase I hypersensitive sites in the 5′-flanking sequences of the tryptophan oxygenase and the tyrosine aminotransferase genes. EMBO J. 3; 2015–20.

Bellard M, Dretzer G, Bellard F, Oudet F, Chambon P (1982). Disruption of the typical chromatin structure in a 2500 base-pair region at the 5′ end of the actively transcribed ovalbumin gene. EMBO J. 1; 223–30.

Bellard M, Dretzer G, Bellard F, Kaye JS, Pratt-Kaye S, Chambon P (1986). Hormonally induced alterations of chromatin structure in the polyadenylation and transcription termination regions of the chicken ovalbumin gene. EMBO J. 5; 567–74.

Bellard M, Gannon F, Chambon P (1977). Nucleosome structure III: The structure and transcriptional activity of the chromatin containing the ovalbumin and globin genes in chick oviduct nuclei. Cold Spring Harbor Symp. Quant. Biol. 42; 779–91.

Berk AT, Sharp PA (1977). Sizing and mapping of early adenovirus mRNA by gel electrophoresis of S1 endonuclease digested hybrids. Cell 12; 721–32.

Bernstein JA, Porter HM, Challberg MD (1986). Template requirements for in vivo replication of adenovirus DNA. Mol. Cell. Biol. 6; 2115–24.

Bloom KS, Anderson JN (1979). Conformation of ovalbumin and globin genes in chromatin during differential gene expression. J. Biol. Chem. 254; 10532–9.

Bogenhagen DF, Shakonju S, Brown DD (1980). A control region in the center of the 5S RNA gene directs specific initiation of transcription II: The 3′ border of the region. Cell 19; 27–35.

Bohmann D, Bos TJ, Admon A, Nishmura T, Vogt PK, Tjian R (1987). Human proto-oncogene c-jun encodes a DNA binding protein with structural and functional properties of transcription factor Ap-1. Science 238; 1386–92.

Brill SJ, DiNardo S, Voelk-Meiman K, Sternglanz R (1987). Need for DNA topoisomerases activity as a swivel for DNA replication for transcription of ribosomal RNA. Nature 326; 414–6.

Burch JBE, Weintraub H (1983). Temporal order of

chromatin structural changes associated with activation of the major chicken vitellogenin gene. Cell. 33; 65–76.

Cartwright IL, Keene MA, Howard GC, Abmayr SM, Fleischmann G (1982). Chromatin structure and gene activity: the role of nonhistone chromosomal proteins. CRC Crit. Rev. Biochem. 13; 1–86.

Choi OR, Engel JD (1986). A 3′ enhancer is required for temporal and tissue-specific transcriptional activation of the chicken adult β-globin gene. Nature 323; 731–4.

Church G, Gilbert W (1984). Genomic sequencing. Proc. Natl. Acad. Sci. USA 81; 1991–6.

Church GM, Ephrussi A, Gilbert W, Tonegawa S (1985). Cell-type-specific contacts to immunoglobulin enhancers in nuclei. Nature 313; 798–801.

Cockerill PN, Garrard WT (1986). Chromosomal loop anchorage of the kappa immunoglobulin gene occurs next to the enhancer in a region containing topoisomerase II sites. Cell 44; 273–82.

Cockerill PN, Yuen M-H, Garrard WT (1987). The enhancer of the immunoglobulin heavy chain locus is flanked by presumptive loop anchorage elements. J Biol Chem 262; 5394–7.

Costlow NA, Simon JA, Lis JT (1985). A hypersensitive site in hsp70 chromatin requires adjacent not internal DNA sequences. Nature 313; 147–9.

Curran T, MacConnell WP, Van Straaten F, Verma IM (1983). Structure of the FBJ murine osteosarcoma virus genome: molecular cloning of its associated helper virus and the cellular homology of the c-fos gene from mouse and human cells. Mol. Cell. Biol. 3; 914–21.

Dynam W, Tjian R (1985). Control of eukaryotic messenger RNA synthesis by sequence specific DNA binding proteins. Nature 316; 774–8.

Earnshaw WC, Halligan B, Cooke CA, Heck MMS, Liu LF (1985). Topoisomerase II is a component of mitotic chromosome scaffolds. J. Cell. Biol. 100; 1706–16.

Eissenberg JC, Cartwright IL, Thomas GH, Elgin SCR (1985). Selected topics in chromatin structure. Ann Rev Genet 19; 485–536.

Elgin SCR (1981). DNase I hypersensitive sites of chromatin. Cell 27; 413–5.

Emerson BM, Felsenfeld G (1984). Specific factor conferring nuclease hypersensitivity at the 5′ end of the chicken adult β-gene globin gene. Proc. Natl. Acad. Sci. USA 81: 95–9.

Emerson BM, Lewis CD, Felsenfeld G (1985). Interactions of specific nuclear factors with the nuclear hypersensitive region of the chicken adult β-gene: Nature of the binding domain. Cell 41; 21–30.

Evans T, Efstratiadis A (1986). Sequence-dependent S1 nuclease hypersensitivity of a heteronomous DNA duplex. J. Biol. Chem. 261; 14771–80.

Fisher LM (1984). DNA supercoiling and gene expression. Nature 307; 686–7.

Fleischmann G, Pflugfelder G, Steiner EK, Javanherian K, Howard GC, Wang JC, Elgin SCR (1984). Drosophila DNA topoisomerase I is associated with transcriptionally active region of the genome. Proc. Natl. Acad. Sci. (USA) 81; 6958–62.

Fritton HP, Igo-Kemenes T, Nowock J, Strech-Jurk U, Theisen M, Sippel AE (1984). Alternative sets of DNase I hypersensitive sites characterize the various functional states of the chicken lysozyme gene. Nature 311; 163–5.

Garel A, Axel R (1976). Selective digestion of transcriptionally active ovalbumin genes from oviduct nuclei. Proc. Natl. Acad. Sci. USA 73; 3966–70.

Gasser SM, Laemmli UK (1987). A glimpse at chromosomal order. Trends Genet 3; 16–22.

Gasser SM, Laroche T, Falquet J, de la Tour EB, Laemmli UK (1986). Metaphase chromosome structure, involvement of topoisomerase II. J Mol Biol 188; 613–29.

Gazit B, Cedar H, Lerer I, Voss R (1982). Active genes are sensitive to deoxyribonuclease I during metaphase. Science 217; 648–50.

Gellert M (1981). DNA topoisomerases. Ann Rev Biochem 50; 879–910.

Gidoni D, Kadonaga JT, Barrera-Saldana H, Takahashi K, Chambon P, Tjian R (1985). Bidirectional SV40 transcription mediated by random Sp1 binding interactions. Science 230; 511–17.

Gilman MZ, Wilson RN, Weinberg RA (1986). Multiple protein-binding sites in the 5′-flanking region regulate c-fos expression. Mol Cell Biol 6; 4305–16.

Gilmour DS, Pflugfelder G, Wang JC, Lis JT (1986). Topoisomerase I interacts with transcribed regions in Drosophila. Cell 44; 401–7.

Green S, Chambon P (1986). A superfamily of potentially oncogenic hormone receptors. Nature 324; 615–7.

Greenberg ME, Siegfried Z, Ziff EB (1987). Mutation of the c-fos gene dyad symmetry element inhibits serum inducibility of transcription in vivo and the nuclease regulatory factor binding in vitro. Mol Cell Biol 7; 1217–25.

Gronostajsk RM, Adhya S, Nagata K, Guggenheimer RA, Hurwitz CG (1985). Site-specific DNA binding of nuclear factor 1: Analyses of cellular binding sites. Mol. Cell. Biol. 5; 964–71.

Groudine M, Weintraub H (1981). Activation of globin genes during chicken developments. Cell. 24; 393–401.

Groudine M, Weintraub H (1982). Propagation of globin DNase I hypersensitive sites in absence of factors required for induction: A possible mechanism for determination. Cell 30; 131–9.

Grummet I, Kuhn A, Bartsch I, Rosenbauer H (1986). A transcription terminator located upstream of the mouse rDNA initiation site affects rRNA synthesis. Cell. 47; 901–11.

Hamer DH (1986). Metallothionein. Annu Rev. Biochem 55; 913–51.

Harland RM, Weintraub H, McKnight SL (1983). Transcription of DNA injected into Xenopus oocytes is influenced by template topology. Nature 302; 38–43.

Henderson S, Sollner-Webb B (1986). A transcriptional termination is a novel element of the promoter of the mouse ribosomal RNA gene. Cell 47; 891–900.

Hesse JE, Nickol JM, Leiber MR, Felsenfeld G (1986). Regulated gene expression in transfected primary chicken erythrocytes. Proc. Natl. Acad. Sci. USA 83; 4312–16.

Hope IA, Struhl K (1986). Functional dissection of a eukaryotic transcriptional activator protein, GCN4 of yeast. Cell 46; 885–94.

Hutchinson N, Weintraub H (1985). Localization of DNase I-sensitive sequences to specific regions of interphase nuclei. Cell 43; 471–82.

Igo-Kemenes T, Zachau HG (1982). Chromatin. Ann. Rev. Biochem 51; 89–121.

Jackson PD, Felsenfeld G (1985). A method for mapping intranuclear protein-DNA interactions and its application to a nuclease hypersensitive site. Proc. Natl. Acad. Sci. (USA) 82; 2296–300.

Johnson PF, Landschulz WH, Graves BJ, McKnight SL (1987). Identification of a rat liver protein that binds to the enhancer core element of three animal viruses. Genes Develop, 1; 133–46.

Jones KA, Kadonaga JT, Rosenfeld PJ, Kelley TJ, Tjian R (1987). A cellular DNA-binding protein that activates eukaryotic transcription and DNA replication. Cell. 48; 79–89.

Jongstra J, Reudelhuber TL, Oudet P, Benoist C, Chae C-B, Jeltsch JM, Mathis DJ, Chambon P (1984). Induction of altered chromatin structures by simian virus 40 enhancer and promoter elements. Nature 307; 708–14.

Kadonaga JT, Jones KA, Tjian R (1986). Promoter specific activation of RNA polymerase II transcription by Sp l. Trends Biochem Sci. 11; 20–3.

Kadonaga JT, Carner KR, Maxiarz FR, Tjian (1987). Isolation of cDNA encoding transcription factor Spl and functional analysis of the DNA binding domain. Cell. 51; 1079–90.

Karin M (1985). Metallothioneins: Proteins in search of function. Cell. 41; 9–10.

Karpov VL, Preobrazhenskaya OV, Mirzabekou AD (1984). Chromatin structure of hsp 70 genes, activated by heat-shock: selective removal of histones from the coding region and their absence from the 5′ region. Cell. 36; 423–31.

Käs E, Chasin LA (1987). Anchorage of the Chinese-hamster dihydrofolate reductase gene to the nuclear scaffold occurs in an intragenic region. J. Mol. Biol. 198; 677–90.

Kaye JS, Bellard M, Dretzer G, Bellard F, Chambon P (1984). A close association between sites of DNase I hypersensitivity and sites of enhanced cleavage by micrococcal nuclease in the 5′ flanking region of the actively transcribed ovalbumin gene. EMBO J. 3; 1137–44.

Kerem B-S, Goitein R, Richler C, Marcus M, Cedar H (1983). In situ nick translation distinguishes between active and inactive X chromosomes. Nature 304; 88–90.

Kmiec EB, Worcel A (1985). The positive transcription factor of the 5S RNA gene induces a 5S DNA-specific gyration in Xenopus oocyte extracts. Cell. 41; 945–53.

Kmiec EB, Ryoji M, Worcel A (1986). Gyration is required for 5S RNA transcription from a chromatin template. Proc. Natl. Acad. Sci. USA 83; 1305–9.

Kuo MT (1981). Preferential damage of active chromatin by bleomycin. Cancer Res. 41; 2439–43.

Kuo MT (1982). Distribution of tightly bound proteins in the chicken ovalbumin gene. Biochem 21; 321–6.

Kuo MT, Plunkett W (1985). Nick translation of metaphase chromosomes; in vitro labeling of nuclease hypersensitive regions in chromosomes. Proc. Natl. Acad. Sci. USA 82; 854–8.

Kuo MT, Mandel JL, Chambon P (1979). DNA methylation: correlation with DNase I sensitivity of chicken ovalbumin and conalbumin chromatin. Nucl. Acids Res. 7; 2015–113.

Kuo MT, Iyer B, Schwartz RJ (1982). Condensation of chromatin into chromosomes preserves an open configuration but alters the DNase I hypersensitive cleavage sites of the transcribed gene. Nucl. Acids Res. 10; 4565–79.

Lawson GM, Knoll BJ, March CJ, Woo SHC, Tsai MJ and O'Malley BW (1982). Definition of 5′ and 3′ structural boundaries of the chromatin domain containing the ovalbumin multigene family. J. Biol. Chem. 256; 1501–7.

Lee W, Haslinger A, Karin M, Tjian R (1987). Activation of transcription by two factors that bind promoter and enhancer sequences of the human metallothionein gene and SV 40. Nature 325; 368–72.

Liu LF (1983). DNA topoisomerases—enzymes that catalyze the breaking and rejoining of DNA. CRC Crit Rev. Biochem 15; 1–24.

Lorch Y, Lapointe JW, Kornberg RD (1987). Nucleosomes inhibit the initiation of transcription but allow chain elongation with the displacement of histones. Cell. 49; 203–10.

Luchnik AN, Bakayev VV, Zbarsky IB, Georgiev GP (1982). Elastic torsional strain in DNA within a fraction of SV 40 minichromosomes: relation to transcriptionally active chromatin. EMBO J. 1; 1353–8.

Ma J, Ptashne M (1987). Deletion analysis of GAL4 defines two transcriptional activating segments. Cell 48; 847–53.

MacArthur CA, Liberman MW (1987). Different types of

hypersensitive sites in the mouse metallothionein gene region. J. Biol. Chem. 262; 2161–5.

Manuelidis L (1984). Different central nervous system cell types display distinct and non random arrangements of satellite DNA sequences. Proc. Natl. Acad. Sci. 81; 3123–7.

Mathis D, Oudet P, Chambon P (1980). Structure of transcribing chromatin. Prog. Nucl. Acid. Res. Mol. Biol. 24; 1–55.

McGhee JD, Felsenfeld G (1980). Nucleosome structure. Ann. Rev. Biochem 49; 1115–56.

McGhee JD, Wood WI, Dolan M, Engel JD, Felsenfeld G (1981). A 200 base pair region at the 5′ end of the chicken adult β-globin gene is accessible to nuclease digestion. Cell 27; 45–55.

McGinnis W, Shermoen AW, Heemskerk J, Beckendorf SK (1983). DNA sequence changes in an upstream DNase I hypersensitive region are correlated with reduced gene expression. Proc. Natl. Acad. Sci. USA 80; 1063–7.

McStay B, Reeder R (1986). A termination site for Xenopus RNA polymerase also acts as an element of an adjacent promoter. Cell. 47; 913–20.

Mirkovitch J, Spierer P, Laemmli UK (1986). Genes and loops in 320,000 base-pairs of the Drosophila melanogaster chromosome. J. Mol. Biol. 190; 255–8.

Morris S, Huang PC (1987). Transient response of amplified metallothionein genes in CHO cells to induction by alpha interferon. Mol. Cell. Biol. 7; 600–5.

Muller BC, Raphael AL, Barton JK (1987). Evidence for altered DNA conformations in the simian virus 40 genomic site-specific DNA cleavage by the chiral complex-A-tris (4,7-diphenyl-1, 10-phenanthroline) cobalt (II). Proc Natl Acad Sci USA 84; 1764–8.

Muller MT, Pfund WP, Mehta VB, Trask DK (1985). Eukaryotic type 1 topoisomerase is enriched in the nucleolus and catalytically active in ribosomal DNA. EMBO J 4; 1237–43.

Muller UR, Wilson CL (1987). The effect of supercoil and temperature on the recognition of palindromic and non-palindromic regions in 6×174 replicative form DNA by S1 and Bal 31. J Biol Chem 262; 3730–8.

Myers RM, Tilly K, Maniatis T (1986). Fine structure genetic analysis of a β-globin promoter. Science 232: 613–18.

Nabel G, Baltimore D (1987). An inducible transcription factor activates expression of human immunodeficiency virus T cells. Nature 326; 711–13.

Nedospasov SA, Georgiev GP (1980). Non random cleavage of SV40 DNA in the compact minichromosome and face solution by micrococcal nuclease. Biochem Biophys Res Commun 92; 532–9.

Ness PJ, Parish RW, Koller TH (1986). Mapping of endogenous nuclease-sensitive regions and of putative topoisomerase sites of action along the chromatin of Dictyostellium ribosomal RNA genes. J. Mol. Biol. 188; 287–300.

Pederson DS, Thoma F, Simpson RT (1986). Core particle, fiber, and transcriptionally active chromatin structure. Ann. Rev. Cell. Biol. 2; 117–47.

Petryniak B, Lutter LC (1987). Topological characterization of the simian virus 40 transcription complex. Cell 48; 289–95.

Prescott DM (1970). The structure and replication of eukaryotic chromosomes. Adv. Cell Biology; 57–117.

Pruijn GJM, Van Driel W, Van der Vliet PC (1986). Nuclear factor III, a novel sequence-specific DNA-binding protein from HeLa cells stimulating adenovirus DNA replication. Nature 322; 656–9.

Prywes R, Roeder RG (1986). Inducible binding of a factor to the c-fos enhancer. Cell 47; 777-84.

Ptashne M (1986). Gene regulation by proteins acting nearby and at a distance. Nature 322; 697–701.

Radice G, Constantine F (1986). Tissue specific DNase I hypersensitive sites in a foreign globin gene in transgenic mice. Nucl. Acids Res. 14; 9765–80.

Reeves R (1984). Transcriptionally active chromatin. Biochem Biophys Acta 782; 343–93.

Renz M, Neuberg M, Kurz C, Bravo R and Miller R (1985). Regulation of c-fos transcription in mouse fibroblasts: identification of DNase I hypersensitive sites and regulatory upstream sequences. EMBO J. 4; 3711–6.

Richmond JJ, Finch JT, Rushton B, Rhodes D, Klug A (1984). Structure of the nucleosome core particle at 7Å resolution. Nature 311; 532–7.

Rosenfeld PJ, Kelley TJ (1986). Purification of NF-1 by DNA recognition site affinity chromatography. J. Biol. Chem. 261; 1398–408.

Ryoji M, Worcel A (1985). Structure of the two distinct types of minichromosomes that are assembled on DNA injected in Xenopus oocytes. Cell 40; 923–32.

Sagami T, Tsai S, Wang H, Tsai M-J, O'Malley BW (1986). Identification of two factors required for transcription of the ovalbumin gene. Mol. Cell. Biol. 4; 4259–67.

Sassone-Corsi P, Borrelli E (1986). Transcriptional regulation by transacting factors. Trends Genet 2; 215–9.

Sen R, Baltimore D (1986a). Multiple nuclear factors interact with the immunoglobulin enhancer sequences. Cell. 46; 705–16.

Sen R. Baltimore D (1986b). Inducibility of k immunoglobulin enhancer-binding protein NF-kB by a posttranslational mechanism. Cell 47; 921–8.

Sen S, Siciliano MJ, Johnston DA, Schwartz RJ, Kuo T (1985). The expression and chromatin structure of the chicken Glyceraldehyde-3-phosphate dehydrogenase gene in mouse cells. J. Biol. Chem. 260; 3071–8.

Senear AW, Palmiter RD (1982). Expression of the mouse

metallothionein-I-gene alters the nuclease hypersensitivity of its 5′ regulatory region. Cold Spring Harbor Symp. Quant. Biol. 47; 539–47.

Serfling E, Jasin M, Schaffner W (1985). Enhancers and eukaryotic gene transcription. Trends Genet. 1; 224–30.

Shuey DJ, Parker CS (1986). Bending of promoter DNA on binding of heat shock transcription factor. Nature 323; 459–61.

Sollner-Webb B, Tower J (1986). Transcription of cloned eukaryotic ribosomal RNA genes. Ann Rev Biochem 55; 801–30.

Stadler J, Larsen A, Engel JD, Dolan M, Groudine M, Weintraub H (1980). Tissue specific DNA cleavages in the globin chromatin domain introduced by DNase I. Cell 20; 451–60.

Struhl K (1987). Promoters, activator proteins, and the mechanism of transcriptional initiation of yeast. Cell 49; 295–7.

Sun YL, Pao CI, Sen S, Kuo T (1987). High resolution mapping of DNase I hypersensitive site in the chicken β tubulin gene transfected into monkey cells. J. Biol. Chem. 262, 7764–9.

Takahashi K, Vigneron M, Mathis H, Wildemann A, Zenke M, Chambon P (1986). Requirement of stereospecific alignments for initiation from the simian virus 40 early promoter. Nature 319; 121–6.

Thoma F, Koller T, Klug A (1979). Involvement of histone Hl in the organization of the nucleosome and of the salt dependent superstructures of chromatin. J Cell Biol 83; 403–27.

Tooze J (1982). DNA tumor viruses. Cold Spring Harbor Laboratory, Cold Spring Harbor, N.Y.

Treisman R (1986). Identification of a protein-binding site that mediates transcriptional response of the c-fos gene to serus factors. Cell 46; 567–74.

Udvardy A, Schedl P, Sander M, Hsieh T (1985). Novel partitioning of DNA cleavage sites for Drosophila topoisomerase II. Cell 40; 933–41.

Uemura T, Yanagida M (1986). Mitotic spindle pulls but fails to separate chromosomes in type II DNA topoisomerase mutants. Uncoordinated mitosis. EMBO J. 5; 1003–10.

Vigneron M, Barrera-Saldana HA, Baty D, Everett RE, Chambon P (1984). Effect of the 21-bp repeat upstream element on in vitro transcription from the early and late SV40 promoters. EMBO J 3; 2373–82.

Wang JC (1985). DNA topoisomerases. Ann Rev Biochem 54; 665–97.

Weintraub H (1983). A dominant role for DNA secondary structure in forming hypersensitive structures in chromatin. Cell. 32; 1191–203.

Weintraub H (1984). Histone Hl dependent chromatin superstructure and the suppression of gene activity. Cell 38; 17–27.

Weintraub H (1985). High resolution mapping of S1 and DNase I hypersensitive sites in chromatin. Mol. Cell. Biol. 5; 1538–9.

Weintraub H, Groudine M (1976). Chromosomal subunits in active genes have an altered conformation. Science 93; 848–56.

Weintraub H, Cheng PF, Conrad K (1986). Expression of transfected DNA depends on DNA topology. Cell. 46; 115–22.

Weisbrod S (1982). Active chromatin. Nature 297: 269–95.

Wiederrecht G, Shuey DJ, Kibbe WA, Parker CS (1987) The *Saccharomyces* and *drosophila* heat shock transcription factors are identical in size and DNA binding properties. Cell 48; 507–15.

Wildeman AG, Zenke M, Schatz C, Wintzerith M, Grundstrom T, Mathis H, Takahashi K, Chambon P (1986). Specific protein binding to the simian virus 40 enhancer in vitro. Mol. Cell. Biol. 6; 2098–105.

Willmann T, Beato M (1986). Steroid-free glucocorticoid receptor binds specifically to mouse mammary tumor virus DNA. Nature 324; 688–91.

Wolffe AP, Andrews MT, Crawford E, Losa R, Brown DD (1987). Negative supercoiling is not required for ssRNA transcription in vitro. Cell 49; 301–3.

Wu C (1980). The 5′ ends of Drosophila heat-shock genes in chromatin are hypersensitive to DNase I. Nature 286; 854–60.

Wu C (1984a). Activating protein factor binds in vitro to upstream control sequences in heat-shock gene chromatin. Nature 311; 81–4.

Wu C (1984b). Two protein-binding sites in chromatin implicated in the activation of heat-shock genes. Nature 309; 229–34.

Wu C, Gilbert W (1981) Tissue-specific expression of chromatin structure at the 5′ terminus of the preproinsulin II gene. Proc. Natl. Acad. Sci USA 78:1577–80.

Wu C, Wilson S, Walker B, David I, Paisley T, Zimarino V, Ueda H (1987). Purification and properties of *Drosophila* heat shock activator protein. Science 238; 1247–54.

Yamamoto KR (1985). Steriod receptor regulated transcription of specific genes and gene networks. Ann Rev Genet 19; 209–52.

Yang L, Rowe TC, Nelson EM, Liu LF (1985). In vivo mapping of DNA topoisomerase II specific cleavage sites on SV40 chromatin. Cell 41; 127–32.

Zaret KS, Yamamoto K (1984). Reversible and persistent changes in chromatin structure accompanying activation of a glucocorticoid-dependent enhancer element. Cell 38; 29–38.

4 Contents of G- and R-bands Defy Contemporary Paradigms

GERALD P. HOLMQUIST

Division of Biology
Beckman Research Institute of the City of Hope
Duarte, CA 91010, USA

Introduction

R-bands and G-bands of mammalian chromosomes differ in replication time and base composition. These two physical differences allow the separation of R- from G-band DNA by means of both base composition fractionation and replication time fractionation. Results from the two techniques are complementary (Bernardi *et al.*, 1985; Goldman *et al.*, 1984; Holmquist and Caston, 1986) as reviewed (Holmquist, 1987, 1988a,b), and summarized herein. A band is a unit of chromosome replication and its DNA is relatively homogeneous in base composition. Bands reflect a compartmentalization of the mammalian genome and at least eight different compartment classes can be identified by their different base compositions. The GC-richest compartment classes correspond to early replicating DNA (R-band DNA) while the AT-richest compartment classes correspond to late replicating DNA (G-band DNA). R-band and G-band DNA are unbelievably different. For example: a gene's codon usage is determined mainly by its band location and not by its function. Q-banding reflects the fact that non-coding DNA is constrained in base composition and implies that this DNA cannot be functionless junk. However, the base composition differences between R- and G-band DNA evolved recently and should have placed an impossible high mutational load on the population if they were functional. New dogma is required to explain these contradictions. The questions are posed in this review to stimulate thought.

What is a Band?

As recently reviewed and referenced (Holmquist, 1988a), a chromosome reflects one long DNA molecule folded and packaged in a linear manner so that the DNA molecule maps 1:1 onto the chromosome at the resolution of the light microscope. There are a total of about 2,000 bands (Yunis, 1981), R-bands and G-bands, in euchromatin; each band contains a stretch of about 1,300 kb of DNA.

A band is a unit of chromosome replication. It is a cluster of about 7–25 contiguous replicons which simultaneously initiate and terminate DNA synthesis during a fraction of the DNA synthesis phase, S-phase, of the cell cycle. This was determined from results of fibre autoradiography (Hand, 1978), premature chromosome condensation during S-phase (Lau and Arrighi, 1981; Hameister and Sperling, 1984) and replication banding. All or almost all autosomal R-bands initiate and terminate replication in the early half of S-phase before G-bands initiate replication (Camargo and Cervenka, 1982; Schmidt, 1980). Thus, R- and G-bands are units of chromosome replication whose replication times are non-overlapping. Kowalski and Cheevers (1976) showed that very long, at least 500–1000 kb, multi-replicon sized stretches of DNA replicate during a limited part of S-phase; DNA labelled in late S-phase was neither interspersed between nor covalently associated with 515 kb DNA segments labelled early in S-phase. Thus, the genome is a mosiac of replicon clusters.

Band DNA Fractionation

Replication time fractionation of these replicon clusters is usually accomplished by using synchronized cells. Mitotic cells are released into G1, labelled in the first half of S-phase with the thymidine analog bromodeoxyuridine (BrdUrd), and this is replaced with thymidine for replication in the last half of S-phase. The subsequent mitotic cells are harvested and some have their chromosomes stained with 33258 Hoechst to insure by replication banding that the BrdUrd has been incorporated into and quenches the Hoechst fluorescence of R-bands. High molecular weight mitotic DNA is extracted from the remaining cells and fractionated in CsCl gradients wherein the denser, early replicating, BrdUrd substituted-DNA is separated from the unsubstituted, late replicating DNA (Holmquist *et al.*, 1982). After digestion with restriction endonucleases, electrophoresis, Southern transfer, and probing of the transfers with radioactive gene probes, one can determine from resulting autoradiographs if the probed signal is stronger in the early or in the late replicating DNA fraction. Since each gene in a population of cells from a cultured cell line replicates during a very short reproducible time interval in S-phase (Brown *et al.*, 1987), the replication time of the gene, early or late, can be determined from the Southern transfer's autoradiographs. Replication time data for various genes is included in Table 1.

Base composition fractionation of R- and G-band DNA is based upon both Q-banding and banding from other base composition sensitive fluorochromes (Schweizer, 1981) being coincident with the replication and trypsin-Giemsa banding patterns. By using base composition sensitive ligands in cesium salt gradients, workers in Georgio Bernardi's laboratory have fractionated mammalian mainband DNA into up to eight different fractions (Salinas *et al.*, 1987). The mammalian genome exhibits a base compositional compartmentalization. It consists of a mosaic of very long (>300 kb) DNA segments called isochores (Bernardi *et al.*, 1985, Bernardi and Bernardi, 1986). Each isochore is fairly homogeneous in base composition along its entire length and at least eight different density classes of isochores can be separated. When the isochore classes were Southern transferred and probed with various genes, the four AT-richest isochore classes contained genes in the late replicating half of the genome while the remaining four GC-rich isochore classes contained genes in the early replicating half of the genome (true for seven of nine doubly tested genes, Table 1).

Another test of the equivalence involved probing the DNA fractions with the interspersed repetitive sequences *Alu* or L1. The half million copies of *Alu* in human or 100,000 copies of B1 in rodents are selectively partitioned into the

TABLE 1

Properties of genes

	1	2	3	4
Heat shock protein	H			I
Vimentin (human)	H		G	
Myosin heavy chain	T	L		
N-myc	T	L		
Nerve growth factor	T			N
alpha-amylase I	T	L		N
Class I histocompatability	H			I
c-src	H			I
alpha-globin	T	E	G	I
c-Ki-ras	H	E		I
c-fos	H	E		I
c-raf, fes, rel		E		
all histones	H	E		I
alpha-l-antitrypsin	T	L		N
beta-casein	T	L		
c-myb	H	E		
Interleukin	T			N
Ribosomal proteins L-30, L-32	H			I
Xenopus somatic 5s RNA	H	E		
Placental lactogen	T	E		I
c-mos	T		G	
beta-actin	H	E		I
gamma-actin	H	E		
c-fms	T	E		
Apolipoprotein	T	E		
beta-globin	T	L	A	N
Triose phosphate isomerase	H			I
Adenosine deaminase	H			I
pre-pro-opiomelanocortin	T		G	I
conalbumin	T		G	
c-Ha-ras	H	E	G	I
alpha 2 (I) collagen	T			I
delta-crystallin	T			N
Phosphoglycerate kinase-l	H	E		I
c-sis	T	E	G	I
Tyrosine amino transferase	H	E		
Myoglobin	T			N
c-myc	H	E	G	I
alpha-l-antichymotrypsin	T	L		
Phenylalanine hydroxylase	T	L		
Metallothionine-I	H	E		I
Ovalbumin (chicken)	T		A	N
Cardiac actin	T		A	
Superoxide dismutase	H			I
Growth hormone	T			I
Enkephalin	T			I
Skeletal muscle actin	T	L	G	
Vimentin (mouse)	H		A	
c-abl	H	E	G	
c-sis	H	E	G	
Immunoglobulin constant (K or gamma)	T	E	A	
Glucose-6-phosphate dehydrogenase	H	E		I
Immunoglobulin Variable	T	L	A	
gamma, delta, epsilon-globins	T		A	N
Hypoxanthine phosphoribosyl transf.	H	E		I
Insulin	T			N
Leukocyte interferon	T			N
Adenine phosphoribosyl transferase	H	E		I
CAD	H	E		
Retinol binding protein	T			I
Thy-l	T	E		I
Class II histocompat.	T	E		I
Dihydrofolate reductase	H	E		I
Arg. succinate synthetase	H	E		
beta-tubulins	H	E		I
Fibronectin	T	L		
HMG CoA reductase	H			I
Glyceraldehyde-3-phosphate dehyd.	H			I
Fibrinogen	T			N
Prothrombin	T			N
Xenopus oocyte 5s RNA	T	L		

Gene data were from human, hamster, or mouse unless specified.

Notes

[1] Genes are (H) housekeeping or (T) tissue specific in expression. Housekeeping genes were defined to include both constitutively active genes and genes such as metallothionine and heat shock which are inducible in all cell types (Goldman *et al.*, 1984).

[2] Replication time fractionation data were tabulated in Goldman *et al.* (1984) with additional data from others as reviewed (Holmquist, 1987; Iqbal *et al.*, 1987): genes are in the early (E) or late (L) replicating halves of the genome.

[3] Base composition fractionation data as listed by Bernardi *et al.* (1985) and Filipski *et al.* (1987): data were summarized by pooling data from isochore fractions representing the AT-richest (A) or the GC-richest (G) halves of the genome respectively.

[4] Clusters or 'islands' of CpG sequences are included in or near the gene (I) or are not (N) near the gene (Bird, 1986; Filipski *et al.*, 1987). Data for brain specific genes and embryo specific genes are unavailable. Most data points are coincident determinations from two or more mammalian species. Only vimentin's data showed a species difference.

Properties 1 and 4 are from the genes and (or) their immediate flanking sequences. Properties 2 and 3 were generally determined from large, > 30 kb, DNA fragments and reflect properties of the domain in which the exon sequences reside.

This list is presented as a puzzle. Each property is digital. Are some property states highly correlated with other property states? The reader may wish to try to rearrange the genes into groups of genes with similar properties. The author's solution is Table 3.

early replicating and GC-rich fractions while the approximately 40,000 L1 family members are selectively partitioned into the late replicating and AT-rich fractions (Holmquist and Caston, 1986; Bernardi *et al.*, 1985). Consistent with these findings, Manuelidis and Ward (1984) showed that biotinylated probes of *Alu* or L1 gave R-banding or G-banding patterns respectively when hybridized *in situ* to human chromosomes. As any biological generalization has exceptions, so all early replicating bands may not be GC-rich but the generalization is good and we can speak of the two fractionation methods as producing band fractionated DNA.

Band Function

While the differences which distinguish R-bands from G-bands are enormous (Table 2), the primary functional relationship coincident with the banding pattern was first suggested by the following. All housekeeping genes, genes constitutively active or activatable in all cell types, are always in the early replicating (R-band) DNA fraction, whereas most tissue specific genes are in the late replicating (G-band) DNA fraction (Goldman *et al.*, 1984), (Tables 1 and 2). We called the R-band and G-band compartment classes the housekeeping sub-genome and ontogenic sub-genome to indicate that these compartments were very different, seemed mutually isolated for the diffusion of mobile repetitive elements, and serviced two different cellular functions (Goldman *et al.*, 1984). Since tissue specific genes can be either active or inactive depending upon the cultured cell used, we began to collect data on the replication time of tissue specific genes in active and inactive cell types. We now know of seven tissue specific genes which replicate late in most cell types but replicate early in cell types in which they are expressed (Holmquist, 1987; Iqbal *et al.*, 1987). The rules we found are: (1) active genes always replicate early, (2) inactive genes usually but not always replicate late, (3) late replicating genes are always inactive (Holmquist, 1987). Thus, late replication is totally coupled to gene repression, a phenomenon known to cytogeneticists for years. G-band genes behave like they are facultatively heterochromatinized (Lyonized) like the inactive, late replicating X of mammalian females or the inactive and late replicating autosomes in male mealy bugs. The tissue dependent replication time of individual genes at about 20 kb resolution coupled with the tissue independent replication banding pattern of whole chromosomes (Camargo and Cervenka, 1982) at about 6000 kb resolution suggested the model in Figure 1. An individual replicon, in that cluster of late replicons which is a G-band, could become early replicating without this difference being detected as altered replication banding by light microscopy.

Does late replication actively determine or passively follow the repressed state? Does the horse pull the cart or does the cart push the horse? The answer is difficult to directly show but circumstantial data favour late replication as an actively determined repression mechan-

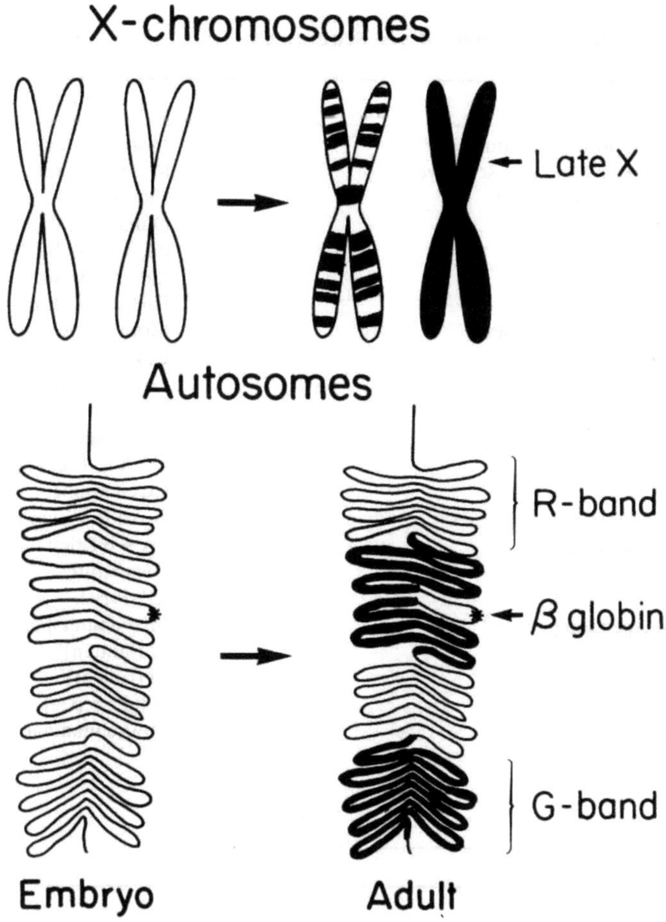

FIGURE 1: A model describing facultative heterochromatinization, acquisition of late-replication (heavy lines), of individual replicons during development. The model incorporates eight points as follows:

1 In anuran embryos, both late-replicating DNA and intrachromosomal patterns of allocyclic replication accumulate gradually during development.

2 In female mammalian embryos, one X becomes late replicating and inactive.

3 In differentiating mammalian cells, the autosomal replication banding pattern at the 500-band (6,000 kb) level of resolution is tissue invariant.

4 All housekeeping genes are in R-bands while most tissue-specific genes are in G-bands.

5 All late-replicating genes are repressed.

6 Only a small fraction of all tissue-specific genes are expressed in any one non-neuronal cell type.

7 In differentiated mammalian cells, the replication time for 50-200-kb DNA segments, which are detected by probing with individual genes, is tissue specific. While the autosomal replication pattern is guided by the G-banding pattern, this guidance is not absolute at the 50-200-kb (replicon) level of resolution.

8 Early replication without expression of beta globin in K562 cells, cells that only express embryonic globin genes in the beta globin gene cluster, suggests that tissue-specific early-replicating domains extend beyond the active gene(s) itself. Together, the data suggest that, like the inactive X, individual G-band replicons become facultatively heterochromatinized during development and that, like the inactive X, their genes become refractory to trans-activation factors.

TABLE 2

Properties of bands and their DNA sequences

R-bands	G-bands
Staining sensitive to protein denaturation[a]	More resistant
DNase II or micrococcyl nuclease sensitive[b]	More resistant
Open to DNase I-nick translation[c]	Closed
Quinacrine bright bands[d]	Quinacrine dull bands
dG + dC rich DNA	dA + dT rich DNA
Early replicating	Late replicating
Pachytene inter-chromomeres[e]	Pachytene chromomeres
Abortions frequent in trisomic state[f]	More viable trisomics
High mitotic chiasmata freq.[g]	Low mitotic chiasmata freq.
Higher X-ray breakage rate[h]	Low X-ray breakage rate
High amplification rate	Low amplification rate
Housekeeping genes (all)[i]	Tissue specific genes (most)
GGGCGG promoters[j]	CAAT, TATA promoters
CCGG rich[k]	CCGG poor
CpG rich islands near sequenced genes[l]	No or few islands
dG + dC rich third coding base	dA + dT rich third base
Slow-sequence divergence rate[m]	Faster-rate
SINE rich[n]	LINE rich
Short-period interspersion pattern[o]	Mixed interspersion pattern
Alu[p]	Ll (human)[p]
Bl[q]	Ll (rodent)[r]
Hepatitis B virus[s]	A36Fc repeat[t]
Bovine leukemia virus[s]	Mouse mammary tumour virus[s]
Epstein-Barr virus[x]	
—— Shared SINEs —— [u]	
—— Processed Pseudogenes —— [v]	
—— SV40 —— [w]	

Notes

The first four properties are for acid-alcohol fixed euchromatin, not heterochromatin. G-, Q-, and replication banding patterns appear identical for more than 95 per cent of all bands; problems of matching the three patterns, often obtained with various degrees of resolution, were addressed (Holmquist, 1987). Since there are at least eight different base composition compartments (Salinas *et al.*, 1987), these may correspond to band sub-types for Q-band intensity and the fluorescent intensity of a Q-band need not reflect its trypsin-Giemsa staining intensity. Most of the properties were referenced in Goldman *et al.* (1984) and Holmquist (1988a).

a. This is a restatement of trypsin induced G-banding. Various protein denaturation agents, trypsin, chymotrypsin, SDS, hot salt, etc., cause R-band DNA to become inaccessible to the cationic thiazine dyes in Giemsa at a rate that is faster than the rate for G-band DNA (Comings, 1972; Evans, 1980).

b. Nuclease digestion followed by Giemsa staining (Sahasrabuddhe *et al.*, 1978).

c. DNase I digestion followed by nick translation with biotinylated precursors. Since the inactive X-chromosome in females G-bands like the active-X but is almost totally refractory to nick translation, nick translation banding is believed to reflect gene activity (Kerem *et al.*, 1984).

d. A variety of base composition sensitive fluorochromes produce the same or reciprocal patterns of bright fluorsecence depending on their base specificity. The most discriminatory of these staining procedures involves energy transfer between two dyes which have alternate base composition specificities (Schweizer, 1981).

e. Meiotic chromomeres correspond to G-bands (Okada

and Comings, 1974; Ambrose and Sumner, 1987) and the inter-chromomeric DNA stains bright with Chromomycin A_3-Distamycin as do R-bands (Ambrose and Sumner, 1987).

f. Hoehn (1975) and

g. Kuhn *et al.* (1985) both review data on the distribution of Bloom-syndrome- mitotic-chiasmata, oncogenes, chromosome breakpoints associated with malignant diseases, and fetal survival with partially trisomic genotypes. Kuhn *et al.* (1985) conclude that the properties associated with R-band chromatin indicates it contains more of the essential genes, 'housekeeping', genes.

h. Holmberg and Jonasson (1973).

i. Table 1.

j. These promoter types (Martini *et al.*, 1986) distinguish constitutively active genes, a subset of housekeeping genes, from all other genes. *Drosophila*'s constitutively active genes, such as GART seem to lack either CAAT or TATA, like mammalian genes; unlike mammalian genes, they seem to lack an extremely GC-rich domain around their promoters.

k–w. Data from base composition fractionated or replication time fractionated DNA.

k. Fig. 2.

l. Table 3.

m. Table 3.

n. Holmquist and Caston (1986). (SINEs are short interspersed repeats and LINEs are long interspersed repeats).

o. From genome fractionation according to percentage renaturation (Strayer *et al.*, 1983).

p–v. From base composition fractionation, replication time fractionation, and *in situ* hybridization (often all three) as described in the text or in Holmquist (1988a).

u. Goldman *et al.* (1984).

v. Data were primarily from replication time of ten processed pseudogenes of argininosuccinate synthetase (Goldman *et al.*, 1984).

w. Based on replication time of SV40 DNA in three different cell lines, each with one copy of the SV40 genome (Marchionni and Roufa, 1981).

x. By *in situ* hybridization of ^{3}H-DNA to chromosomes (Shiraishi *et al.*, 1985).

ism; the replicon must be early replicating before classical trans-activation factors which bind to gene promoters can turn on a gene in the replicon (Holmquist, 1987).

If late replication is included in the repertoire of mammalian gene repression mechanisms and is reflected by the G-banding pattern of chromosome organization, then what other organisms utilize this repression mechanism and banding pattern? *Drosophila* DNA, by fibre autoradiography, lacks the temporal clusters of replicons (Stineman, 1981) that are present in mammalian DNA (Hand, 1978) and that identify mammalian bands as units of chromosome replication. (A polytene band is one-seventh of a replicon in length and bears no similarity to a mammalian band). Thus, *Drosophila* chromosomes lack the prerequisites of the mammalian banded organization and lack the ability to facultatively heterochromatinize individual replicons (Holmquist, 1987). Since a variety of insects and other non-chordates have the ability to facultatively heterochromatinize whole chromosomes, I suggested that vertebrates have G-bands because they have acquired the ability to adopt a common but crude repression mechanism, facultative heterochromatinization of large chromosome segments, which vertebrates have delicately applied to individual replicons.

Band Evolution

When did the banded pre-pattern evolve? What extant creatures most distant from mammals show evidence of banding? Since the chromosomes of teleost fish and frogs show replication banding and trypsin G-banding (Wiberg, 1983; Delany and Bloom, 1984; Schempp and Schmid, 1981; Stock and Mengden, 1975), the banded pre-pattern evolved before bony fish diverged from sharks and rays. Since the function of the banded pre-pattern involves facultative heterochromatinization of visibly large chromosome replication units and there is no known a priori reason why the Giemsa basophilia of these units should be differentially sensitive to protein denaturing agents, I would suggest that an astute cytotaxonomist should use replication banding techniques on lower chordates and deuterostomes to discover when, since the divergence of insects and chordates, the banded pre-pattern probably evolved.

Frog chromosomes do not band with quinacrine or mithramycin/chromomycin A_3 (Schmid, 1980). Also, as opposed to avian and mammalian DNA, amphibian (*Xenopus* and *Pleurodeles*) DNA lacks the base composition isochores, long DNA stretches of homogeneous base composition, and have little DNA in the 'GC-rich' region of the density gradient where mammalian R-band DNA is found (Macaya *et al.*, 1976; Bernardi, *et al.*, 1985). Thus, the quinacrine bandability of mammalian chromosomes evolved since mammals diverged from amphibians, it is directional, it is due to an increased dG and dC content of R-band DNA, and it may be a directional reponse to homeothermia (Bernardi and Bernardi, 1986; Bernardi *et al.*, 1985). The homeothermia cause implies Q-bandability evolved twice, independently in birds and mammals, and would require showing that reptiles do not have the isochores required for Q-bandability of their chromosomes.

The student of cytogenetics should step back and note the logic of the aforementioned arguments. First, a species' chromosomes may be reported unbandable. But this negative data may be due to shortcomings of the investigator's technique and not reflect a property of the chromosomes. However, if the molecular basis of a bandedness, chromosome replication units in *Drosophila* or GC-rich isochores in frogs, is known and these molecular properties can be unambiguously ascertained, then the bandedness of a species' chromosomes can be ascertained without positive cytogenetic data. Second, the words 'chromosome band' describe a simple morphology like the words 'ball shaped'. As all ball shaped biological structures are not evolutionarily or functionally related, so also are the bands of *Drosophila* polytene, human diploid, and *Platyhelminthes* diploid (Ekaterina de Vaio, personal communications) chromosomes not related. When someone reports trypsin G-bands in red pine (Drewry, 1982), *Aedes albopictus* (Steiniger and Mukherjee, 1975), or *Triatomine* bugs (Maudlin, 1974), this does not imply that these structures have the same function or share an evolutionary continuum with mammalian bands. While bands were discovered by light microscopists, light microscopy alone is insufficient to really describe bands.

The New Importance of Bands to Molecular Biology

While banding was a Rosetta Stone for cytogenetics, the stone's translations have been limited to descriptive morphology. All the while, molecular biologists considered themselves uniquely anointed to discuss their findings with Allah. Their productive but myopic reductionism is about to change as lengths of DNA greater than 20 kb are inspected. That gene sequences can be determined by where the gene lives, instead of what it does, and that mobile elements are concentrated in certain band types, shatter reductionist dogma and require new paradigms.

Bernardi and Bernardi (1986) describe a genomic phenotype for all organisms which is characterized by the average base composition of the entire genome. They found from the sequences of many genes in many organisms including bacterial phages and metazoa, that the average base composition of coding and noncoding sequences is positively correlated with the average base composition of the genome. The finding of importance to us is that various mammalian chromosomes' compartments have different base compositional strategies (Bernardi and Bernardi, 1986). When a variety of gene sequences in cold-blooded vertebrates are compared with those of the same gene in warm-blooded vertebrates, a clear directional change is seen. The comparisons show genes which are in GC-rich isochores (R-bands) in mammals not only are flanked by GC-richer DNA, but also their coding sequences are GC-richer than the corresponding amphibian gene. The differences are not only in non-coding DNA, introns and interspersed repeat sequences, but also in silent (third codon bases) DNA and even in first and second codon positions (Bernardi and Bernardi, 1986). The recent directional increase in GC-richness of R-bands is evident throughout the R-band compartment classes. The most illustrative individual example (Filipski *et al.*, 1987) involves alpha globin, an R-band gene, and beta globin, a G-band gene (Table 3). The human alpha globin gene has a GC content of 67 per cent and a GC usage in the third codon of 88 per cent compared with 43.5 and 46.5 per cent respectively in the *Xenopus* gene. By contrast, the base composition and third codon usage of the human beta globin gene are only 40 and 67 per cent respectively.

Such a directional and compartmental strategy has earth-shaking theoretical implications. Kimura (1983) argued that most amino acid substitutions are selectively neutral and are fixed by random drift. Bernardi and Bernardi (1986) can now show from DNA sequence data that first and second coding position changes, changes which alter an amino acid, are directional and in accord with the genome's base compositional strategy. This is not random drift but a direct response to a selective force which operates in a particular direction; the kinds of the amino acid changes can be determined by knowing the average base composition of the genome or, in the case of mammals, the base composition of the genomic compartment.

While silent third coding position changes do not alter an amino acid sequence, they do alter the transfer RNA which will be used to translate the gene. These 'silent codon' mutations were thought to be selectively fixed according to a strategy which optimized translational efficiency through the adaptation of codons to the actual population sizes of the various isoaccepting tRNAs in the cytoplasm. For example, the alpha globin protein is translated at a very high rate and should utilize those isoaccepting tRNAs which are present in high concentrations so that any one isoaccepting species does not become rate limiting during rapid translation in hematopoietic cells. Housekeeping proteins, on the other hand, are generally translated slowly and should be less constrained, more free to use any isoaccepting tRNA species. Since the different genomic compartments of a mammalian genome utilize different isoaccepting tRNA frequencies, and since translation via mRNA and tRNA occurs in the cytoplasm wherein all these RNAs are in a common pool, then silent codon usage is determined by the compartment or band wherein the gene is located and not by a property related to the gene's function. Now one may argue that housekeeping genes are in R-bands and, being translated at a slower rate than most active tissue specific genes, require a different codon strategy. While this may be true and hint at one selective advantage to the organism for partitioning different kinds of genes in different genomic compartments, still, alpha globin, while tissue specific, has adopted the strategy of a housekeeping compartment not because of what the alpha globin protein does but because of where the alpha globin gene sequence is located.

CpG islands are several hundred base pair long tracts of DNA wherein the frequency of the dinucleotide CpG is ten times its concentration in bulk DNA (Bird, 1986). The islands extend into coding sequences (Bird, 1986) but do not occur in amphibian genomes (Bernardi and Bernardi, 1986). Table 1 has been rearranged to show two natural groupings of gene properties; these two groups correspond to R-band genes and G-band genes (Table 3). Some tissue specific genes are known to be uncharacteristically located in R-bands because they are early

TABLE 3

Properties of genes

	R-band genes					
	1	2	3	4	5	6
Hypoxanthine phosphoribosyl transf.	H	E		I	+	+
Adenine phosphoribosyl transferase	H	E		I	–	–
CAD	H	E			+	+
Dihydrofolate reductase	H	E		I	–	+
Arg. succinate synthetase	H	E			–	+
Glucose-6-phosphate dehydrogenase	H	E		I	–	-
beta-tubulins	H	E		I	+	+
Phosphoglycerate kinase-l	H	E		I	+	+
Tyrosine amino transferase	H	E			+	+
beta-actin	H	E		I	+	+
gamma-actin	H	E			+	+
Metallothionine-I	H	E		I		
Superoxide dismutase	H			I		
Heat shock protein	H			I		
Vimentin (human)	H		G		+	
HMG CoA reductase	H			I		
Glyceraldehyde-3-phosphate dehyd.	H			I		
Triose phosphate isomerase	H			I		
Adenosine deaminase	H			I		
Class I histocompatability	H			I		
c-myc	H	E	G	I	+	+
c-Ha-ras	H	E	G	I	+	+
c-abl	H	E	G			
c-src	H			I		
c-Ki-ras	H	E		I		
c-fos	H	E		I		
c-myb	H	E				
c-raf, fes, rel		E				
all histones	H	E		I	+	+
Ribosomal proteins L-30, L-32	H			I		
Xenopus somatic 5s RNA	H	E				
alpha-globin	T	E	G	I	–	
Placental lactogen	T	E		I	–	
c-mos	T		G			
c-sis	T	E	G	I		
c-fms	T	E				
Apolipoprotein	T	E				
pre-pro-opiomelanocortin	T		G	I		

	R-band genes					
	1	2	3	4	5	6
conalbumin	T		G			
alpha 2 (I) collagen	T			I		
Retinol binding protein	T			I		
Thy-l	T	E		I		
Class II histocompat.	T	E		I		
Growth hormone	T			I		
Enkephalin	T			I		
			??-band			
Skeletal muscle actin	T	L	G		+	+
Immunoglobulin constant (K or gamma)	T	E	A			
	G-band genes					
beta-globin	T	L	A	N	–	
Immunoglobulin Variable	T	L	A			
gamma, delta, epsilon-globins	T		A	N		
Ovalbumin (chicken)	T		A	N		
Cardiac actin	T		A		+	+
Vimentin (mouse)	T		A		+	
alpha-l-antichymotrypsin	T	L				–
alpha-l-antitrypsin	T	L		N	–	
beta-casein	T	L			–	–
Phenylalanine hydroxylase	T	L			–	+
Fibronectin	T	L				
Myosin heavy chain	T	L		N		
N-myc	T	L		N		
Nerve growth factor	T	-		N		
alpha-amylase I	T	L		N		
Insulin	T			N		
Leukocyte interferon	T			N		
Fibrinogen	T			N		
delta-crystallin	T			N		
Myoglobin	T			N		
Interleukin	T			N		
Prothrombin	T			N		
Xenopus oocyte 5s RNA	T	L				

Notes

Gene properties from Table 1 grouped into R-band genes and G-band genes. The added properties are (5) the gene's cDNA sequence cross hybridized rodent–human (+) at high stringency or it did not (–) cross hybridize (Goldman *et al.*, 1984); and (6) Ibid., but at low stringency conditions.

From Table 1, if one were to plot the data in five dimensional digital phase space, one sees two major data clusters, HEGI and TLAN, with some nebulously situated data points in between. Some tissue specific genes do not utilize late replication repression (Holmquist, 1987) and these are always early replicating. Since housekeeping genes are always in R-bands but tissue specific genes can be in either R- or G-bands, the nebulous data almost disappears and two groupings are evident; these are (H/T)EGI and TLAN corresponding to R- and G-band genes respectively. The TEG genes, tissue specific genes in R-bands, are most informative. One can assess from these if a new property, $?^+$, is more highly correlated with either band location (TEG$?^+$, HEG$?^+$, TLA$?^-$), or with gene function (TEG$?^+$, HEG$?^-$, TLA$?^+$). Obviously, inclusion of a CpG island is correlated with band position, ie, TEGI, HEGI, TLAN and not TEGN, HEGI, TLAN.

Base sequence divergence data, properties 5 and 6, are insufficient to distinguish if the divergence rate is more highly correlated with band type or gene function.

Note that a gene can be assisgned to a band type not by locating it there cytologically, but by knowing its properties. For a gene such as enkephalin (T), only some of its properties are known. Due to its CpG island, Baysian statistics would assisgn it to an R-band with a fairly high probability. As more of its properties or even the properties of other genes become known, the confidence of enkephalin's assisgnment would vary. For example, if several TLAI genes were found, this would reduce the probability that (T--I) indicates an R-band location. On the other hand, if enkephalin was found to be TE-I, then this would almost assure assignment to an R-band.

replicating in all tested tissues and are in GC-rich isochores (Holmquist, 1987). Such tissue specific genes have CpG islands as do housekeeping genes. Thus, the presence of CpG islands in coding sequences is more highly correlated with the gene's position than with the gene's function.

The final dogma shattering property of bands involves the force which makes G-band DNA so different from R-band DNA. I will summarize the force by analogy. When one makes a CsCl equilibrium density gradient, one starts with a centrifuge tube containing a homogeneous distribution of CsCl. During centrifugation, the high centripetal force acts to concentrate the CsCl away from the centre of rotation. At equilibrium, this is balanced by a diffusive force tending to randomize the CsCl molecules; the net result is an equilibrium gradient. The exact mathematical treatment is an application of Fick's laws of diffusion. Kimura (1983) used Fick's laws to argue that most fixed mutations are neutral. As a new mutant allele diffuses through the population, the probability functions describing both its frequency in the gene pool over time and its eventual probability of fixation are random functions as is that of the aforementioned diffusion function. Those mutant alleles which confer a selective advantage to the individual spread throughout the population as if driven by a directional force mathematically identical to the centripetal force.

Now consider an R-band and a neighbouring G-band. They differ in their concentration of dA, dT, CpG, CpCpGpG, Alu, Ll, and a variety of other sequences (Table 2). These concentration gradients reflect a force which prevents the concentration of each base, oligonucleotide, or mobile genetic element from becoming homogeneous across the band boundaries. On comparing a histone H4 gene from *Drosophila* and man, we would find most of the base sequence, the constrained bases, to be identical. Kimura (1983) would argue that these bases are essential to the function of a histone protein and that a force, natural selection, has acted to preserve these sequences from the randomizing effect of spontaneous mutation. Now we question the kind of force which sets up and maintains the concentration differences along the chromosome. Either this force is unknown to biology or it is, as suggested by Bernardi and Bernardi (1986), natural selection acting to mould a genomic phenotype (Q-banding for example being a visualization of one phenotypic characteristic of the genotype). If the force is natural selection, then the DNA which responds to this force has a function upon which selection acts. The DNA most responsive to this force is non-coding DNA and therefore non-coding

DNA must have a function.

We can estimate the Haldane-mutational load due to Q-bandability. For an advantageous new mutant DNA sequence or 'allele' to become fixed in the population, it must confer a selective advantage upon organisms harbouring the new allele. These more fit individuals survive at the expense of less fit individuals carrying the old allele. For a population of size n to fix a functional and advantageous DNA sequence, Haldane's theory shows that $30n$ less fit individuals must undergo genetic death to purge the population of the old allelic sequence (MacArthur and Connell, 1967). The genetic death incurred while eliminating these less fit individuals is the mutational load associated with fixing an advantageous mutation. Q-bandability was derived from a directionally increased dG + dC content of R-band DNA which occurred by a variety of mechanisms acting on introns, interspersed repeats, and even coding sequences (Bernardi and Bernardi, 1986) and amounts to an increased GC content of 3.24 per cent (Holmquist *et al.*, 1982). That is, 4.4×10^7 times an A or T was directionally exchanged for a G or C and, we assume, then fixed in the population. This is only 0.121 bases fixed per year in the 360 million years since frogs diverged from mammals. If, by Haldane's calculations, $30n$ genetic deaths must occur for each selectively advantageous base fixed in the population, and we conservatively assume the outbreeding population size of frog or small mammal populations is 200,000, then an average of at least one million genetic deaths per year have resulted from the fixation of mutations responsible for Q-bandability. This is an intolerable genetic load.

Ohno (1972) used Haldane's mutational load theory as the basis for calling non-coding DNA functionless junk. Knowing the average mutation rate per base pair of DNA, Ohno argued that if all the DNA in a mammal was functional and therefore that mutations in it were subject to natural selection, then the genetic load upon a mammalian population would be intolerable. Therefore, most of a mammal's DNA, the non-coding DNA, must be functionless junk.

Conclusion

Eighteen years after the discovery of euchromatic bands, their properties are finally being realized. The banding patterns so dear to our hearts now threaten the dogma of population genetics and molecular biology by revealing contradictions. The sequences of non-coding DNA are constrained to a non-random distribution along the banding pattern and therefore must be functional by neutral gene theory. However, they must also be functionless by mutational load theory. Silent coding mutations and possibly even some amino acid altering mutations are determined not by the protein's function nor by fixation of neutral mutations but by directional fixation, the direction being determined by the gene's band of residence.

The ordered DNA sequence distribution of the banded chromosome's pattern reveals a force operating to determine chromosome organization. We begin to comprehend the force's strength by its ability to measurably direct codon uasage and possibly even amino acid substitutions. However, the force cannot be identical to that which moulds and constrains amino acid sequences because its mutational load would be impossibly high.

References

Ambrose PF, Sumner AT (1987). Correlation of pachytene chromomeres and metaphase bands of human chromosomes and distinctive properties of telomeric regions. Cytogenet Cell Genet 44; 223–8.

Bernardi G, Bernardi G (1986). Compositional constraints in genome evolution. J. Mol. Evol. 24; 1–11.

Bernardi G, Olofsson B, Filipski J, Zerial M, Salinas J, Cuny G, Meunier-Rotival M, Rodier F (1985). The mosaic genome of warm-blooded vertebrates. Science 228; 953–8.

Bird AP (1986). CpG-rich islands and the function of DNA methylation. Nature 321; 209–13.

Brown EH, Iqbal MA, Stuart S, Hatton KS, Valinsky J, Schildkraut CL (1987). Rate of replication of the murine immunoglobulin heavy-chain locus: evidence that the region is part of a single replicon. Mol. Cell Biol. 7; 450–7.

Camargo M, Cervenka J (1982). Patterns of DNA replication of human chromosomes. II. Replication map and replication model. Am J. Hum. Genet 34; 757–80.

Comings D (1972). The structure and function of heterochromatin Adv. Hum. Genet 3; 237–431.

Delany ME, Bloom SE (1984). Replication banding patterns in the chromosomes of the rainbow trout. J. Hered 75; 431–4.

Drewry A (1982). G-banded chromosomes in *Pinus resinosa*. J. Heredity 73; 305–6.

Evans HJ (1980). Some facts and fancies relating to chromosome structure in man. Adv. Hum. Genet 8; 347–438.

Filipski J, Salinas J, Rodier F (1987). Two distinct compositional classes of vertebrate gene-bearing DNA stretches, their structures and possible evolutionary origin. DNA 6; 109–18.

Goldman MA, Holmquist GP, Gray MC, Caston LA, Nag A (1984). Replication timing of mammalian genes and middle repetitive sequences. Science 224; 686–92.

Hameister H, Sperling K (1984). Description of a chromosome replication unit in individual prematurely condensed human S-phase chromosomes. Chromosoma 90; 389–93.

Hand R (1978). Eukaryotic DNA: organization of the genome for replication. Cell 15; 317–25.

Hoehn H (1975). Functional implications of differential chromosome banding. Am J. Hum. Genet 27; 676–86.

Holmberg M, Jonasson J (1973). Preferential location of X-ray induced chromosome breakage in the R-bands of human chromosomes. Hereditas 74; 57–68.

Holmquist GP (1987). Role of replication time in the control of tissue-specific gene expression. Am J. Hum. Genet 40; 151–73.

Holmquist GP (1988a). DNA sequences in G-bands and R-bands. In Chromosomes and Chromatin Vol. II (ed. Adolph KW). CRC Press, Boca Raton, Florida; 76–121.

Holmquist GP (1988b). Mobile genetic elements in G-band and R-band DNA. In The cytogenetics of mammalian autosomal rearrangements (ed. Daniel A). Alan R. Liss, Inc. New York; 803–34.

Holmquist GP, Caston La (1986). Replication time of interspersed repetitive DNA sequences in hamsters. Biochimica et Biophysica Acta 868; 164–77.

Holmquist GP, Gray M, Porter T, Jordan J (1982). Characterization of Giemsa dark- and light-band DNA. Cell 31; 121–9.

Iqbal MA, Chinsky J, Didamo V, Schildkraut CL (1987). Replication of proto-oncogenes early during the S phase in mammalian cell lines. Nucleic Acids Res. 15; 87–103.

Kerem B-S, Goitein R, Diamond G, Cedar H, Marcus M (1984). Mapping of DNAse I sensitive regions on mitotic chromosomes. Cell 38; 493–9.

Kimura M (1983). The neutral theory of molecular evolution. Cambridge University Press, Cambridge, England.

Kuhn EM, Therman E, Denniston C (1985). Mitotic chiasmata, gene density, and oncogenes. Hum. Genet 70; 1–5.

Kowalski J, Cheevers WP (1976). Synthesis of high molecular weight DNA strands during S phase. J. Mol. Biol. 104; 603–15.

Lau Y-F, Arrighi FE (1981). Studies of mammalian chromosome replication. II. Evidence for the existence of defined chromosome replication units. Chromosoma 83; 721–41.

MacArthur RH, Connell JH (1967). The biology of populations. John Wiley & Sons, New York, 94.

Macaya G, Thiery J-P, Bernardi G (1976). An approach to the organization of eukaryotic genomes at a macromolecular level. J. Mol. Biol. 108; 237–54.

Manuelidis L, Ward DC (1984). Chromosomal and nuclear distribution of the FHind III 1.9 kb human DNA repeat segment. Chromosoma 91; 28–38.

Marchionni MA, Roufa DJ (1981). Replication of viral DNA sequences integrated within the chromatin of SV 40 transformed Chinese hamster lung cells. Cell 26; 245–58.

Martini G, Toniolo D, Vulliamy T, Luzzatto L, Dono R, Viglietto G, Paonessa G, D'Urso MD, Persico MG (1986). Structural analysis of the X-linked gene encoding human glucose 6-phosphate dehydrogenase. EMBO J. 5; 1849–55.

Maudlin I (1974). Giemsa banding of metaphase chromosomes in triatomine bugs. Nature 252; 392–3.

Ohno S (1972). Evolutional reason for having so much junk DNA. In Modern aspects of cytogenetics: constitutive heterochromatin in man (ed. Pfeiffer RA). F.K. Schattauer Verlag, Stuttgart.

Okada TA, Comings DE (1974). Mechanisms of chromosome banding III. Similarity between G-bands of mitotic chromosomes and chromomeres of meiotic chromosomes. Chromosoma 48; 65–71.

Sahasrabuddhe CG, Pathak S, Hsu TC (1978). Responses of mammalian metaphase chromosomes to endonuclease digestion. Chromosoma 69; 331–8.

Salinas J, Zerial M, Filipski J, Crepin M, Bernardi G (1987). Nonrandom distribution of MMTV proviral sequences in the mouse genome. Nucleic Acids Res. 15; 3009–22.

Schempp W, Schmid M (1981). Chromosome banding in amphibia VI. BrdU-replication patterns in Anura and demonstration of XX/XY sex chromosomes in *Rana esculenta*. Chromosoma 83; 697–710.

Schmid M (1980). Chromosome banding in amphibia. IV. Differentiation of GC- and AT-rich chromosome regions in Anura. Chromosoma 77; 83–93.

Schmidt M (1980). Two phases of DNA replication in human cells. Chromosoma 76; 101–10.

Schweizer D (1981). Counterstain-enhanced chromosome banding. Hum. Genet 57; 1–14.

Shiraishi Y, Takahiro T, Ohta Y, Hirai K (1985). Chromosomal localization of the Epstein-Barr virus (EBV) genome in Bloom's syndrome B-lymphoblastoid cell lines transformed with EBV. Chromosoma 93; 157–64.

Soriano PM, Meunier-Rotival, Bernardi G (1983). The distribution of interspersed repeats is nonuniform and conserved in the mouse and human genomes. Proc. Natl. Acad. Sci. USA 80; 1816–20.

Steiniger GE, Mukherjee AB (1975). Insect chromosome banding: technique for G- and Q-banding pattern in the mosquito *Aedes albopictus*. Can J. Genet Cytol 17; 241–4.

Stineman M (1981). Chromosomal replication in *Drosophila virilis*. III. Organization of active origins in the highly polytene salivary gland cells. Chromosoma 82; 289–307.

Stock AD, Mengden GA (1975). Chromosome banding pattern conservatism in birds and nonhomology of chromosome banding patterns between birds, turtles, snakes, and amphibians. Chromosoma 50; 69–77.

Strayer D, Heintz N, Roeder R, Gillespie D (1983). Three organizations of human DNA. Proc. Natl. Acad. Sci. USA 80; 4770–4.

Wiberg UH (1983). Sex determination in the European eel (*Anguilla anguilla*, L.) Cytogenet Cell Genet 36; 589–98.

Yunis JJ (1981). Mid-prophase human chromosomes: the attainment of 2,000 bands. Hum. Genet 56; 293–8.

5 Examination of the Structure and Function of the Mammalian Centromere

ELTON STUBBLEFIELD AND LORRAINE M. CHERRY

Department of Molecular Genetics,
The University of Texas M.D. Anderson Cancer Center,
Houston, Texas 77030, USA

Introduction

In most eukaryotic organisms, and especially in mammals, the centromere is a distinctive feature of metaphase chromosomes. It appears as a definite constriction where the two chromatids are most firmly bound together in metaphase. This region may be located anywhere along the length of the chromosome, not just near the centre. The term centromere always applies, even when the centromere is located at one end of a chromosome, as it is in all the chromosomes of the laboratory mouse. The centromere allows the chromosome to interact with the mitotic spindle in mitosis, so its presence and adequate function are essential to the chromosome's persistence in the genome of a cell. When a chromosome breaks and creates an acentric fragment, the result is usually the loss of the fragment from the cell, since it can no longer attach to the division spindle and participate in mitotic segregation.

In this review we wish to examine current views on centromere structure and function, especially as they apply to mammalian cells. Several recent reviews have dealt with this subject in depth, and some have compared the ultrastructure of the centromere in a wide variety of species (Rieder, 1982; Blackburn and Szostak, 1984; Godward, 1985; Nicklas, 1987). We will not attempt to duplicate their efforts. Instead, we will focus on a variety of available techniques and models that have appeared in the literature and attempt to make some sense from the many conflicting concepts in circulation about this very interesting feature of chromosomes.

Before launching into the details of our topic, however, we must define two terms: the *centromere* and the *kinetochore*. In the literature these terms were used interchangeably until about 1965. After electron microscopy revealed the structural detail associated with the centromere, the term kinetochore was usually reserved for the specific molecular-attachment apparatus that linked spindle fibres (microtubules) to the chromosome (Brinkley and Stubblefield, 1966). The term centromere is now usually applied to the region of the chromosome containing the kinetochore. Because ultrastructural detail is

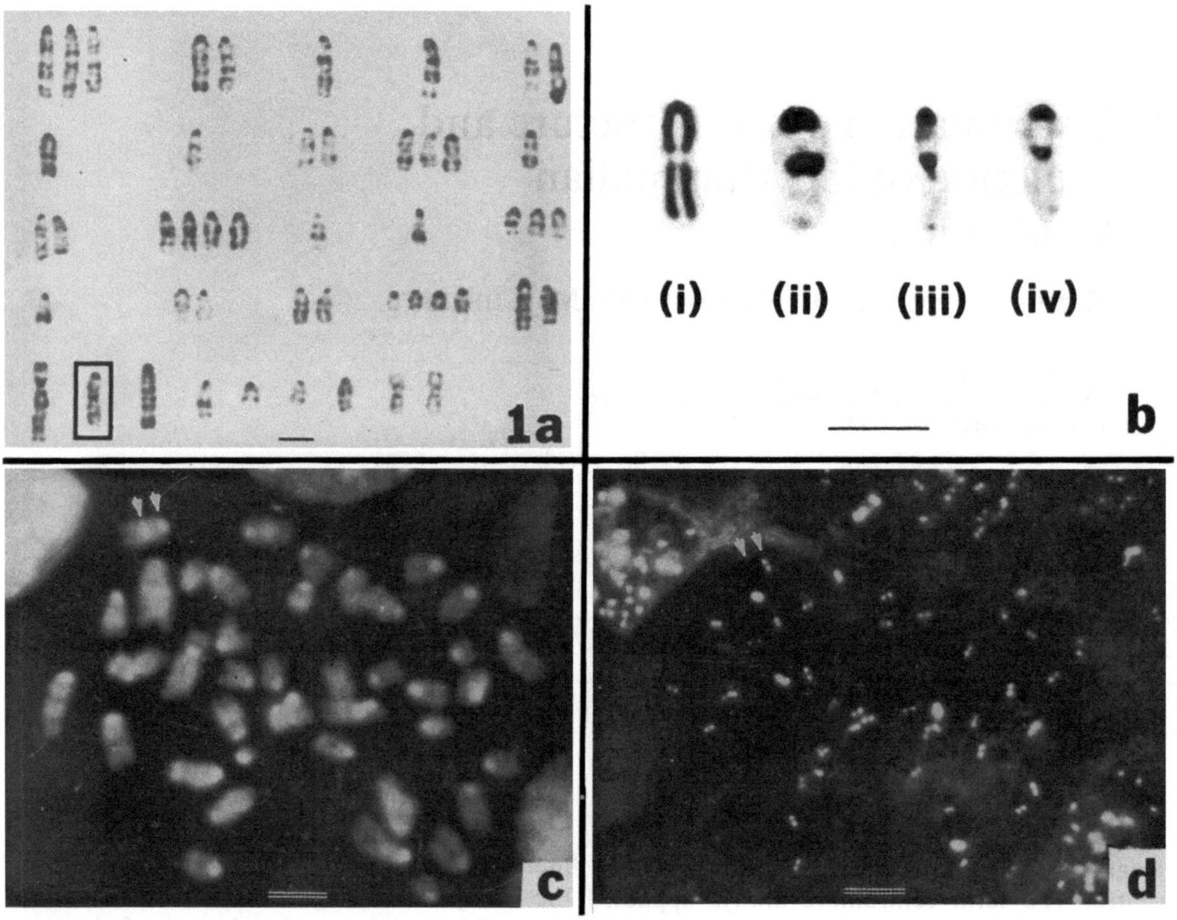

FIGURE 1: Centromere staining in a compound chromosome of the murine sarcoma cell line, SEWA. (a) G-banded karyotype of a SEWA cell in metaphase. The marker chromosome (box) at the bottom is a tandem fusion product of chromosome No. 19 (upper) and chromosome No. 18 (lower). (b) Staining techniques used to identify centromeric areas of the compound chromosome (box) in Fig. 1a: (i) Giemsa staining, (ii) C-banding, (iii) Cd staining, and (iv) Kt staining. (c) Chromosomes of a metaphase SEWA cell stained with DAPI. (d) The same cell as in Fig. 1c stained with rhodamine-labelled antikinetochore antibodies. The arrowheads at the upper left point to the two centromere areas in the compound tandem fusion chromosome found in these cells. Note that the inactive centromere fails to stain with the rhodamine antibodies in Fig. 1d. Bars are all 5 μm long.

lacking, this term is used extensively in light microscope studies. We will adhere to this convention in the following review.

Visualizing the Centromere with Light Microscope

From the beginning of cytogenetics, the location of the centromere has been a factor in the description of chromosomes. Chromosomes are described as metacentric, telocentric, or acrocentric, depending on whether the centromere is located near the centre, toward the end, or at the end of the chromosome. As we have learned more about the importance of compound chromosomes in neoplastic and evolutionary change, our interest in the ability to observe the location of the centromere has increased. Specific banding techniques developed in the 1970s (Hsu and Arrighi, 1971; Seabright, 1971) allowed us for the first time to identify unequivocally each chromosome in a complement. It soon became apparent that the fusion of whole chromosome into various configurations with no obvious loss of material was rather common. Interestingly, these compound chromosomes behaved as if they had only one centromere. These findings spurred the development of new techniques to examine the centromere, and these in turn have enriched our knowledge of centromere structure and function. In this section, we will examine the development and use of techniques to observe the major components of the centromere: the primary constriction, the pericentromeric heterochromatin, and the kinetochore.

The primary constriction

When mammalian chromosomes are examined with a light microscope, it is immediately apparent that a major constriction occurs somewhere along the length of each chromosome. This constriction appears to be a point of very tight connection between the two chromatids and is, in fact, the last point to separate as the cells move into anaphase. When two centromeres occur in one chromosome, a second constriction may be seen (Figure 1a,b), but it is usually not as obvious as the first (Schwartz *et al.*, 1983). In addition, the second constriction often tends to disappear as the cell progresses toward anaphase (Vig and Zinkowski, 1986). In some instances, compound chromosomes are observed in which both centromeres appear to be active, but these tend to show a high frequency of breakage, as we would expect (Niebuhr, 1972; Dewald *et al.*, 1979). Commonly, one centromere or the other becomes dominant, but we cannot predict which one. For instance, a chromosome translocation (5;13) reported by Niebuhr (1972) showed both centromeres active in 36 per cent of the cells, the No. 5 centromere active in 53 per cent, and the No. 13 centromere active in 11 per cent. A translocation (Y;13) studied by Ing and Smith (1983) revealed 22 per cent of the cells with Y centromere activity and 78 per cent with centromere 13 activity. It seems, then, that the inactivation of a centromere in a compound chromosome is a cell-by-cell event, and that particular centromeres are not dominant for some a priori reason.

Pericentromeric heterochromatin

In the course of experiments aimed at *in situ* localization of mouse satellite DNA, Pardue and Gall (1970) found that specific segments of the mouse chromosomes immediately adjacent to the centromere tended to stain intensely. These deeply staining areas resulted from sequential treatment of the chromosomes with hydrochloric acid, an alkaline solution such as sodium or barium hydroxide, a 60°C salt solution, and pH 6.8 phosphate-buffered Giemsa stain. Subsequently, Arrighi and Hsu (1971) reported that this treatment also resulted in dark-staining pericentromeric regions in human chromosomes. The short arms of the acrocentric chromosomes, the secondary constrictions of chromosomes 1, 9, and 16, and the distal end of the Y chromosome were also stained. These intensely staining regions represent constitutive heterochromatin, a highly condensed chromatin that is a constitutive component of the cell. This contrasts with the facultative heterochromatin such as that found in the inactive X Chromosome of female mammals. Chromosomal regions containing constitutive heterochromatin are referred to as C-bands (Paris Conference, 1971).

The name C-band has been incorrectly translated to mean *centromere* band because of the propensity in many species for this type of heterochromatin to localize around centromeres. Because of this tendency, the presence of C-band material has often been used as a marker to detect the presence of a second centromere in multicentric chromosomes (Figure 1b). In some cases this approach has been successful (Nakagome *et al.*, 1986; Ward et al. 1981; Ghosh and Chaudhuri, 1984), but it is subject to several major problems. To begin with, although the presence of pericentromeric C-bands is widespread in mammals, there are species in which some or all of the chromosomes lack C-band areas (Pathak and Wurster-Hill, 1977). Further, even in those species with easily detected C-bands, there can be a marked lack of correlation between C-bands and other centromere locators in multicentric chromosomes (Daniel, 1979; Lambiase *et al.*, 1984; Maraschio *et al.*, 1980). These two observations indicate that constitutive heterochromatin is probably not always associated with centromere function. We will discuss this topic again in a later section.

The kinetochore

The kinetochore is a protein and nucleic acid structure that serves as the point of chromosomal attachment for spindle microtubules. It can be easily recognized as an electron-dense structure with electron microscopy. Until recently, however, it was visible in light microscope preparations only by indirect means that reflected kinetochore activity. When they are first formed, secondary kinetochores in multicentric chromosomes retain full or partial activity for unpredictable periods of time (Zinkowski *et al.*, 1986). During this period of activity, the chance of anaphase abnormality can be very high, with one kinetochore going to one mitotic pole and the second to the other pole. The resulting anaphase bridge that forms, or the fragments that result from subsequent stretching and breakage, are clear indicators of multicentric activity in careful preparations of anaphase and telophase cells (Schwartz *et al.*, 1983; Ying and Ives, 1971).

Beginning in the early 1970s, several staining techniques were developed that purported to stain active kinetochores. One of these, developed by Eiberg (1974), was based on hot salt incubation and buffered Giemsa staining. This technique, called Cd staining (Figure 1b), produced two identical dots, one on each chromatid arm, at the site of the centromere in normal chromosomes. Eiberg thought that these dots might represent a protein-complexed subfamily of the repetitive DNA found in constitutive heterochromatin. Other researchers, however, postulated that the dots were actually the kinetochores and their associated proteins (Evans and Ross, 1974). The subsequent reports examining this idea generated a fair amount of confusion about the distinction between centromere activity and centromere structure, especially kinetochore structure. Early reports (Daniel, 1979; Maraschio *et al.*, 1980) examined compound chromosomes that were C-band positive and Cd negative. In these reports the C-band-positive staining was taken to mean that no material had been lost; the Cd staining seemed to serve as a measure of activity rather than an indicator of actual structural presence. In a later paper Lambiase *et al.*, (1984) reported a stable compound chromosome in which the second centromere was C-band negative and Cd positive. Although the interpretation of Lambiase *et al.* is somewhat different from ours, it would seem that in this case the Cd staining marked an actual structure rather than indicated an activity. We note that in all three cases the chromosomes involved are reported to be stable in culture, which means there was no kinetochore activity in any of the second centromere areas; if there had been, the chromosomes would have been involved in anaphase bridge formation and would have been unstable in culture.

An alternative technique thought to stain kinetochores involves incubation with aqueous silver nitrate (Denton *et al.*, 1977; Brat *et al.*, 1979). This technique, known as Kt staining (Figure 1b), is essentially the same as that used routinely to stain nucleolar organizer regions (NORs) in mammalian chromosomes, and so it has been suggested that the target molecules are acidic proteins common to the NORs and the centromeric regions (Rocchi *et al.*, 1982). As with Cd staining, however, it is not really known with any certainty whether the kinetochore itself is being stained. It is possible, in fact, that the Cd and Kt methods are not even staining the same structures. In mouse melanoma cell lines we have found Cd-positive and Kt-negative presumptive kinetochore areas in compound chromosomes.

A key problem with both these techniques is that the timing involved in the staining is very critical, but an optimal time is difficult to establish. Overstaining can lead to marking of C-bands, NORs, or even G-bands. Even when this is not a problem, the 'centromeric dots' do not correspond in size or location to kinetochores as detected by electron microscopy, according to Roos (1975).

In 1980, Moroi *et al.* reported that individuals with the autoimmune disease CREST scleroderma manufacture antibodies against the kinetochores to their own chromosomes. The antibodies bind to kinetochore-specific antigens (Earnshaw and Migeon, 1985; Valdivia and Brinkley, 1985) and react with both the inner and outer layers of the kinetochore (Brenner *et al.*, 1981). Indirect immunofluorescence can be used to mark kinetochores for observation with a fluorescence microscope (Figure 1c,d). Merry *et al.* (1985) used this technique to examine the inactive kinetochore in a human 9;11 chromosome translocation and found that it showed a significant decrease in fluorescence compared with the active kinetochore. Subsequent studies by Earnshaw and Migeon (1985), Peretti *et al.* (1986), and Vig and Zinkowski (1986) have reported everything from full fluorescence to total lack of fluorescence in the inactive kinetochore. Cherry and Johnston (1987) have recently used anti-kinetochore indirect immunofluorescence to demonstrate an apparent size heterogeneity among the kinetochores of human chromosomes. This size heterogeneity may reflect the evolutionary history of certain chromosomes and may be an influencing factor in the preferential gain or loss of these chromosomes in cell division.

Visualizing the Centromere with the Electron Microscope

As expected, the electron microscope reveals details about chromosomes that are not apparent with light microscopy. The first views of chromosome ultrastructure came with studies of embedded, sectioned cells. Since the chromosome is a fibrous organelle, it came as no surprise that the basic stuff of chromosomes was very similar to that found in the cell nucleus, a filamentous material called chromatin. Some chromatin in the interphase nucleus appeared more densely packaged than the rest and so was called heterochromatin in agreement with the lower resolution picture of the light microscope. The loosely organized chromatin was termed euchromatin, and modern concepts indicate that this is the material containing the active genetic elements in a living cell. Mitotic chromosomes contain chromatin of the heterochromatin class, which suggests that mitotic chromosomes are not transcriptionally active, a conclusion that has been repeatedly demonstrated (Crippa, 1966). In sectioned chromosomes only two regions contained chromatin different from all the rest: one was in the centromere, and the other was the nucleolus organizer where ribosomal RNA is made in interphase. In the latter case chromatin fibres were thinner and less dense than most chromatin (Brinkley and Hsu, 1966); otherwise the packing was quite similar to the rest of the chromosome. The centromere, on the other hand, contained material with a distinctly different organization.

Kinetochore ultrastructure

Once fixation technology allowed the adequate preservation of cytoplasmic structure with glutaraldehyde, sectioned mitotic material was seen to contain microtubules, the major structural components of the mitotic spindle. In fortuitous sections through chromosome centromeres with proper orientation, a distinct organelle was seen where the spindle fibres attached to the chromosome (Robbins and Gonatas, 1964; Krishan and Buck, 1965). The term kinetochore was appropriated to apply to this specific structure, nestled in the primary constriction of the centromere. Figure 2 demonstrates how this structure appears in sectioned Chinese hamster chromosomes.

The first impression is that the kinetochore is a trilayered structure, with alternating light-dark-light regions (Brinkley and Stubblefield, 1966). The microtubules from the spindle usually seem to penetrate the outer light layer, called the corona, and to terminate in the middle dark layer. The corona is most obvious in prophase or in cells blocked by Colcemid (Figure 3); as mitosis proceeds, the corona thins and disappears. This has led to some confusion in the description of kinetochores, since many researchers have described the three layers as dark-light-dark, with the inner dark layer closely applied to the dark chromosomal surface. The real problem comes in imagining what the unsectioned kinetochore looks like, because the actual organization is not easily extrapolated from the thin-section images. Most of the published images are highly selected orientations that reveal the most spectacular profiles; very few workers have actually taken the time to carefully reconstruct kinetochores from complete serial sections in a variety of orientations.

When technological advances permitted the study of isolated whole chromosomes, it was something of a disappointment to find that nothing like the kinetochore structure could be seen in the primary constriction of whole chromosomes (Stubblefield and Wray, 1971). It seemed that the kinetochore did not survive isolation. Although there was continuity across the centromere near the chromosome axis, peripherally the primary constriction was empty. Our own early studies of chromosomes isolated at neutral pH frequently demonstrated a few chromatin fibres spanning the primary constriction, but we were never able to find cases in which microtubles were attached to these fibres or anything else. Ris and Witt (1981) solved that problem by stabilizing the microtubules with duterium oxide (heavy water), which enabled them to demonstrate that microtubules were directly attached to chromatin fibres. They concluded that kinetochore microtubules are nucleated within the fibrous corona and subsequently become attached to the outer layer. Our earliest impression had been that the dark layer under the corona was fibrous, and we had tentatively suggested that it might be chromatin that was

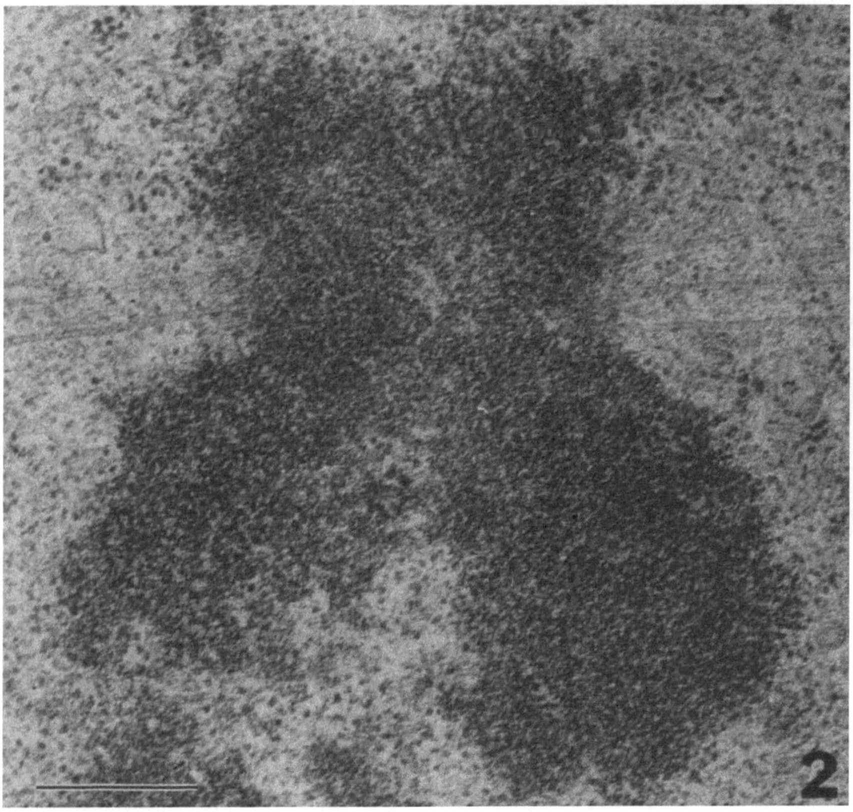

FIGURE 2: Chinese hamster chromosome showing microtubule attachment to kinetochores. The metaphase cell was fixed in glutaraldehyde, embedded in Epon, sectioned, stained with lead citrate, and photographed with an electron microscope. The termination of the microtubules in the dense layer is best shown in the right-hand kinetochore. Bar is 0.5 μm long.

somehow specialized to organize microtubules in mitosis (Brinkley and Stubblefield, 1966, 1970).

The heterochromatin adjacent to the centromere was visible in isolated chromosomes, even if the kinetochore itself was not. Relaxation of the chromosome structure in isolated chromosomes by removal of divalent ions revealed the constitutive heterochromatin on either side of the primary constriction and in both chromatids. Figure 4 shows the usual image in Chinese hamster chromosomes, in which such structures are relalitvely small. The interconnections often seen between heterochromatin masses lends support to the concept that they are involved in holding the two chromatids together until the moment of anaphase (Stubblefield, 1973). The recent finding that there is a brief rise in the concentration of calcium ions at the onset of anaphase (Poenie *et al.*, 1986) suggests a simple mechanism for the synchronization of chromatid separation at this point in mitosis. The rise in calcium ions may alter the heterochromatin strand interaction between chromatids by lower-

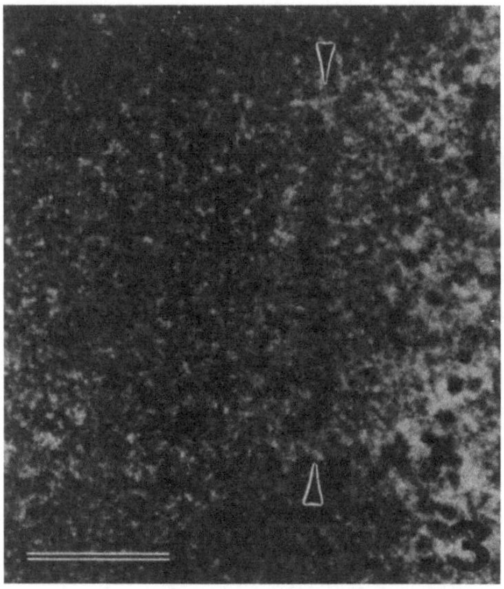

FIGURE 3: Kinetochore of a Chinese hamster chromosome prepared as in Fig. 2, but in this case treated with Colcemid, which blocks microtubule formation. The corona fibres are especially clear in this specimen, as they pass from the chromatid through the dense layer (arrow heads) into the corona area as loops. Bar is 0.25 μm long.

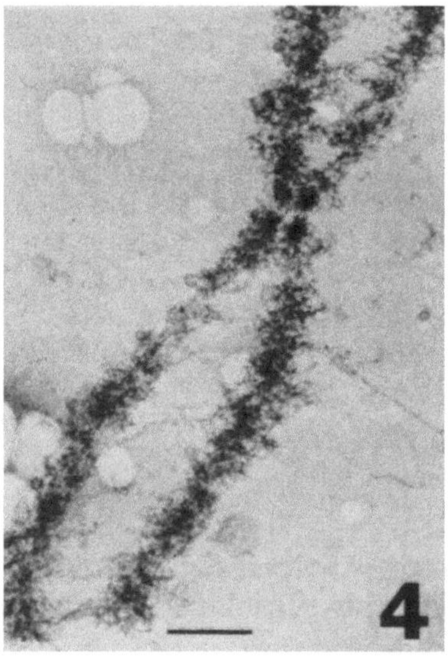

FIGURE 4: Isolated Chinese hamster chromosome prepared by critical point drying and viewed unstained by electron microscopy. The four dense heterochromatin clumps are seen at the primary constriction, but the typical kinetochore structure is not preserved. Bar is 2 μm long.

ing the affinity of a DNA binding protein, thereby releasing the connection holding the two daughter chromatids together.

Many published works have described the kinetochore as a proteinaceous disc that becomes attached to the chromosome in mitosis (Jokelainen, 1965, 1967). Although everyone seemed to agree that the innermost dark layer was chromatin, our view was that the outer dark layer also was chromatin, even though this was not easily demonstrated in isolated chromosomes. Demonstration of microtubule nucleation in broken cell preparations went a long way toward solving that particular controversy (Pepper and

Brinkley, 1979). Those studies clearly indicated that microtubules could be formed with attachment to chromatin fibres spanning the primary constriction, even when a typical kinetochore structure could not be demonstrated. One might argue that this was an artificial system and not characteristic of live kinetochores, but more recent evidence makes this unlikely.

Although this review considers mainly the kinetochore as seen in mammalian cells, studies have revealed similarities in the kinetochores of many other organisms. Perhaps the most informative case is that of yeast, in which the chromosomes appear to consist of single long chromatin strands, with no condensation into compact chromosomes. Ultrastructure studies reveal no typical kinetochore structure. Each daughter chromatin strand attaches to single microtubule of the mitotic apparatus (Peterson and Ris, 1976). This seems to be the limiting case for eukaryotes, and it confirms the concept of specific chromatin structures uniquely and directly attaching to the mitotic spindle fibres. In the typical case of an organism with large compact chromosomes, as many as a dozen to perhaps a hundred microtubules attach to the kinetochore with fairly uniform spacing. Considering the uniformity of such mechanisms in all living cells, we can anticipate that the more complex kinetochore is a scaled-up version of the simple system found in yeast. With this interpretation, we can cut through the uncertainties about what has been described by others to arrive at a simple concept: the microtubules attach directly to chromatin, and the kinetochore is a genetic element that maps to the primary constriction of condensed chromosomes.

Kinetochore cytochemistry

It is not surprising that the kinetochore contains DNA, and evidence suggests that DNA is found in the dense outer layer where chromatin fibres can be seen (Valdivia and Brinkley, 1985). RNA has also been detected in the region of the kinetochore (Rieder, 1979; Pepper and Brinkley, 1980). Whether this RNA is a product of gene transcription in the kinetochore or an association of RNA from another chromosomal site that selectively binds to the kinetochore is not yet known. From the sequence data for yeast kinetochore DNA, however, it seems clear that no protein would be translated from such an RNA (Blackburn and Szostak, 1984).

Several proteins that are probably specific components of the kinetochore are known. One of these is tubulin, the major structural protein of microtubules (Pepper and Brinkley, 1977). Although tubulin does not bind to DNA, some microtubule-associated-proteins (MAPS) do have DNA-binding capacity. In addition, several antigens that selectively bind to the antibodies are found in the sera of scleroderma patients (Brenner *et al.*,1981); the picture of the protein composition of kinetochores that has emerged with the use of these sera is a confusing one, but it is still helpful. Several kinetochore-specific proteins have been identified with these antibodies, and it is tempting to assume that the variations in observations by researchers from different laboratories are due to minor differences in technique. In this view, we find at least two major antigens reported by most groups, one protein in the 17–25 kD range, and another in the 70–95 kD range.

As shown in Table 1, however, there are still some discrepancies remaining. In the 17–25 kD antigen, for example, the acidic protein reported by Cox *et al.* (1983), Guldner *et al.* (1984), and McNeilage *et al.* (1986) is clearly different from the histone-like protein reported by Palmer *et al.* (1987). The Palmer protein, on the other hand, may be the same as that reported by Earnshaw and Rothfield (1985), and by Valdivia and Brinkley (1985), and its histone-like nature has been confirmed by Balczon and Brinkley (1986). In the 70–95 kD class there is more confusion. The 95 kD acidic protein observed by Guldner *et al.* (1986) is

TABLE 1

Kinetochore-specific antigens

Protein Size Range 17-25 kD	70-95 kD	Reference
20 (highly acidic)	–	Cox *et al.*, 1983
19.5 (non-histone)	95 (identified as DNA Topoisomerase I)	Guldner *et al.*, 1984 Guldner *et al.*, 1986
17 (shares epitopes with 70 kD, but product of a different gene)	70 (acidic, may be associated with centromeric heterochromatin)	Earnshaw and Rothfield, 1985 Earnshaw *et al.*, 1987
18 (may be associated with centromeric DNA; only in cycling cells; not required for microtubule binding or nucleation)	80 (in both cycling and noncycling cells; associated with tubulin)	Valdivia and Brinkley, 1985 Balczon and Brinkley, 1986, 1987
19.5 (does not share epitopes with 72)	72 (highly conserved across species)	McNeilage *et al.*, 1986
17 (like nuclear core histones; human sera not reactive in rat or chicken)	80 (human form; 50-kD protein in rat)	Palmer *et al.*, 1987
25 (extracted from nuclei with 0.8 M NaCl)	70 (resistant to extraction with 1.6 M NaCl; associated with nuclear lamina)	Pankov and Hancock, 1987

DNA topoisomerase I. Earnshaw *et al.* (1987) found a 70 kD protein that may be associated with constitutive heterochromatin. The McNeilage protein is highly conserved in organisms as disparate as humans and protozoans, whereas the Palmer protein appears to be quite different even in another mammal. It is possible that the 70 kD protein of Earnshaw and Rothman is the same as that of Valdivia and Brinkley. If Earnshaw *et al.* (1987) are correct in their assumption that this protein is not involved in microtubule binding, then agreement would be found in Balczon and Brinkley's (1986) observation that this antigen is found in both growing and non-cycling cells, whereas the smaller protein is found only in cycling cells. However, Balczon and Brinkley subsequently observed that the 80 kD protein is closely associated with tubulin in the cell, whereas the smaller protein does not seem to be required for microtubule binding or nucleation.

Several other kinetochore-specific antigens have been identified using scleroderma antisera. Balczon and Brinkley (1987) observed 21 kD and 52 kD proteins which, along with the 70 kD antigen, appear to be tightly associated with tubulin. Pankov and Hancock (1987) reported a 38 kD antigen that is associated with the nuclear matrix in nuclear fractionation experiments. The poten-

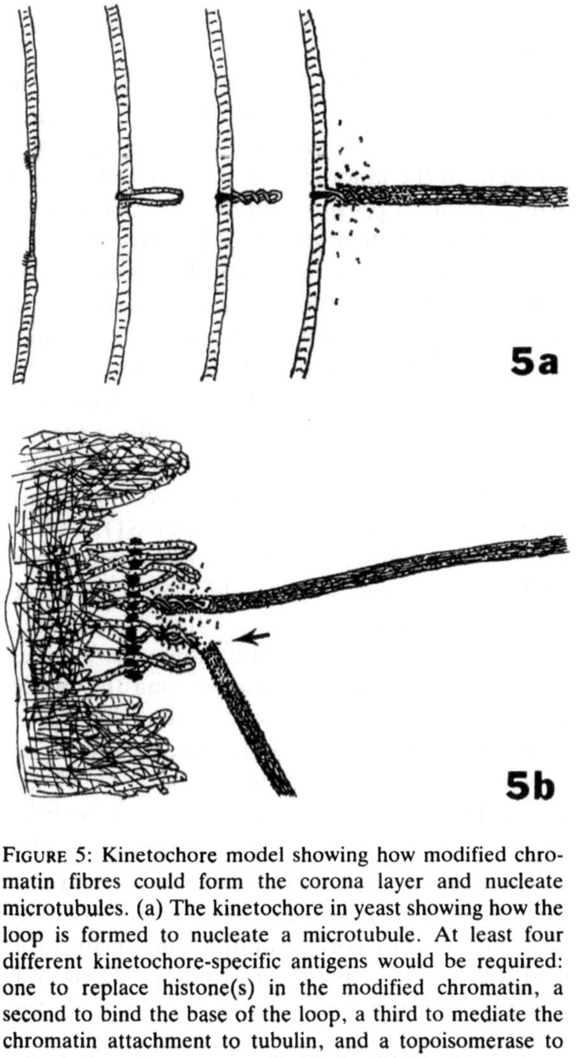

5a

5b

FIGURE 5: Kinetochore model showing how modified chromatin fibres could form the corona layer and nucleate microtubules. (a) The kinetochore in yeast showing how the loop is formed to nucleate a microtubule. At least four different kinetochore-specific antigens would be required: one to replace histone(s) in the modified chromatin, a second to bind the base of the loop, a third to mediate the chromatin attachment to tubulin, and a topoisomerase to twist the loop into a supercoiled helix. (b) The kinetochore in a mammalian cell. Four antigens are also required here, and the multiple-loop structure would function as it does in yeast. One microtubule is shown as it would appear if nucleated by a corona loop, and a second (arrow) in the process of being captured by a loop. The binding sites between the chromatin and microtubule would be within the lumen of the microtubule. As the microtubule disassembles in anaphase, these binding sites would pull the chromatin loop deeper within the remaining microtubule structure, generating the force that pulls the chromatid to the mitotic pole.

tial functions of these proteins are less clear than those of the major antigens.

This survey reveals that there are at least three, and possibly four or five, different kinetochore-specific proteins. One is very similar to a histone and may alter the structure of chromatin in the kinetochore. A second is a DNA topoisomerase. Some of the others may bind to DNA or chromatin in the kinetochore. This provides an interesting array of components for assembly into a model.

A Structural Model

The core and loop architecture of chromosomes (Stubblefield, 1973) provides the basis for a general model of the kinetochore. The studies of Ris and Witt (1981) are the clearest demonstration of the chromatin loop as a component of this structure. If each chromatin loop is a kinetochore unit of the type described in yeast by Peterson and Ris (1976), then the compound structure that we see in mammalian cells would be composed of as many loops as there are microtubules eventually attached to the kinetochore in mitosis. The basic idea is diagrammed in Figure 5. Before any microtubules are attached, the tips of the chromatin loops are evident as the corona. One of the kinetochore-specific antigens would then bind to the lateral sides of the loops to form the base of the corona, the outer dense layer. Brinkley and Stubblefield (1966) thought this layer was a chromatin fibre, since it had the dimensions and density of the condensed chromatin found in the rest of the chromosome. It seems clear now, however, that chromatin loops penetrate this layer vertically in prophase. As this layer forms, it holds all the loops together in one plane, whereas earlier they could have been oriented in almost any direction. The loops of chromatin in the corona are of finer diameter than elsewhere in the chromosome, which may reflect the substitution of another kinetochore-specific antigen for one or more of the histones in the chromatin and perhaps the elimination of the more typical nucleosome architecture in these loops (Palmer and Margolis, 1985; Palmer *et al.*, 1987). As the loops capture or nucleate microtubules they seem to disappear, and the corona is no longer an obvious component of the kinetochore. The organization described by Rattner (1986) in which chromatin fibres collapsed into the dense layer, is then one possibility. In this model, the variation in kinetochore size seen in specific chromosomes would correspond with the number of loops located in the centromere and the number of microtubules bound to the metaphase chromosome. The microtubule binding site would be on the corona loop imbedded in the outer layer and not on the antigen of that layer itself.

A Functional Model

The kinetochore functions in several specific ways. First, it must provide a binding site for the microtubules of the mitotic spindle to attach to the chromosome. In addition, it must provide for oriented binding of the kinetochores of the two chromatids to opposite poles of the spindle. We also think it must provide for capture of existing microtubules (at least initially) and must nucleate microtubule assembly and growth as well. Finally, the kinetochore somehow provides the motive force for chromatid migration toward the mitotic pole at anaphase. The centromere region must allow for the adhesion of chromatids until anaphase and then allow simultaneous release when anaphase migration starts. These functions must be accounted for in any model that we propose.

The unique nature of the chromatin strands of the corona may derive from the substitution of a specific kinetochore antigen for one or more of the components of nucleosomes in ordinary chromatin, as described earlier. This DNA-protein affinity must be determined by the base sequence of the DNA, as it appears to be in yeast centromeres. If this altered chromatin provides a unique binding site for the other antigens of the kinetochore, then we can explain why spindle fibres bind only to the kinetochore and not to other regions of the chromosome. One of these antigens would ligate the corona loops laterally, and another antigen would be a microtubule-associated protein that links tubulin to the corona chromatin. This would raise the tubulin concentration in the corona layer and provide nucleation sites for microtubule formation from tubulin dimers. Initially, newly formed microtubules would probably be randomly oriented, as seen by Ris and Witt (1981); as they grow longer, however, charge distributions on such linear structures would probably force them into parallel arrays so that they would eventually lie vertical to the kinetochore outer layer. The linkage of the corona loops to microtubules is dynamic rather than fixed, because both nucleated growth and disassembly of the microtubule can occur without loss of the connection to the chromatin (Gorbsky *et al.*, 1987). Existing microtubules can also be captured by the kinetochore (Mitchison and Kirschner, 1985); this can be accomplished by incorporation of corona-bound tubulin into the end of a microtubule or binding of tubulin already incorporated in a microtubule to the coronal antigen, as indicated by the arrow in Figure 5b.

The forces responsible for chromosome movement are present throughout mitosis, not just at anaphase. The force that moves chromosomes onto the mitotic spindle during prometaphase (congression) is thought to be the same as that which is detected later during anaphase (Nicklas and Kubai, 1985). The well-organized anaphase motion is easier to describe, although we are uncertain about the source of the forces propelling the chromatids toward the mitotic poles. The synchronous anaphase motion is the result of a release of the binding affinity holding together the heterochromatin masses near the centromere. This would be initiated by the brief rise in calcium ion concentration, as described earlier. The poleward migration has been shown to correlate with a corresponding shortening of the microtubules bound to the kinetochore *from the chromosome end* (Gorbsky *et al.*, 1987). If, as it appears, the chromosome pulls itself along the microtubule bundle as the microtubules shorten, then the bundle must be firmly anchored near the spindle pole. The interaction of kinetochore microtubules and pole-to-pole microtubules generated by the centrioles may serve to anchor the spindle fibre; this had earlier been thought to be a fibre interaction of some kind and the source of the motive power.

This still does not reveal why the chromatids move. How can a force be generated to pull the chromatid as the microtubules are disassembling *at the kinetochore*? In one possibility offered by Hill (1985), the kinetochore is modelled as a perforated layer through which the microtubules penetrate; disassembly would occur below the dense layer, allowing the kinetochore plate to maintain a loose attachment to the spindle fibre.

The model which we favour (Stubblefield and Cherry, 1987) is illustrated in Figure 5. In this proposal, the corona loops nucleate the formation of microtubules from within and are inserted into the microtubule lumen. In order for the loop to fit inside the microtubule it must be supercoiled into a tightly twisted helix. This is the role of the DNA topoisomerase I, and the action requires that the loops be firmly anchored in the dense layer. In this model the microtubule forms a sleeve over the twisted loop of chromatin, allowing intimate binding internally, while permitting disassembly at the microtubule end around the base of the loop. This idea also accounts for the disappearance of the corona layer when microtubules attach to the kinetochore. As the microtubule shortens, the chromatin loop would shift its binding sites to positions deeper in the lumen, and this molecular repositioning would be the force that pulls the chromatid. This model has the additional value of explaining why hollow microtubules are used as spindle fibres instead of solid microfilaments. It also provides a mechanism that would work in yeast, in which a single microtubule attaches directly to the chromatin fibre that comprises the chromatid, as in Figure 5a. With this mechanism we can also understand why alteration of chromatin structure is necessary in the corona loops; a thinner, more flexible structure is needed to make the twisted helix which is inserted into the confines of the microtubule lumen.

Summary

Studies of the ultrastructure and cytochemistry of the centromere region of chromosomes has revealed a wealth of information. Many laboratories have contributed to this body of knowledge. Although all of the details are not yet available, enough is already known to allow us to outline the structure and composition of the kinetochore and to begin to understand how it functions in mitosis. The anaphase motion of chromatids appears to be a direct interaction of modified chromatin loops with the ends of microtubules as they disassemble. A model for this function has been presented in this review. We do not yet know which of these components are missing in inactive centromeres of compound chromosomes, but at least we now have the tools for detailed immunological studies and can formulate meaningful questions for further research.

References

Arrighi FE, Hsu TC (1971). Localization of heterochromatin in human chromosomes. Cytogenetics 10; 81–6.

Balczon RD, Brinkley BR (1986). Kinetochore associated proteins of mammalian chromosomes: Analysis with scleroderma CREST autoantibodies. J Cell Biol 103; 413a.

Balczon RD, Brinkley BR (1987). Tubulin interaction with kinetochore proteins; analysis using a bifunctional chemical crosslinker. J Cell Biol 105; 282a.

Blackburn EH, Szostak JW (1984). The molecular structure of centromeres and telomeres. Ann Rev Biochem 53; 163–94.

Brat SV, Verma RS, Dosik H (1979). A simplified technique for simultaneous staining of nucleolar organizer regions and kinetochores. Stain Tech 54; 107–8.

Brenner S, Pepper D, Berns MW, Tan E, Brinkley BR (1981). Kinetochore structure, duplication and distribution in mammalian cells: Analysis by human autoantibodies from scleroderma patients. J Cell Biol 91; 95–102.

Brinkley BR, Hsu TC (1966). Structure and behavior of specialized regions of mammalian chromosomes—kinetochores, telomeres, and nucleolar organizers. J Cell Biol 31; 16a.

Brinkley BR, Stubblefield E (1966). The fine structure of the kinetochore of a mammalian cell in vitro. Chromosoma 19; 28–43.

Brinkley BR, Stubblefield E (1970). Ultrastructure and interaction of the kinetochore and centriole in mitosis and meiosis. Adv in Cell Biol 1; 119–85.

Cherry LM, Johnston DA (1987). Size variation in kinetochores of human chromosomes. Hum. Genet 75; 155–8.

Cox JV, Schenk EA, Olmsted JB (1983). Human anticentromeric antibodies: Distribution, characterization of antigens, and effect on microtubule organization. Cell 35; 331–9.

Crippa M (1966). Rate of ribonucleic acid synthesis during the cell cycle. Exptl Cell Res 42; 371–4.

Daniel A (1979). Single Cd band in dicentric translocations with one suppressed centromere. Hum. Genet 48; 85–92.

Denton TE, Brooke WR, Howell WM (1977). A technique for the simultaneous staining of both nucleolar organizer regions and kinetochores of human chromosomes with silver. Stain Tech 52; 311–13.

Dewald GW, Boros SJ, Conroy MM, Dahl RJ, Spurbeck JL, Vitek HA (1979). A translocation dicentric (5;15) (p31;p11) chromosome showing variation for constriction in the centromeric regions in a patient with the cri du chat syndrome. Cytogenet Cell Genet 24; 15–26.

Earnshaw WC, Migeon BR (1985). Three related centromere proteins are absent from the inactive centromere of a stable isodicentric chromosome. Chromosoma 92; 260–96.

Earnshaw WC, Rothfield N (1985). Identification of a family of human centromere proteins using autoimmune sera from patients with scleroderma. Chromosoma 91; 313–21.

Earnshaw WC, Sullivan KF, Machlin PS, Cooke CA, Kaiser DA, Pollard TD, Rothfield NF, Cleveland DW (1987). Molecular cloning of cDNA for CENP-B, the major human centromere autoantigen. J Cell Biol 104; 817–29.

Eiberg H (1974). New selective Giemsa technique for human chromosomes, Cd staining. Nature 248; 55.

Evans HJ, Ross A (1974). Spotted centromeres in human chromosomes. Nature 249; 861–2.

Ghosh S, Chaudhuri A (1984). Analysis of three whole-arm translocations in a mouse sarcoma cell line. Cytogenet Cell Genet 38; 161–4.

Godward MBE (1985). The kinetochore. Internatl Rev Cytol 94; 77–105.

Gorbsky GJ, Sammak PJ, Borisy GG (1987). Chromosomes move poleward in anaphase along stationary microtubules that coordinately disassemble from their kinetochore ends. J Cell Biol 104; 9–18.

Guldner H-H, Lakomek H-J, Bautz FA (1984). Human anticentromere sera recognize a 19.5 kD nonhistone chromosomal protein from HeLa cells. Clin Exptl Immunol 58; 13–20.

Guldner H-H, Szostechi C, Vosberg HP, Lakomek H-J, Penner E, Bautz FA (1986). Scleroderma 70 autoantibodies from scleroderma patients recognize a 95 kD protein identified as DNA Topoisomerase I. Chromosoma 94; 132–8.

Hill TL (1985). Theoretical problems related to the attachment of microtubules to kinetochores. Proc Natl Acad Sci USA 82; 4404–8.

Hsu TC, Arrighi FE (1971). Distribution of constitutive heterochromatin in mammalian chromosomes. Chromosoma 34; 243–53.

Ing PS, Smith SD (1983). Cytogenetic studies of a patient with mosaicism of isochromosome 13q and a dicentric (Y;13) translocation showing differential centromeric activity. Clin Genet 24; 194–9.

Jokelainen PT (1965). Fine structure aspects of kinetochores in dividing cells from fetal rat kidney. Anat Rec 151; 367.

Jokelainen PT (1967) The ultrastructure and spatial organization of the metaphase kinetochore in mitotic rat cells. J Ultrastr Res 19; 19–44.

Krishan A, Buck RC (1965). Structure of mitotic spindle in L strain fibroblasts. J Cell Biol 24; 433–44.

Lambiase S, Maraschio P, Zuffardi O (1984). The Cd technique identifies a specific structure related to cen-

tromeric function. Hum Genet 67; 214–5.
Maraschio P, Zuffardi O, LoCurto F (1980). Cd bands and centromeric function in dicentric chromosomes. Hum Genet 54; 265–7.
McNeilage LJ, Whittingham S, McHugh N, Barnett AJ (1986). A highly conserved 72,000 Dalton centromeric antigen reactive with autoantibodies from a patient with progressive systemic sclerosis. J Immunol 137; 2541–7.
Merry DE, Pathak S, Hsu TC, Brinkley BR (1985). Antikinetochore antibodies: Use as probes for inactive centromeres. Am J Hum Genet 37; 425–30.
Mitchison TJ, Kirschner MW (1985). Properties of the kinetochore in vitro. II. Microtubule capture and ATP-dependent translocation. J Cell Biol 101; 766–77.
Moroi Y, Peebles C, Fritzler MJ, Steigerwald J, Tan EM (1980) Autoantibody to centromere (kinetochore) in scleroderma sera. Proc Natl Acad Sci USA 77; 1627–31.
Nakagome Y, Nakahori Y, Mitani K, Matsumoto M (1986). The loss of centromeric heterochromatin from an inactivated centromere of a dicentric chromosome. Jpn J Hum. Genet 31; 21–6.
Nicklas RB (1987). Chromosomes and kinetochores do more in mitosis than previously thought. In Chromosome structure and function: impacts of new concepts (eds Gustafson JP, Appels R, Kaufman RJ), Plenum Press, New York; 53–74.
Nicklas RB, Kubai DF (1985). Microtubules, chromosome movement, and reorientation after chromosomes are detached from the spindle by micromanipulation. Chromosoma 92; 313–24.
Niebuhr E (1972). Dicentric and monocentric Robertsonian translocations in man. Humangenetik 16; 217–26.
Palmer DK, Margolis RL (1985). Kinetochore components recognized by human autoantibodies are present on mononucleosomes. Mol Cell Biol 5; 173–86.
Palmer DK, O'Day K, Wener MH, Andrews BS, Margolis RL (1987). A 17 kD centromere protein (CENP-A) copurifies with nucleosome core particles and with histones. J Cell Biol 104; 805–15.
Pankov R, Hancock R (1987). Polypeptides of the kinetochore of Chinese hamster ovary cells. J Cell Biol 105; 240a.
Pardue ML, Gall JG (1970). Chromosomal localization of mouse satellite DNA. Science 168; 1336–8.
Paris Conference (1971). Standardization in human cytogenetics. The National Foundation, Birth Defects: Original Article Series. Vol. XI, No. 9 (1972).
Pathak S, Wurster-Hill D (1971). Distribution of constitutive heterochromatin in carnivores. Cytogenet Cell Genet 18; 245–54.
Pepper DA, Brinkley BR (1977). Localization of tubulin in mitotic apparatus of mammalian cells by immunofluorescence and immunoelectron microscopy. Chromosoma 60; 223–35.
Pepper DA, Brinkley BR (1979). Microtubule initiation at kinetochores and centromers in lysed mitotic cells: Inhibition of site-specific nucleation by tubulin antibody. J Cell Biol 82; 585–91.
Pepper DA, Brinkley BR (1980). Tubulin nucleation and assembly in mitotic cells: Evidence for nucleic acids in kinetochores and centrosomes. Cell Motil 1; 1–15.
Peretti D, Maraschio P, Lambiase S, LoCurto F, Zuffardi O (1986). Indirect immunofluorescence of inactive centromeres as indicator of centromere function. Hum Genet 73; 12–16.
Peterson JB, Ris H (1976). Electron-microsopic study of the spindle and chromosome movement in the yeast *Saccharomyces cerevisiae*. J Cell Sci 22; 219–42.
Poenie M, Alderton J, Steinhardt R, Tsien R (1986). Calcium rises abruptly and briefly throughout the cell at the onset of anaphase. Science 233; 886–9.
Rattner JB (1986). Organization within the mammalian kinetochore. Chromosoma 93; 515–20.
Rieder CL (1979). Localization of ribonucleoprotein in the trilaminar kinetochore of PTK1. J Ultrastr Res 66; 109–19.
Rieder CL (1982). The formation, structure, and composition of the mammalian kinetochore and kinetochore fiber. Internatl Rev Cytol 79; 1–58.
Ris H, Witt PL (1981). Structure of the mammalian kinetochore. Chromosoma 82; 153–70.
Robbins E. Gonatas NK (1964). Ultrastructure of mammalian cell during mitotic cycle. J Cell Biol 21; 429–63.
Rocchi A, Prantera G, DiCastro M, Capanna E (1982). Silver staining of the centromeric area of acrocentric and Robertsonian metacentric mouse chromosomes. Genetica 60; 65–9.
Roos U-P (1975). Are centromeric dots kinetochores? Nature 254; 463.
Seabright M (1971). A rapid banding technique for human chromosomes. Lancet 2; 971–2.
Schwartz S, Palmer CG, Weaver DD, Priest J (1983). Dicentric chromosome 13 and centromere inactivation. Hum Genet 63; 331–7.
Stubblefield E (1973). The structure of mammalian chromosomes. Internatl Rev Cytol 35; 1–60.
Stubblefield E, Cherry LM (1987). Kinetochore model for mitotic functions: Microtubule interactions and chromatid mobility. J Cell Biol 105; 280a.
Stubblefield E, Wray W (1971). Architecture of the Chinese hamster metaphase chromosome. Chromosoma 32; 262–94.
Valdivia MM, Brinkley BR (1985). Fractionation and initial characterization of the kinetochore from mammalian

metaphase chromosomes. J Cell Biol 101; 1124–34.
Vig BK, Zinkowski RP (1986). Sequence of centromere separation: A mechanism for orderly separation of dicentrics. Cancer Genet Cytogenet 22; 347–59.
Ward BE, Bradley CM, Cooper JB, Robinson A (1981). Homodicentric chromosomes: A distinctive type of dicentric chromosome. J Medical Genet 18; 54–8.
Ying KL, Ives EJ (1971). Mitotic behavior of a human dicentric Y chromosome. Cytogenet 10; 208–18.
Zinkowski RP, Vig BK, Broccoli D (1986). Characterization of kinetochores in multicentric chromosomes. Chromosoma 94; 243–8.

6 Comparative Aspects of Chromosome Replication in *Drosophila* and Mammals

RAJIVA RAMAN AND S.C. LAKHOTIA

Cytogenetics Laboratory
Centre of Advanced Study in Zoology
Banaras Hindu University
Varanasi 221005, India

The double helical model for DNA structure provides a self-replicating property to the molecule. Essentially, DNA replication requires polymerization of complementary chain of nucleotides on a template. Fulfilment of this requirement, however, involves an elaborate enzymatic machinery even for the simplest of organisms where the genetic material could be a naked DNA molecule, only a few kilobases long (Kornberg, 1980). Although the basic process has been conserved through the evolutionary time scale, the increasing complexity of the genome has correspondingly resulted in a more intricate mechanism of replication in higher organisms. The genetic material in eukaryotes is not only compartmentalized in the nucleus, but is also distributed in different linkage groups and is complexed with a variety of proteins, some permanent others transient (see Sen and Kuo, this volume). The genetic material thus exists as chromatin and becomes visible as discrete individual units, the chromosomes, in certain stages of the cell reproduction cycle. Unlike in prokaryotes, cell division does not immediately follow replication nor does replication necessarily follow division. Two periods of growth respectively span cell division and DNA replication (G_1), and vice-versa (G_2). This, however, leaves a well defined period in the cell division cycle, the S-phase, in which replication is effected.

Individual chromosomes comprise large amounts of DNA and while their number and sizes vary within and between taxa, an average-sized chromosome has several-fold more DNA than in *E. coli*. Yet the ability of a cell to replicate so much DNA within a few hours indicates that replication proceeds in different chromosomes and in different regions of the same chromosome at the same time. However, as will be discussed later, not all the DNA is replicated simultaneously and segments of DNA are programmed to replicate at different but generally predetermined periods within the S-phase. Thus the increase in the amount of DNA, its compaction into chromatin and confinement of replication to a limited period in the cell cycle have made the process of replication more elaborate in eukaryotes than in prokaryotes.

Studies on chromosome replication were initially handicapped by the fact that chromatin is generally cytologically invisible at the time of replication (S-phase). The advent of the cellular autoradiography technique and the commercial availability of radiolabelled nucleic acid precursors brought in their wake a volume of data on chromosomal replication. In the simplest form of technique, mitotically cycling cells are grown in a medium containing ^{3}H-thymidine for desired periods so that the cells that are in S-phase take up the label at the sites of ongoing DNA synthesis and these cells are then allowed to progress to metaphase before the cells are harvested for cell autoradiographic studies. The slides containing chromosome preparations are coated with a fine-grained photographic emulsion and exposed in darkness for a certain period of time before photographic development. ^{3}H-thymidine, wherever incorporated during the S-phase, is revealed on the chromosomes as dark silver grains. Extremely close proximity of chromosomes with the photographic emulsion ensures a reasonably accurate representation of the sites replicating at the time of ^{3}H-thymidine incorporation. There are cell types where chromosomes are visible during their synthetic phase itself, e.g. polytene cells in diptera and certain other organisms. Obviously, in these cases, ^{3}H-thymidine incorporation is not required to be chased up to metaphase, instead it can be visualized directly. In the case of mitotically cycling cells also, a comparable condition can be experimentally simulated by premature chromosome condensation of interphase chromatin by fusion with mitotic cells (Sperling and Rao, 1974).

A major advance in the study of chromosome replication, especially in mammalian cells, was the development of a variety of chromosome banding (C, G, Q, R-bands, etc.) techniques and the use of 5-bromodeoxyuridine as the label for newly synthesized DNA While the various banding techniques allowed a better resolution of structure of mitotic chromosomes which otherwise looked monotonously uniform, the bromodeoxyuridine label bypassed the requirement of the autoradiographic step since special methods of fluorescent or Giemsa staining directly allowed a precise visualization of sites of incorporation of the label with a much higher degree of resolution than that obtained with the autoradiographic technique (see Zakharov and Egolina, 1972; Latt, 1973).

The present chapter describes some aspects of replication at chromosomal level, especially features that regulate the S-phase and the temporal order of chromosomal replication in relation to chromatin activity and cell differentiation. Since our own areas of research are confined to mammals and *Drosophila,* the discussion will mainly be confined to these two systems with an attempt to bring out similarities and differences in their strategies for chromosomal replication.

Different Strategies of Chromosomal Replication in Mammals and *Drosophila*

In a majority of organisms, including mammals, cell proliferation is the main source of growth and differentiation. In a typical cell proliferation cycle, the products of a mitotic division undergo a variable period of cell growth (the G_1-period) preparatory to the following S- or the DNA replication phase. The G_2-period which follows the S-phase prepares the cell for entry into mitosis which typically involves chromosome condensation, an orderly separation of sister chromatids to two poles, reformation of nuclear envelopes around the two daughter nuclei and finally cytokinesis to complete the generation of two daughter cells (Prescott, 1976). In several groups of eukaryotes, however, certain cell types short circuit the regular mitotic cell cycle in various ways to achieve somatic growth and differentiation (see Rudkin, 1972, Nägl, 1978).

For example, in a number of tissues of *Drosophila,* the post-S-phase cells, instead of entering the cell division cycle, resume a fresh cycle of DNA synthesis following a pause of varying duration. Thus in organisms like *Drosophila*, somatic growth is achieved largely by endoreplication and cell growth following an initial phase of cell multiplication. While the early embryonic and those larval cells that are destined to form adult structures proliferate mitotically, a majority of larval and some of the adult cell types result from a well regulated pattern of endoreplication cycles. It is also notable that in the adult stage of *Drosophila*, DNA synthesis occurs only in gonadal cells.

General Aspects of Replication and Chromosome Structure in *Drosophila*

Drosophila provides examples of mitotic and endoreplication cycles in its different cell types and at different stages of development. The occurrence of endoreplication cycles in somatic cells of *Drosophila* (and other similar organisms) can lead to at least two types of nuclear organizations, viz., the well known polytene types with distinct band-interband organization of chromosomes in interphase nucleus, and the polyneme type (see Rieger *et al.*, 1976 for definitions) in which all the daughter chromatids of different chromosomes remain less tightly synapsed so that the nucleus looks more like a typical interphase nucleus but much larger in size and DNA content. An interesting aspect of endoreplication cycles, *Drosophila* is the occurrence of specific patterns of unequal replication of different DNA sequences on the same chromosome. This under- and/or over-replication raises intriguing possibilities about chromosome structure in these endoreplicating polytene or polyneme cell types.

Among the non-polytene tissues, the brain ganglia of *Drosophila* larvae have received considerable attention for studies on chromosomal structure and replication since this tissue provides the most common source for studies on mitotic chromosomes in *Drosophila*. Although not so commonly appreciated, the brain ganglia of *Drosophila* larvae also contain endoreplicating cells. Evidence for endoreplication in larval brain cells was first provided by Berendes and Keyl (1967) who found by cytophotometric measurements that the DNA content in different interphase cells in brain ganglia of *D. hydei* larvae was not in geometric multiples of the 2C values; rather the hetero- and euchromatin components appeared to endoreplicate to independently varying levels. Occurrence of such independent endoreplication cycles of hetero- and euchromatin regions in different brain cells in *D. nasuta* larvae was also shown by autoradiographic and fluorescence studies by Lakhotia and Kumar (1980). Lakhotia (1982) further raised the possibility that some of these endoreplicating brain cells are triggered to enter mitotic division cycles and since these mitotic cells display typical 2N metaphase figures, the metaphase chromosomes in these cell types could be polynemic as suggested by several cytophotometric studies earlier (Rudkin, 1963; Swift, 1965; Gay *et al.*, 1970). However, another cytophotometric study has contradicted the polynemy concept (van de Flierdt, 1975). Since the patterns of segregation of newly synthesized DNA in daughter chromatids can be followed and since these patterns provide information about the uninemic or polynemic structure of mitotic chromosomes (see Prescott, 1970), a detailed analysis of cell cycle and chromosomal labelling patterns in mitotically dividing cells of *Drosophila* becomes specially important for understanding their metaphase chromosome structure.

Unfortunately, studies of chromosomal replication and cell cycle patterns in these mitotically dividing larval cell types are rather few in the published literature (Pimpinelli *et al.*, 1976; Wienberg, 1977; Tsuji and Tobari, 1980; Steinemann, 1980) and these, too, have been based on equivocal assumptions. Steinemann (1980), using the conventional ^{3}H-thymidine pulse-chase method but scoring from a small metaphase sample, estimated the neuroblast cell cycle in *D. virilis* larvae to be about 23h long (S = 12h; G_2 = 7h). However, a basic assumption in Steinemann's (1980) study that the replicating cells in larval brain ganglia constitute an exponentially proliferating population of cells is also not valid in view of the known presence of a significant number of endoreplicating cells in the larval brain (see above). Lakhotia *et al.* (1979) found at least some of the larval brain cells in *D. nasuta* to complete cell cycle within 1–2h. In another more extensive study, Roy (1983) noted that the percentage of labelled metaphases in brain ganglia of *D. nasuta* larvae pulse labelled with ^{3}H-thymidine for 10–15 min and chased for 15 min to 48h, remained greater than 50 per cent in nearly all samples and thus the cell cycle parameters could not be estimated conventionally. Study of cell cycle parameters in the brain ganglia of *Drosophila* larvae is interesting but difficult to interpret since this tissue includes both mitotically dividing and endoreplicating cells. If the mitotic cells in larval brain ganglia are indeed polynemic as some of the evidence may suggest, it implies that the metaphase chromosome organization in early cleavage division cells and in larval brain cells differs and that the patterns of chromosomal replication cycles bring about this difference. A difference in the mitotic chromosome organization between embryonic and larval brain cells was also noted in studies on the sensitivity of these chromosomes to exposure to certain DNA-ligands like Hoechst 33258, Distamycin A, etc. (Lakhotia and Roy, 1981, 1983; Dolfini and Razzini, 1983; Dolfini, 1987). Thus the replication cycles in larval brain ganglia continue to raise intriguing questions.

A very striking feature of endoreplication cycles in *Drosophila* and several other organisms is a tissue- and developmental stage-specific regulation of participation of different DNA sequences in endoreplication. Classical observations of Heitz (1934) and numerous other subsequent studies (Rudkin, 1964; Gall *et al.*, 1971; Lakhotia, 1974, 1984; Hammond and Laird, 1985a) have shown that the bulk of pericentric heterochromatin regions does not endoreplicate during polytenization in salivary glands of *Drosophila* larvae. Notwithstanding some recent claims to the contrary (e.g. see Lamb, 1982; Dennhoffer, 1982a, b), it is now well established that in nearly all polytene cell types in *Drosophila*, the pericentric heterochromatin regions, including the entire Y-chromosome in males, remain grossly underreplicated although the extent of underreplication may vary in a narrow range in relation to developmental stage and tissue type (see Lakhotia, 1984; Hammond and Laird, 1985a, b; Zacharias, 1986; Spradling and Orr-Weaver, 1987; P.K. Tiwari and Lakhotia, unpublished).

A number of other specific sequences are also known to be underreplicated in salivary gland polytene nuclei of *Drosophila* larvae. These include ribosomal sequences, histone genes, nearly all intercalary heterochromatin sites, the bithorax locus, etc. (for a detailed discussion and references, see Spradling and Orr-Weaver, 1987). In view of these instances of underreplication and several other considerations, Laird (1980) proposed an interesting model of polytene chromosome organization in *Drosophila* larvae where it was suggested that the level of polyteny varies along the length of a chromosome arm such that the interbands are generally less polytenized than the bands (also see Lakhotia and Sinha, 1983). Although Spierer and Spierer (1984) did not find any difference in the levels of polyteny in interbands and bands in a specific region of the 3rd chromosome, a variable polytenization along the euchromatic arms seems to be a reality (Zhimulev *et al.*, 1982; Lamb and Laird, 1987). This may not, however, be as extensive as proposed by Laird (1980). These instances of under-replication and consequent differences in polyteny levels along the length of a chromosome arm, require the presence of cascades of arrested replication forks on either sides of the under-replication domains.

The patterns of differential replication in other endo replicating cell types of *Drosophila* have been studied less extensively. Endow and Gall (1975) found the relative amounts of satellite sequences in different polytene and non-polytene cell types of *D. virilis* to vary, which suggested a tissue-specific pattern of differential replication of these sequences during endoreplication cycles. The larval brain ganglia have generally been taken as representatives of diploid cell types because they show typical diploid mitotic figures and also because the relative amounts of satellite sequences in this tissue were found to be typical for the species. However, the progression of endoreplication cycles in many of the larval brain cells is well documented cytologically and since the heterochromatin may be under- or over-replicated in different brain cells (Berendes and Keyl, 1967; Lakhotia and Kumar, 1980), it is possible that biochemical estimations of satellite DNA content in this tissue would fail to detect any change due to the averaging effect (Lakhotia, 1982). Grimm and Kunz (1980) and Grimm *et al.* (1984) reported over-replication of rDNA sequences in brain ganglia and/or thoracic muscles in certain specific genotypes of *D. hydei.* Starting with the presumption that the brain tissue is essentially diploid, these authors suggested that this over-replication of rDNA in certain genotypic combinations in 'diploid' cells is distinct from the regulated under-replication, magnification, etc. that occur in other polytene cell types. However, in view of the above, the over-replication of rDNA sequences in brain ganglia of *D. hydei* (Grimm *et al.*, 1984) could also be linked to independent endo-replication cycles of euchromatin and heterochromatin (within which the rDNA sequences are located) in a significant proportion of brain cells.

A different category of differential replication in *Drosophila* cells is the amplification of chorion gene clusters in follicle cells at specific stages of ovarian follicle development. This aspect has been reviewed periodically (for details, see Kafatos *et al.*, 1985a, b; Kalfayan *et al.*, 1985; Spradling and Orr-Weaver, 1987). So far no other instances of gene amplification are known in *Drosophila*, although DNA puffs, which reflect gene amplification in polytene chromosomes, are known in some other dipterans like *Rhynchosciara* (Gabrusewycz-Garcia, 1964; Amabis and Amabis, 1984; Spradling and Orr-Weaver, 1987).

In summary, the control of replication, mitotic vs endoreplication cycles, is a very important aspect of cell differentiation in *Drosophila.* Moreover, in cells that are channellized to enter endoreplication pathways, the *Drosophila* genome can identify specific DNA sequences and regulate their participation in replication such that they show characteristic levels of under- or over-replication. The precise mechanisms that effect this intricate regulation are not known, but as may be expected, a rather large number of gene loci affecting replication patterns of specific cell types and/or DNA sequences have been identified in *D. melanogaster* (reviewed in Spradling and Orr Weaver, 1987). Hormonal and developmental conditions also modulate replication patterns of specific cell types. From this point of view, cell differentiation in a typical mammalian system shows certain very basic differences from that in *Drosophila.* Control of replication, i.e., the number and timing of mitotic division cycles are of course also important in mammalian cells but these replication cycles do not, in general, bring about such extensive qualitative and quantitative changes in the genomic content of differentiating cells as in *Drosophila* and similar other systems.

Temporal Sequence of Chromosomal Replication

Initial studies on chromosomal replication in mitotic (Lima-de-Faria, 1959; Taylor, 1960) and polytene cells (Plaut, 1963; Keyl and Pelling, 1963) demonstrated that different chromosomes and regions thereof replicated at different times in the S-phase. One of the earliest (and more or less global) correlations between the time of replication and chromatin structure/organization was the replication of heterochromatin during the late S-phase (Lima-de-Faria and Jaworska, 1968). The literature is replete with observations in both plants and animals that heterochromatin replicates late in the S-phase. The correlation between heterochromatization, genetic inactivity and replication during the late S-phase (Lima-de-Faria and Jaworska, 1968; also see Shah *et al.*, 1974 for review of earlier literature) was regardless of whether the heterochromatin was constitutive or facultative (Brown, 1966) like the inactive or lyonized X-chromosome in somatic cells of female mammals (Lyon, 1961, 1974).

With the advent of metaphase chromosome banding techniques, it was shown by Ganner and Evans (1971) and Sharma and Dhaliwal (1974) in humans and the barking deer, *Muntiacus muntjak*, respectively, that besides heterochromatin, the G-band regions also replicated during the later half of the S-phase. The use of BrdU-Giemsa staining technique has since provided much more detailed and better resolved information on the temporal and regional order of replication in individual mitotic chromosomes in mammalian cells (Stubblefield, 1975; Dutrillaux *et al.*, 1976; Das and Savage, 1978; Popescue and Dipaolo, 1979; Camargo and Cervenka, 1982; Dubey, 1985).

A major aspect of the large number of autoradiographic studies on polytene chromosome replication was to ascertain the temporal order of replication of different chromosomal regions during a given polytene S-period. However, the absence of real time marker in these endoreplication cycles and a varying degree of asynchrony in different nuclei have been limiting factors which necessitated application of other strategies for analysis (see Rudkin, 1972 for review). It is now well established that during the later polytene replication cycles, a given S-phase is initiated by replication at disperse regions like interbands and puffs (designated variously as Early Discontinuous, Interband or Disperse Discontinuous patterns); this early phase is quickly followed by a phase when all disperse and dense chromosome regions are more or less uniformly labelled (the Continuous patterns) and finally the denser bands, intercalary and chromocentric heterochromatin regions continue replication till the very late S-phase (the Discontinuous patterns, see Rudkin, 1972, Hägele, 1973; Roy and Lakhotia, 1979; Mukherjee *et al.*, 1980). In addition to correlating these gross patterns of ^{3}H-thymidine labelling of polytene nuclei with their early or late S-phase, such studies also provided a wealth of data concerning the temporal order of replication of specific interband, band, puff sites, etc. (see later). These studies also showed that besides the heterochromatin regions (chromocentric and intercalary), many of the denser bands continued to replicate in late S in a reproducible manner (Plaut *et al.*, 1966; Lakhotia and Mukherjee, 1970; Rudkin, 1972) to generate 'replication bands' analogous to those in the mitotic chromosomes of mammals, etc. Curiously, however, when mitotic chromosomes of *Drosophila* were examined for ^{3}H-thymidine labelling or BrdU-Giemsa staining patterns to analyse the temporal sequence of chromosomal replication, no segmental patterns or 'replication bands' were seen: even with a brief pulse of 10–15 min, the labelled metaphases were uniformly labelled either throughout euchromatin or throughout heterochromatin or throughout both regions (Barigozzi *et al.*, 1966; Kumar, 1979; Steinemann, 1980; Lakhotia 1982; Roy, 1983). Thus, unlike in the polytene nuclei, all the euchromatin regions in mitotic nuclei of *Drosophila* appear to replicate synchronously during the early and mid S period while during the late S, only the pericentromeric heterochromatin blocks replicate. It is possible that some of the later replicating sites in the euchromatin part of *Drosophila* mitotic nuclei may be very small in proportion to the much smaller genome size in *Drosophila* and may thus escape detection in the usual autoradiographic preparations. However, it is significant to note that the mitotic chromosomes of *Drosophila* do not show any G-bands etc. (see Holmquist, 1987 and this volume). Implications of these differences in chromosomal replication of *Drosophila* and mammalian cells are discussed later.

Patterns of Chromosomal Replication in Differentiated Tissues

The segmental replication patterns of mitotic chromosomes in mammals and of the polytene chromosomes of *Drosophila* encouraged comparative studies on patterns of chromosomal replication of different tissue types of a species. A rather painstaking study in different human tissues was initiated by Prokofieva-Belgovskaya and her group. Employing autoradiography and observing generally the late S, they and certain other groups detected subtle tissue-specific variations in temporal orders of replication among chromosomes (Slesinger and Prokofieva-Belgovskaya, 1968; Slesinger *et al.*, 1974; Prokofieva-Belgovskaya *et al.*, 1976; German and Aronian, 1971; Farber and Davidson, 1977). It was suggested that the basic replication pattern in human cells is that of lymphocytes and the variant patterns superimpose on it as a function of differentiation (Farber and Davidson, 1978). However, autoradiographic studies of Govosto *et al.* (1968) in different hemopoetic cell types in human, and studies on several other mammals (Martin, 1966; Pfluger and Yunis, 1966) could not detect tissue-related differences in the replication order of individual chromosomes. Sheldon and Nichols (1981a,b) employed both autoradiography and BrdU labelling methods to study replication in human as well as avian tissues: they concluded that the minor differences observed in replication within and among different cell types fell within the range of statistical insignificance. Likewise, studies by Epplen *et al.* (1975), employing high resolution BrdU techniques also failed to reveal marked differences among different tissues.

Earlier studies on mitotic chromosome replication generally suffered from the fact that they were mostly carried out during the late S-phase without giving due attention to the duration of the S-phase. A systematic study on the chromosomal replication coupled with measurement of duration of the S-phase in different tissues of two mole rats, *Bandicota bengalensis* and *Nesokia indica*, was carried out in our laboratory (Dubey, 1985). While the length of the S-phase varied among tissues (bone marrow and PHA-stimulated lymphocytes: 13–15h; skin fibroblasts and cells from new born litters: 7.5–9h), the regional order of replication of individual chromosome regions was generally conserved. Nevertheless, due to differences in the length of S-phase more chromosomes replicated at a given time in tissues with a shorter S-phase. It is possible that the differences observed by earlier authors were more a manifestation of differences in the length of S-phase rather than actual differences in the order of replication.

Another approach that has unambiguously demonstrated the permanency of replication patterns has been the study of somatic cell hybrids in which chromosomes of one parental cell line are preferentially lost. A large body of data confirmed that chromosomes in these hybrids retain their parental pattern and that this programme was not altered even if all but one chromosome from one of the parental lines survived in the hybrid (Sonnenschein, 1970; Graves, 1972; Marin and Colletta, 1974; Lin and Davidson, 1975; Farber and Davidson, 1977, 1978).

The above suggests that by and large the replication pattern of each chromosome is determined and regulated autonomously: it is affected neither by the state of differentiation nor by its genomic surroundings. This conclusion, however, needs to be qualified in reference to the late replicating regions of the genome, especially the constitutive heterochromatin and the lyonized X-chromosome in female mammals. Studies on a variety of mammals, including humans, confirm asynchrony of replication between the two X's in female somatic cells. One of them, the genetically active homologue, replicates early in the S-phase, while the other, lyonized and genetically inert, replicates during the later half (Lyon, 1961, 1974). This asynchrony is an ontogenic phenomenon since the two X's replicate synchronously during early S in pre-implantation embryo and the onset of late replication of one X occurs in a differentiation specific manner later in embryogeny. This temporal switch coincides with the formation of sex-chromatin body and genetic quiescence (Lyon, 1974). Recent evidence suggests that an analogous, but not identical, pattern may follow in constitutive heterochromatin in the mouse. Selig *et al.* (1988) have shown that while always replicating in the later quarter of the late S-phase in all body cells of mice, the constitutive heterochromatin replicates during the earlier quarter of the late S in teratocarcinoma of embryonal origin.

In spite of the temporal asynchrony, the regional order of replication in the two X's is generally alike (Schmidt *et al.*, 1982; Camargo and Cervenga, 1982; Reddy *et al.*, 1988). However, whereas the order is almost inviolate in the early-replicating X, it shows considerable heterogeneity in the late-replicating X (Schmidt *et al.*, 1982; Popescu and Dipaolo, 1979) in a tissue-specific manner (Willard, 1977; Cawood, 1981; Reddy *et al.*, 1988). In aneuploid cell lines of female origin, having multiple X's, the heterogeneity among the late-replicating X's is easily marked (Willard and Breg, 1980; Raman, 1984).

Like the facultative X-heterochromatin (lyonized X chromosome), the constitutive heterochromatin on X chromosome also has been found to replicate asynchronously in several rodents. Sen and Sharma (1982) reported that in *Mus dunni*, the constitutive heterochromatin on the active X chromosome in female cells replicates earlier than its homologue on the lyonized X. A comparable situation was noted in the majority of cells of *B. bengalensis* and *N. indica* (Dubey, 1985). Although conclusively demonstrated in only a few species, asynchronous replication of homologous heterochromatin regions appears to be a characteristic feature of X-linked heterochromatin. Temporal heterogeneity in the autosomal heterochromatin has also been recently demonstrated in the mouse (Selig *et al.*, 1988).

The elaborate banding pattern of polytene chromosomes has permitted extensive studies on the temporal order of replication of specific bands, puffs, interbands, etc. in relation to the influence of sex, chromosomal rearrangements and other genotypic factors, developmental conditions and cell types. An important revelation of the early autoradiographic studies on polytene chromosome replication was that the temporal order of replication, particularly during the late S, of specific band regions was remarkably constant in different wild type strains of *D. melanogaster* and in different laboratories across the world (Rudkin, 1972). However, certain genotypic factors like chromosomal rearrangements, etc., particularly those involving heterochromatin or other late-replicating sites were found to locally alter the replication timings of adjacent regions (Arcos-Terán and Beermann, 1968; Kalisch and Hägele, 1973; Mukherjee *et al.*, 1980; Wargent and Hartmann-Goldstein, 1976; Lakhotia and Mishra, 1982). Developmental conditions or certain mutants that increase the polyteny levels of salivary gland nuclei of late 3rd instar larvae (Hartmann-Goldstein and Goldstein, 1979) also affected the temporal order of replication mainly by prolonging the normally brief early S of a polytene replication cycle (Mishra and Lakhotia, 1982).

Unlike in mammals, the two X chromosomes in polytene cells of *Drosophila* females replicate synchronously with each other as well as with autosomes while the single X in male polytene cells was found to complete its replication earlier than autosomes in the S-phase (Berendes, 1966; Lakhotia and Mukherjee, 1970). This early completion of replication of the X in male polytene cells was correlated with its hyperactivity to achieve dosage compensation (Mukherjee and Beermann, 1965; Lakhotia and Mukherjee, 1969, 1970). It was also shown that parallel with its transcriptional status, the polytene X chromosome in male cells continued its normal faster replication without any effect of or on autosomal segment in autosome-X insertion genotypes (Lakhotia, 1970; Ghosh and Mukherjee, 1986) or in XX/XO mosaic salivary glands (Chatterjee and Mukherjee, 1977). Intriguingly, however, the X chromosome in mitotic cells of male or female *Drosophila* does not show any kind of asynchrony comparable to that noted in polytene nuclei; this aspect will be discussed later (also see Lakhotia and Sinha, 1983).

In spite of the enormously increased resolution offered by polytene chromosome banding pattern, there have been relatively fewer studies on the temporal order of replication of specific chromosome segments in different tissue types. This paucity of studies has been mainly due to the fact that except for the salivary glands of late 3rd instar larvae, the polytene chromosomes found in other cell types in *Drosophila* are of poorer cytological quality and thus unsuitable for such studies. However, in species of *Drosophila* or in certain mutant genotypes of *D. melanogaster*, analysable polytene chromosomes are found in other tissues also and some of these situations have been utilized for comparative studies on temporal orders of replication in different polytene cell types. Lakhotia and Tiwari (1984) and Tiwari and Lakhotia (1984) compared replication of specific chromosome segments in polytene chromosomes of larval salivary glands and gastric caeca in *D. nasuta* and *D. hydei* and found that the independently labelled replication units identifiable at the chromosomal level (see Plaut *et al.*, 1966; Lakhotia and Mukherjee, 1970; Rudkin, 1972) were cytologically comparable in all cases and the units that were late-replicating in one cell type continued to be late-replicating in the other as well. However, within the late S, certain sites displayed characteristically different temporal order of completion of their replication. Recently, Sinha *et al.*, (1987), taking advantage of *otu* mutant alleles in *D. melanogaster* that alter the normally polyneme type of ovarian nurse cells into cells of polytene type (see King and Storto, 1988), compared the replication in larval salivary gland polytene nuclei with that in these nurse cell polytenes. Their results were in general similar to those noted earlier. In a study with polytene chromosomes of the mosquito, *Anopheles stephensi*, Redfern (1981) had also reached a similar conclusion.

The above considerations lead to the following general conclusions: (i) each chromosome has a characteristic temporal and regional order of replication which is autonomously regulated and (ii) whereas this order is rigidly maintained in the early replicating segments, those replicating in the late S tolerate a certain degree of flexibility both in terms of their temporal and regional orders. It is interesting to note that even among different mammals, chromosomal regions of homologous functions (regions of evolutionary homology) show comparable early S replication patterns (von Kiel *et al.*, 1985).

Factors Affecting S-Phase—Organization of Active Replicons

It is now well known that the nuclear DNA in eukaryotes replicates in units, the replicons, much smaller than the long DNA molecule that spans even the smallest chromosome. In general, each replicon starts replication bidirectionally from an origin point giving rise to two more or less symmetrically placed replication forks which progress in opposite directions (Hubermann and Riggs, 1968) till they meet forks advancing towards them from the neighbouring replicons or else they encounter a fixed termination site; fusions of replication forks of adjacent replicons leads ultimately to production of two daughter DNA molecules (for reviews see Callan, 1972; Hand, 1978; Taylor, 1978, 1984; Van't Hof and Bjerkens, 1979). There are several possible variables in the organization of eukaryotic replicons: (1) distance between adjacent origin points (replicon size); (2) unidirectional vs bidirectional migration of replication fork; (3) rate of progression of replication fork (rate of replication); (4) fixed termini vs fusion of adjacent forks when they meet; (5) numbers and spatial locations of replicons that are active at a given moment of time in the S-phase (synchrony between replicons of same and different chromosomes in a nucleus). It is these different variables and their subtypes that influence the characteristic patterns of chromosomal replication and duration of S-phase in different species and in different cell types of a species.

Much of the information on these determinants of the S-phase has been obtained using the technique of DNA-fibre autoradiography (Hubermann and Riggs, 1968; Callan, 1972). Some aspects of these will be briefly considered here. Pioneering studies by Callan (1972) and Blumenthal *et al.* (1973) had suggested that the major factor determining the duration of S-phase in a particular cell type of a species was the size of active replicons or the distance between adjacent origin points that are activated to initiate replication since cleavage cells of amphibians and *Drosophila* with very short cell cycles had very closely spaced active origin points while their differentiated cells with longer S-periods utilized far fewer of the potential origin points (also see Taylor and Hozier, 1976; Spradling and Orr-Weaver, 1987). Other studies using human and mammalian cells (Hand and German, 1977; Jagiello *et al.*, 1983; Sung *et al.*, 1986) suggested that the rate of fork migration rather then replicon size determined the length of the S-phase. On the other hand, in the case of plant cells, evidence was presented that a modulation of timing of initiation of replicon families was the critical factor affecting the S-phase duration (Van't Hof, 1975).

Studies in our laboratories, using *Drosophila* as well as mammalian cells (Lakhotia and Sinha, 1983; Lakhotia and Tiwari, 1985; Dubey and Raman, 1983, 1987) have revealed that more than one aspect of the replicon organization is developmentally regulated by different cell types of an organism to attain their characteristic S-phase patterns. In the mole rat, *B. bengalensis*, cell types with S-phase varying from 8h to 14h showed differences both in the rate of fork migration and in replicon size: new-born cells with shorter S-phase had more closely spaced origin points as well as faster migrating forks (Dubey and Raman, 1987). Interestingly, this study also showed that cells with similar

S-phase may also differ in their replicon parameters (size and rate). In an earlier study, Dubey and Raman (1983) had shown that treatment of mole rat cells with Hoechst 33258 affected yet another parameter of replicon activity: the initiation of neighbouring replicon was more asynchronous in treated than in untreated cells.

Studies on replicon organization in polytene (salivary glands) and non-polytene (brain and imaginal disk) cells of *Drosophila* larvae (Lakhotia and Sinha, 1983; Lakhotia and Tiwari, 1985) revealed a novel feature: two very different classes of replicons (designated as Types I and II), differing in their size as well as rate of fork migration, were active in the same cell type during early and late phases of the S, respectively. Moreover, the size and the rate of fork migration in Type I replicons (longer and faster) were found to differ between the polytene and non-polytene cell types analysed (Lakhotia and Sinha, 1983; Lakhotia and Tiwari, 1985).

An important difference in the replicon organization in mammalian and *Drosophila* type systems is the presence of replicon clusters in the former (Edenberg and Hubermann, 1975; Hand, 1978) and their absence in *Drosophila* (Steinemann, 1981a,b; Lakhotia and Sinha, 1983; Lakhotia and Tiwari, 1985). Neighbouring replicons which are activated more or less synchronously constitute the replicon clusters and a mammalian chromosome is constituted of many such clusters. On the other hand, in *Drosophila* chromosomes, entire euchromatin or entire heterochromatin regions appear to constitute single clusters of replicons. The synchrony of replication within clusters seems to be an important aspect of mammalian chromosome organization as reflected in their various banding patterns, etc. (for a more detailed discussion, see Holmquist, this volume). In this context, it is interesting to emphasize differences in the nature of bands of mitotic chromosomes of mammals and polytene chromosomes of *Drosophila*, etc. While their darker staining, species specificity and generally later replication may suggest apparent similarity, a deeper analysis reveals fundamental differences, e.g., each G-band of a metaphase chromosome contains several replicons (replicon cluster), while in *Drosophila* several polytene bands may comprise one replicon. The apparent late replication of many of the polytene bands, giving rise to 'replication bands,' is also a reflection of the complexities associated with replication of the large number of laterally associated sister chromatids of both homologues since as mentioned earlier, mitotic chromosomes of *Drosophila* do not show any later replicating segments or 'replication bands' in their euchromatic regions. To account for the ^{3}H-thymidine autoradiographic labelling patterns seen at polytene chromosome as well as at their DNA fibre level, Lakhotia and Sinha (1983) proposed that the polytene chromosome organization imposes a temporal and spatial asynchrony in the activation of origin points on homologous replicons of the large number of lateral sister chromatids. This asynchrony leads to continued incorporation of ^{3}H-thymidine in some of the chromatids while the corresponding replicons in other sister chromatids had already finished and this in turn generates the various segmental or discontinuous labelling patterns, so characteristic of polytene chromosomes but never seen in mitotic chromosomes of *Drosophila*. This asynchrony in activity of homologous replicons on lateral polytene strands also accounts for the overall S-phase being considerably longer than expected on the basis of replicon size and rate of fork migration found in these cells (Lakhotia and Sinha, 1983).

Thus the different possible variables in replicon organization have been utilized in diverse organisms in their various cell types to regulate their S-phases and temporal as well as regional orders of replication of different chromosomes during the S-phase.

Time of Replication and Transcriptional Activity

Since both the processes of replication and transcription require the same template on which specific polymerases progress to make the respective products, they are expected to influence each other. While the process of replication during a typical S-period makes a global use of nuclear DNA as template, the process of transcription has to be necessarily selective in template usage in a developmentally regulated tissue-specific manner. The specificity of regulated transcription requires a large variety of transient factors binding to the DNA sequences that are to be or not to be transcribed. The process of replication fork migration during the S-period may dislocate these transcription regulatory factors (Brewer, 1988). This obviously necessitates a very intricate organization of transcriptional and replicational activities so that the cell functions proceed unhindered and in orderly fashion. The temporal and regional patterns of chromosome replication are a reflection of this order.

The genes which are required to make their products available in a cell may be expected to replicate early in the S so that the particular DNA sequence is available, without further interruption till the mitosis, for RNA polymerase to act upon. The general temporal exclusion of eu- and hetero-chromatin replication provides a gross compartmentalization of the above type. A finer resolution is seen in the metaphase chromosome band specific replication differences. Stubblefield (1975) in Chinese hamster and others in human cells (Camargo and Cervenka, 1982; Schmidt *et al.*, 1982) have convincingly reinforced the observations obtained from autoradiography of human and muntjac chromosomes that the structural chromosome bands are also distinct in their time of replication in that the early replicating chromosome regions correspond to the R-bands and the late replicating ones to the G-bands. Thus the early and late replicating chromatin appear to be structurally distinct (Korenberg and Rykowski, 1988). Holmquist *et al.* (1982) succeeded in isolating R- and G-band DNA from CHO cells and showed that active and inactive constituents of chromatin were spatially distinct with R-bands harbouring the former, and G-bands the latter. Extending this study further to replication of individual genes, they found a number of house-keeping genes (those active in nearly all body cells) replicating in the early S in different cell types. Among the tissue-specific genes, a strong correlation between their activity and time of replication was established (Goldman *et al.*, 1984). In a series of papers on replication of multigene families, viz., immunoglobulin, globin and histones, Schildkraut and his group established their replication in early S in the tissues of their expression (Furst *et al.*, 1981; Braunstein *et al.*, 1982; Iqbal *et al.*, 1984; Brown *et al.*, 1987). Comparing the replication of functional and pseudogene alleles of globin genes in mice, they found that while the active locus replicated in early S, most of the pseudogenes were late replicating (Calza *et al.*, 1984), thus suggesting that not sequence homology but the expression competence of genes and their location in the genome determines the time of replication in S-phase. Goldman (1988) has chronicled replication timings of all the genes so far studied, and it shows that though the evidence for correlation between early S replication and the expression of a gene is strong, it is not absolute: no gene products has been identified so far for a number of early replicating DNA segments and transcripts from a few late replicating regions have also been obtained.

In the frog, *Xenopus*, there are two sets of 5S-rRNA genes, a cluster of about 800 copies that is expressed in somatic cells (s5S-rRNA) and a much larger cluster expressed in oocytes (o5S-rRNA) but not in somatic cells. In somatic cells, the s5S-rRNA genes replicate in early S while the o5S ones in late S (Gilbert, 1986; Guinta *et al.*, 1986). In the oocytes also, the o5S cluster is late replicating but is expressed. Hatton *et al.*, (1988) recorded genes in the heavy chain immunoglobulin complex in murine cell lines that were early replicating but were not expressed. However, when expressed, their time of replication was switched to even earlier in the S. Eul *et al.* (1988) fused a lymphoblastoid cell line, in which c-myc expressed and replicated early in S-phase, with lymphocytes and with fibroblasts. The lymphocytic hybrids were tumorigenic while those fused with fibroblasts reverted to a nontumorigenic phenotype. C-myc continued to express and replicate early in the former while in the fibroblastic hybrid, c-myc was unexpressed and late replicating. Thus the correlation between the time of replication and the expression of genes seems to be strong and it may be safe to conclude that in general, replication in early S is a necessary if not essential condition for an expressed gene (Goldman *et al.*, 1984).

The relation between gene activity and time of replication has also been examined in polytene chromosomes. In these chromosomes, the transcriptionally more active regions are distinctly visible as puffs and interbands (Beermann, 1972), and during polytene replication these regions are known to be replicated early (Hägele, 1973; Lakhotia and Roy, 1979; Mukherjee *et al.*, 1980). The early replication of the single X-chromosome in polytene nuclei of male larvae of *Drosophila* has also been correlated with its transcriptional hyperactivity (Berendes, 1966; Lakhotia and Mukherjee, 1970). While these results may suggest a correlation between gene activity and replication time as in mammalian and other cell types, a more critical analysis reveals this correlation between active genes and their early replication in polytene nuclei to be due to certain aspects of polytene chromosome organization rather than this being a general feature of the *Drosophila* genome. As discussed earlier, an interpretation of early and late replication in polytene chromosomes of *Drosophila* is not straightforward in view of the constraints imposed by their unique structural organization. Therefore, it would be proper to look for correlation between early replication and gene expression in non-polytene diploid cell types rather than in endoreplicating ones. The endoreplicated nuclei have multiple copies of most genes on the multitudes of daughter chromatids generated and thus may not perhaps be under the same constraints to regulate time of replication and transcription as the diploid cells. The lateral multiplicity of gene copies allows that, while some gene copies are being replicated, their other homologues may continue to transcribe if the cell requires their transcripts. The model of polytene chromosome replication proposed by Lakhotia and Sinha (1983) provides for such temporal asynchrony in replication of homologous DNA sequences on the lateral strands of polytene chromosomes. If the processes of replication and transcription were mutually exclusive at a given chromosomal location in polytene nuclei, the autoradiographic labelling patterns in ^{3}H-uridine labelled nuclei could be expected to be different in nuclei that were replicating and those that were not replicating at the time of pulse labelling. Although this aspect has not been specifically examined, experience shows that progression of replication in a polytene nucleus does not affect its labelling patterns in transcription autoradiograms. From this point of view, the apparent early completion of replication by the X-chromosome in polytene nuclei of male *Drosophila* is not directly and causally related to this chromosome being hyperactive for dosage compensation. As suggested by Lakhotia and Sinha (1983), the earlier completion of the male X-chromosome in polytene cells appears to be due to a greater synchrony in replication of the lateral chromatids of this more loosely organized and hemizygous chromosome: both these conditions presumably facilitate a greater and quicker availability of the replication machinery to the different lateral subunits. This interpretation finds support in the fact that in mitotic cells of *Drosophila*, the single X in male is not faster replicating. In an interesting study, Hägele and Kalisch (1980) compared replication time of a specific polytene chromosome band before and after puffing (i.e., before and after transcriptional activation) and found that in both situations it continued to be late replicating. Indeed in view of the fact that nearly all replicons in euchromatic regions of the *Drosophila* genome are fired more or less synchronously, the question of correlation between 'early' and 'late' replication of a gene and its expressivity in a tissue becomes a moot point.

Concluding Remarks

The fidelity and precision of chromosomal DNA replication in eukaryotic nuclei is remarkable considering that the DNA is several thousand-fold packed in a typical eukaryotic nucleus and that this highly packed DNA is very orderly and faithfully replicated once and only once every cell cycle. Some of the intriguing questions that arise in this context are: (1) what provides the force and what is the mechanism that permits a rather very rapid unwinding of the double-stranded DNA during replication (and transcription) while it remains as condensed chromatin; (2) what controls the fairly well conserved temporal order of replication of different chromosome regions during an S-phase; (3) what constitutes an origin point of a replicon and how different cell types are able to modulate the use of these origin points with reference to an origin being operative or inoperative in a given cell, and if operative, its time of firing; and (4) how transcription or transcriptional competence affects the replication time of a replicon or replicon cluster. Although clearcut answers to these and related questions are still not available, some understanding of these processes is beginning to emerge with successful development of cell free *in vitro* DNA replication systems and with a better understanding of nuclear matrix and its relation to chromosome structure and organization (Nelson *et al.*, 1986; Gasser and Laemmli, 1987; Blow and Laskey, 1986, 1988; Blow, 1987, 1988; Hubermann, 1987; Bravo *et al.*, 1987; Prelich *et al.*, 1987; Sheehan *et al.*, 1988). It is becoming increasingly clear that the nuclear matrix (or scaffold for metaphase chromosomes) provides a structural framework for attachment of specific chromosomes and chromosome regions. These attachment points not only provide a structural framework for DNA unwinding during replication (and transcription) but also facilitate chromosome and chromatid movements during cell division. In addition, this spatial sub-compartmentalization of chromatin within the nucleus may influence its synthetic activities in other ways also, e.g., by restricting specific chromosomes or chromosome regions to certain functionally distinct spatial domains within the nucleus so that they are or are not available for replication (and/or transcription) at specific times in the cell cycle or in specific cell types. Although not documented conclusively, the nuclear matrix is expected to play a crucial role in defining the early and late replicating chromosomal domains. Thus in the case of late replicating, genetically inert X-chromosomes in female mammals, it is suggested that sometimes during early development, the relegation of one of the two X's to a spatially distinct nuclear location leads to its heterochromatization followed by methylation of its cytosine residues so that the inactive state is cell-inherited (Gartler *et al.*, 1985). It appears that this distinct spatial location is responsible for this X-chromosome becoming late-replicating which in turn leads to its heterochromatization and methylation since in early cleavage embryos of mammals, late replication of one X-chromosome precedes by a few division cycles the formation of heterochromatic sex-chromatin (Issa *et al.*, 1969; Mukherjee, 1976). The constitutive heterochromatin in *Drosophila* is also known to occupy a distinct nuclear location (Lakhotia, 1974; Foe and Alberts, 1985; Hammond and Laird, 1985a,b). An analysis of the 3-dimensional organization of chromosomes in intact polytene nuclei of *Drosophila* also revealed a remarkable order (Hochstrasser *et al.*, 1986; Hochstrasser and Sedat, 1987a,b). It is very likely that the specific under-replication of heterochromatin and other sequences in diverse endoreplicating cell types of *Drosophila* is governed by their specific spatial intra-nuclear locations due to interactions with nuclear matrix. An orderly arrangement of chromosomes in a nucleus was suggested more than 100 years ago by Rabl, and a good knowledge of this order and its basis will be essential for understanding chromosomal replication in its totality.

Studies on replication of multigene families (30–250 kb long) in vertebrates showed that the contiguously aligned genes replicate as one replicon (Brown *et al.*, 1987) so that replication is regulated in domains larger than individual genes. These larger domains seem to coincide with binding of chromatin at specific points with nuclear matrix through topoisomerase (Gasser and Laemmli, 1987) and with Bal31 cut CHO chromosome fragments of ~90kb length which were considered to represent individual replicons (Clark *et al.*, 1987). In this context, it is interesting to note that Dubey and Raman (1987) reported asynchronous replication of contiguous replicons in DNA fibre autoradiograms and which they suggested to be related to active or inactive state of those genes. Thus within a functional domain, a finer regulation of replication is still possible.

In recent years, some progress has also been made in identifying the replicon origins, although the results have often been rather contradictory. Striking similarities were found in origin sequences (*ori* region) of eukaryotic viruses like simian (SV40), polyoma (Py), adeno and Epstein Bar (EPB) viruses. In all of them, the *ori* region comprises a core DNA sequence, the ori sequence, flanked by small DNA motifs; a functional origin requires the entire complex and it is important to realize that the flanking sequences have transcription promoter and enhancer sites which bind with transcriptional and other nuclear factors for initiation and maintenance of replication and transcription (for brief reviews, see DePemphilis and Wassermann, 1987; DePemphilis, 1988). However, the above structure of eukaryotic viral origins is not necessarily ubiquitous for all viral, and even less likely for mammalian replicons. However, the combined organization for regulation of transcriptional and replicational activities does raise interesting possibilities. Significantly, one of the ori sequences recently analysed in HeLa cells, shows some features comparable to those of the viruses, especially in the occurrence of transcription and other nuclear factor binding sites (Triboli *et al.*, 1987). Another approach in the study of the origin sequences has been to isolate the autonomously replicating sequences (ARS) from eukaryotic genomes. There are several reports claiming isolation of such sequences. However, none of them have been convincingly shown to be identifiable as a true replicon origin. Thus Thireos *et al.* (1980) identified a region in one of the domains of amplified chorion genes in *Drosophila* as showing ARS activity in yeast; however, this does not coincide with the amplification control element identified for this domain (see Spradling and Orr-Weaver, 1987).

Acknowledgements

Research on chromosomal replication in *Drosophila* and mammalian cells in our laboratories has been supported at various times by grants from the University Grants Commission, the Council of Scientific & Industrial Research, Indian National Science Academy, the Department of Science & Technology (Govt. of India) and the Department of Atomic Energy (Govt. of India).

References

Amabis DC, Amabis JM (1984). Effects of ecdysterone in polytene chromosomes of *Trichosia pubescens*. Dev Biol 102; 1–9.

Arcos-Terán L, Beermann W (1968). Changes of DNA replication behaviour associated with intragenic changes of the *white* region in *Drosophila melanogaster*. Chromosoma 25; 377–91.

Barigozzi C, Dolfini S, Fraccaro M, Rezzonico RG, Tiepolo L (1966). *In vitro* study of DNA replication patterns of somatic chromosomes of *Drosophila melanogaster*. Exp Cell Res 43; 231–4.

Beermann W (1972). Chromomeres and genes. In (ed. Beermann W). Results and Problems in Cell Differentiation, vol. 4. Springer-Verlag, Berlin, Heidelberg, New York: 1–33.

Berendes HD (1966). Differential replication of male and female X-chromosomes in *Drosophila*. Chromosoma 20; 32–43.

Berendes HD, Keyl HG (1967). Distribution of DNA in heterochromatin and euchromatin of polytene nuclei of *D. hydei*. Genetics 57; 1–13.

Blow JJ (1987). DNA replication: many strands converge. Nature 326;441–2.

Blow JJ (1988). Eukaryotic DNA replication reconstituted outside the cell. Bioessays 8; 149–52.

Blow JJ, Laskey RA (1986). Initiation of DNA replication in nuclei and purified DNA by a cell free extract of *Xenopus* eggs. Cell 47; 577–87.

Blow JJ, Laskey RA (1988). A role for the nuclear envelope in controlling DNA replication within the cell cycle. Nature 332; 546–8.

Blumenthal AB, Kriegstein HJ, Hogness DS (1973). The units of DNA replication in *Drosophila melanogaster* chromosomes. Cold Spring Harb Symp Quant Biol 38; 205–33.

Braunstein JD, Schulze D, Del Giudice T, Furst A, Schildkraut L (1982). The temporal order of replication of murine immunoglobulin heavy chain constant region sequences corresponds to their linear order in the genome. Nucleic Acids Res 10; 6887–902.

Bravo R, Frank R, Blundell A, Macdonald-Bravo H (1987). Cyclin/PCNA is the auxillary protein of DNA polymerase-delta. Nature 326; 515–17.

Brewer BJ (1988). When polymerases collide: replication and the transcriptional organization of the *E.coli* chromosome. Cell 53; 679–86.

Brown EH, Iqbal MA, Stuart S, Hatton KS, Valinsky J, Schildkraut CL (1987). Rate of replication of the murine immunoglobulin heavy chain locus. Evidence that the region is part of a single replicon. Mol Cell Biol 7; 450–7.

Brown S (1966). Heterochromatin. Science 151; 417–25.

Callan HG (1972). Replication of DNA in the chromosomes of eukaryotes. Proc Roy Soc Lond Ser B 181; 19–41.

Calza RE, Eckhardt LA, Del Guidice T, Schildkraut CL (1984). Changes in gene position are accompanied by a change in time of replication. Cell 36; 689–96.

Camargo M, Cervenka J (1982). Patterns of DNA replication of human chromosomes. II. Replication map and replication model. Am J Human Genet 34; 757–80.

Cawood AH (1981). Cell specific variation of X-chromosome replication in the Syrian hamster (*Mesocricetus auratus*). Chromosoma 84; 365–72.

Chatterjee RN, Mukherjee AS (1977). Chromosomal basis of dosage compensation in *Drosophila*. IX. Cellular autonomy of the faster replication of the X-chromosome in haplo-X cells of *D. melanogaster* and synchronous initiation. J Cell Biol 74; 168–80.

Clark RW, Tseng P, Lechuga JM (1987). A nuclease-derived fragment of metaphase DNA and its relationship to the replicon. Exp Cell Res 169; 296–310.

Das RK, Savage JRK (1978). Chromosome replication patterns in the Djungarian hamster (*Phodopus sungarus*). Chromosoma 67; 165–76.

Dennhoffer L (1982a). Cytophotometric DNA determinations and autoradiographic studies in salivary gland nuclei from larvae with different karyotypes in *Drosophila melanogaster*. Chromosoma 86; 123–47.

Dennhoffer L (1982b). Underreplication during polytenization? Theor Appl genet 63; 193–9.

DePemphilis ML (1988). Transcriptional elements as components of eukaryotic origins of DNA replication. Cell 52; 635–8.

DePemphilis ML, Wasserman PM (1987). Regulation of chromosomal replication and transcription during early mammalian development. Bioessays 7; 265–71.

Dolfini SF (1987). The effect of distamycin A on heterochromatin condensation of *Drosophila* chromosomes. Chromosoma 95; 57–62.

Dolfini SF, Razzini AB (1983). High resolution of heterochromatin of *Drosophila melanogaster* by distamycin A. Experientia 39; 1402–4.

Dubey DD (1985). Studies on synthetic activity in mammalian genome: Replication in the rodents *Bandicota bengalensis* and *Nesokia indica*. Ph.D. thesis, Banaras Hindu University, Varanasi, India.

Dubey, DD, Raman R (1983). Effects of Hoechst 33258 on different cell cycle events. I. Inhibition of synthetic activities in bone marrow cells of the mole rat, *Bandicota bengalensis*. Exp Cell Res 149; 419–32.

Dubey DD, Raman R (1987). Factors influencing replicon organization in tissues having different S phase durations in the mole rat, *Bandicota bengalensis*. Chromosoma 95; 285–9.

Dutrillaux B, Courturier J, Richter C-L, Viegas-Pequinot E (1976). Sequence of DNA replication in 277 R- and Q-bands of human chromosomes using a BrdU treatment. Chromosoma 58; 51–61.

Edenberg HJ, Huberman JA (1975). Eukaryotic chromosome replication. Ann Rev Genet 9; 245–84.

Endow SA, Gall JG (1975). Differential replication of satellite DNA in polyploid tissues of *Drosophila virilis*. Chromosoma 50; 175–92.

Epplen JT, Siebers J-W, Vogel W (1975). DNA replication pattern of human chromosomes from fibroblasts and amniotic fluid cells revealed by a Giemsa staining technique. Cytogenet Cell Genet 15; 177–85.

Eul J, Gronemeyer H, Adolph S, Hameister H (1988). Suppression of tumorigenicity in T-cell lymphoma hybrids is correlated with changes in *myc* expression and DNA replication of the *myc* chromosomal domain. Chromosoma 96; 248–54.

Farber RA, Davidson RL (1977). Differences in the order of termination of DNA replication in human chromosomes in peripheral blood lymphocytes and skin fibroblasts from the same individual. Cytogenet Cell Genet 18; 349–63.

Farber RA, Davidson RL (1978). Altered pattern of replication of human chromosomes in a human fibroblast-mouse cell hybrid. Proc Nat Acad Sci 75; 1470–4.

Farber RA, Davidson RL (1979). Maintenance of replication patterns in human-mouse hybrids retaining only one human chromosome. Somat Cell Genet 5; 519–28.

Foe VE, Alberts BM (1985). Reversible chromosome condensation induced in *Drosophila* embryos by anoxia: visualization of interphase nuclear organization. J Cell Biol 100; 1623–36.

Furst A, Brown EH, Braunstein JD, Schildkraut CL (1981). Alpha-globin sequences are located in a region of early replicating DNA in murine erythroleukemia cell. Proc Nat Acad Sci 78; 1023–7.

Gabrusewycz-Garcia N (1964). Cytological and autoradiographic studies in *Sciara coprophila* salivary gland chromosomes. Chromosoma 15; 312–44.

Gall JG, Cohen EH, Polan ML (1971). Repetitive DNA sequences in *Drosophila*. Chromosoma 33; 319–44.

Ganner E, Evans HJ (1971). The relationship between patterns of DNA replication and of quinacrine fluorescence in the human chromosome complement. Chromosoma 35; 326–41.

Gartler SM, Dyer KA, Graves JAM. Rocchi M (1985). A two-step model for mammalian X-chromosome inactivation. In Biochemistry and Biology of DNA Methylation (eds. Cantoni GL, Razini A). Alan R. Liss, New York; 223–35.

Gartler SM, Riggs AD (1983). Mammalian X-chromosome inactivation. Ann Rev Genet 17; 155–90.

Gasser SM, Laemmli UK (1987). The organization of chromatin loops: characterization of a scaffold attachment site. EMBO J 5; 511–18.

Gay H, Das CC, Forward K, Kaufmann BP (1970). DNA content of mitotically active condensed chromosomes of *Drosophila melanogaster*. Chromosoma 32; 212–23.

German J, Aronian D (1971). Autoradiographic studies of human chromosomes. VI. Comparisons of the end of S patterns in lymphocytic and amniotic epithelial cells. Chromosoma 35; 99–110.

Ghosh M, Mukherjee AS (1986). Transcriptive and replicative activity of the X-chromosome in an autosomal segmental hyperdiploid in *Drosophila* and its significance. J Cell Sci 81; 267–81.

Gilbert DM (1986). Temporal order of replications of *Xenopus laevis* 5S ribosomal RNA genes in somatic cells. Proc Nat Acad Sci 83; 2924–8.

Goldman MA (1988). The chromatin domain as a unit of gene regulation. Bioessays 9; 50–5.

Goldman MA, Holmquist GP, Gray MC, Caston LA, Nag A (1984). Replication timing of genes and middle repetitive sequences. Science 224; 688–92.

Govosto F, Pegararo L, Maserà P, Povera G (1968). Late replication patterns in human haemopoetic cells. A comparative investigation using high resolution autoradiography. Exp Cell Res 49; 340–58.

Graves JAM (1972). Cell cycle and chromosome replication patterns in interspecific somatic cell hybrids. Exp Cell Res 73; 81–94.

Grimm C, Kunz W (1980). Disproportionate rDNA replication does occur in diploid tissue in *Drosophila hydei*. Molec Gen Genet 180; 23–6.

Grimm C, Kunz W, Franz G (1984). The organ specific rRNA gene number in *Drosophila hydei* is controlled by sex heterochromatin. Chromosoma 89; 48–54.

Guinta DR, Tso JY, Narayanswami S, Hamkalo BA, Korn LJ (1986). Early replication and expression of oocyte type 5S RNA genes in a *Xenopus* somatic cell line carrying a translocation. Proc Nat Acad Sci 83; 5150–4.

Hägele K (1973). Komplementäre DNA-Replikationsmuster bei *Drosophila melanogaster*. Chromosoma 41; 231–6.

Hägele K, Kalisch WE (1980). DNA replication of a polytene chromosome section in *Drosophila* prior to and during puffing. Chromosoma 79; 75–83.

Hammond MP, Laird CD (1985a). Control of DNA replication and spatial distribution of defined DNA sequences in salivary gland cells of *Drosophila melanogaster*. Chromosoma 91; 267–78.

Hammond MP, Laird CD (1985b). Chromosome structure and DNA replication in nurse and follicle cells of *Drosophila melanogaster*. Chromosoma 91; 279–86.

Hand R (1978). Eucaryotic DNA: organization of the genome for replication. Cell 15: 317–25.

Hand R, German J (1977). Bloom's syndrome. DNA replication in cultured fibroblasts and lymphocytes. Hum Genet 38; 297–306.

Hartmann-Goldstein IJ, Goldstein DJ (1979). Effect of temperature on morphology and DNA content of polytene chromosomes of *Drosophila*. Chromosoma 71; 333–46.

Hatton KS, Dhar V, Brown EH, Iqbal MA, Stuart S, Didamo VT, Schildkraut CL (1988). The replication programme of active and inactive multigene families in mammalian cells. Mol Cell Biol 8; 2149–58.

Heitz E (1934). Über alpha- unde beta-Heterochromatin sowie Konstanz und Bau der Chromomeren bei *Drosophila*. Biol Zbl 54; 588–609.

Hochstrasser M, Mathog D, Gruenbaum Y, Saumweber H, Sedat JW (1986). Spatial organization of chromosomes in the salivary gland nuclei of *Drosophila melanogaster*. J Cell Biol 102; 112–23.

Hochstrasser M, Sedat JW (1987a). Three-dimensional organization of *Drosophila melanogaster* interphase nuclei. I. Tissue-specific aspects of polytene nuclear architecture. J Cell Biol 104; 1455–70.

Hochstrasser M, Sedat JW (1987b). Three-dimensional organization of *Drosophila melanogaster* interphase nuclei. II. Chromosome spatial organization and gene regulation. J Cell Biol 104; 1471–83.

Holmquist GP (1987). Role of replication time in the control of tissue-specific gene expression. Am J Hum Genet 40; 151–73.

Holmquist GP, Gray M, Porter T, Jordan T (1982). Characterization of Giemsa-dark and light band DNA. Cell 31; 121–9.

Hubermann JA (1987). Eukaryotic DNA replication: A complex picture partially clarified. Cell 48; 7–8.

Hubermann JA, Riggs AD (1968). On the mechanism of DNA replication in mammalian chromosomes. J Mol Biol 32; 327–41.

Iqbal MA, Plumb M, Stein J, Stein G, Schildkraut CL (1984). Coordinate replication of members of the multigene family of core and H1 human histone genes. Proc Nat Acad Sci 81; 7723–7.

Issa M, Blank CE, Atherton GW (1969). The temporal appearance of sex chromatin and of the late replicating X-chromosome in blastocysts of the domestic rabbit. Cytogenetics 8; 219–37.

Jagiello G, Sung WK, Van't Hof J (1983). Fibre DNA studies of premeiotic spermatogenesis. Exp Cell Res 146; 281–7.

Kafatos FC, Mitsialis SA, Spaereb N, Mariam B, Lingappa R, Delidakis C (1985a). Studies on the developmentally regulated expression and amplification of insect chorion genes. Cold Spring Harb Symp Quant Biol 50; 537–48.

Kafatos FC, Orr W, Delidakis C (1985b). Developmentally regulated gene amplification. Trends Genet 1; 301–6.

Kalfayan L, Levine J, Orr-Weaver T, Park S, Wakimoto B, de Cicco D, Spradling A. (1985). Localization of sequences regulating *Drosophila* chorion gene amplification and expression. Cold Spring Harb Symp Quant Biol 50; 527–35.

Kalisch WE, Hägele K (1973). Different DNA replication behaviour of a tandem duplication in *Drosophila melanogaster*. Chromosoma 44; 265–83.

Keyl HG, Pelling C (1963). Differentielle DNS-Replikation in den Speicheldrüssen Chromosomen von *Chironomus thummi*. Chromosoma 14; 347–59.

King RC, Storto PD (1988). The role of the *otu* gene in *Drosophila* oogenesis. Bioessays 8; 18–24.

Kornberg A (1980). DNA Replication. Freeman, New York.

Korenberg JR, Rykowski MC (1988). Human genome organization: Alu, Lines and the molecular structure of metaphase chromosome bands. Cell 53; 391–400.

Kumar M (1979). Studies on hetero and eu-chromatin in *Drosophila:* Organization of heterochromatin in *D. nasuta* brain cells. Ph.D. Thesis, Banaras Hindu University, Varanasi, India.

Laird CD (1980). Structural paradox of polytene chromosome. Cell 22; 869–74.

Lakhotia SC (1970). Chromosomal basis of dosage compensation in *Drosophila*. II. The DNA replication patterns of the male X-chromosome in an autosome-X insertion in *D. melanogaster*. Genet Res 15; 301–7.

Lakhotia SC (1974). EM autoradiographic studies on polytene nuclei of *Drosophila melanogaster*. III. Localization of non-replicating chromatin in the chromocentre heterochromatin. Chromosoma 46; 145–59.

Lakhotia SC (1982). Independent endoreplication cycles of hetero- and euchromatin in mitotic chromosomes in somatic cells of *Drosophila*. In Cellular Control Mechanisms. Dept Atomic Energy, Govt of India, Bombay; 289–302.

Lakhotia SC (1984). Replication in *Drosophila* chromosomes. XII. Reconfirmation of underreplication of heterochromatin in polytene nuclei by cytofluorometry. Chromosoma 89; 63–7.

Lakhotia SC, Kumar M (1980). Replication in *Drosophila* chromosomes. Part VI. Relative heterochromatin content and autonomous endoreplication cycles of hetero- and eu-chromatin in brain cells of *D. nasuta*. Ind J Exp Biol 18; 1066–71.

Lakhotia SC, Mishra A (1982). Functional organization of polytene X-chromosome in two X-chromosome inversion carrying larvae of *Drosophila melanogaster* reared at 24°C. or at 10°C. Ind J Exp Biol 20; 643–51.

Lakhotia SC, Mukherjee AS (1969). Chromosomal basis of dosage compensation in *Drosophila*. I. Cellular autonomy of hyperactivity of the male X-chromosome in salivary glands and sex differentiation. Genet Res 14; 137–50.

Lakhotia SC, Mukherjee AS (1970). Chromosomal basis of dosage compensation in *Drosophila*. III. Early completion of replication by the polytene X-chromosome in male. Further evidence and its implications. J Cell Biol 47; 18–33.

Lakhotia SC, Roy JK (1981). Effects of Hoechst 33258 on condensation patterns of hetero- and euchromatin in mitotic and interphase nuclei of *Drosophila nasuta*. Exp Cell Res 132; 423–31.

Lakhotia SC, Roy JK (1983). Effects of distamycin A and netropsin on condensation of mitotic chromosomes in early embryos and larval brain cells of *Drosophila nasuta*. Ind J Exp Biol 21; 357–62.

Lakhotia SC, Roy JK, Kumar M (1979). A study of heterochromatin in *Drosohpila nasuta* by the 5-bromodeoxyuridine-Giemsa staining technique. Chromosoma 89; 249–55.

Lakhotia SC, Roy S (1979). Replication in *Drosophila* chromosomes. I. Replication of intranucleolar DNA in polytene cells of *D. nasuta*. J Cell Sci 36; 185–97.

Lakhotia SC, Sinha P (1983). Replication in *Drosophila* chromosomes. X. Two kinds of active replicons in salivary gland polytene nuclei and their relation to chromosomal replication patterns. Chromosoma 88; 265–76.

Lakhotia SC, Tiwari PK (1984). Replication in *Drosophila* chromosomes. XI. Differences in temporal order of replication in two polytene cell types in *D. nasuta*. Chromosoma 89; 212–17.

Lakhotia SC, Tiwari PK (1985). Replication in *Drosophila* chromosomes. Part XV. Organization of active replicons in brain ganglia and wing imaginal disks of *D. nasuta* larvae. Ind J Exp Biol 23; 413–20.

Lamb MJ (1982). The DNA content of polytene nuclei in midgut and malpighian tubule cells of adult *Drosophila melanogaster*. Wilhelm Roux Arch 191; 381–4.

Lamb MM, Laird CD (1987). Three euchromatic DNA sequences under-replicated in polytene chromosomes of *Drosophila* are localized in constrictions and ectopic fibres. Chromosoma 95; 227–35.

Latt SA (1973). Microfluorometric detection of deoxyribonucleic acid replication in human metaphase chromosomes. Proc Nat Acad Sci 70; 3395–9.

Lima-de-Faria A (1959). Differential uptake of tritiated thymidine into hetero- and eu-chromatin in *Melanoplus* and *Secale*. J Biophys Biochem Cytol 6; 457–6.

Lima-de-Faria A, Jaworska H (1968). Late DNA synthesis in heterochromatin. Nature 217; 138–42.

Lin MS, Davidson RL (1975). Replication of human chromosomes in human-mouse hybrids: evidence that the timing of DNA synthesis is determined independently in each chromosome. Somat Cell Genet 1; 111–22.

Lyon MF (1961). Gene action in the X-chromosome of the mouse (*Mus musculus*). Nature 190; 372–3.

Lyon MF (1974). Gene and chromosome inactivation. Genetics 78; 305–9.

Marin G, Colletta G (1974). Patterns of late chromosomal DNA replication in unbalanced Chinese hamster-mouse somatic cell hybrids. Exp Cell Res 89; 368–76.

Martin PG (1966). The pattern of autosomal DNA replication in four tissues of Chinese hamster. Exp Cell Res 45; 85–95.

Mishra A, Lakhotia SC (1982). Replication in *Drosophila* chromosomes. VII. Influence of prolonged larval life on patterns of replication in polytene chromosomes of *D. melanogaster*. Chromosoma 85; 221–36.

Mukherjee AB (1976). Cell cycle analysis and X-chromosome inactivation in the developing mouse. Proc Nat Acad Sci 16; 202–9.

Mukherjee AS, Beermann W (1965). Synthesis of rebonucleic acid by the X-chromosome of *Drosophila melanogaster* and the problem of dosage compensation. Nature 207; 785–6.

Mukherjee AS, Duttagupta AK, Chatterjee SN, Chatterjee RN, Majumdar D, Chatterjee C, Ghosh M, Achary PM, Dey A (1980). Regulation of DNA replication in *Drosophila*. In Development and Neurobiology of *Drosophila* (eds. Siddiqi O, Babu P, Hall LM, Hall JC). Plenum Publ Co, New York; 57–83.

Nägl W (1978). Endopolyploidy and polyteny in differentiation and evolution. North-Holland Publ Co, Amsterdam.

Nelson GW, Pienta KJ, Barrack ER, Coffey DS (1986). The role of the nuclear matrix in the organization and function of DNA. Ann Rev Biophys Biophys Chem 15; 457–75.

Pfluger OH, Yunis JJ (1966). Late replication patterns of chromosomal DNA in somatic tissues of the Chinese hamster. Exp Cell Res 44; 413–20.

Pimpinelli S, Pignone D, Gatti M, Olivieri G (1976). X-ray induction of the chromatid exchanges in somatic cells of *Drosophila melanogaster*: variations through the cell cycle of the pattern of rejoining. Mutat Res 35; 101–10.

Plaut W (1963). On the replicative organization of DNA in the polytene chromosomes of *Drosophila melanogaster*. J Mol Biol 7; 632–5.

Plaut W, Nash D, Fanning T (1966). Ordered replication of DNA in polytene chromosomes of *Drosophila melanogaster*. J Mol Biol 16; 85–93.

Popescu NC, Diapaolo JA (1979). Heterogeneity of constitutive heterochromatin in somatic Syrian hamster chromosomes. Cytogenet Cell Genet 24; 53–60.

Prelich G, Kostura M, Marshak DR, Mathews MB, Stillman B (1987). The cell-cycle regulated proliferating cell nuclear antigen is required for SV40 DNA replication *in vitro*. Nature 326; 471–6.

Prescott DM (1970). The structure and replication of eukaryotic chromosomes. Adv Cell Biol 1; 57–117.

Prescott DM (1976). Reproduction of eukaryotic cells. Acad Press Inc, New York.

Prokofieva-Belgovskaya AA, Slesinger SI, Lozovskaya ER, Atayeva JM (1976). Comparative study of human chromosome replication in primary cultures of embryonic fibroblasts and in cultures of peripheral blood lymphocytes. II. Replication of centromeric regions of chromosomes at the termination of the S-period. Chromosoma 57; 261–7.

Raman R (1984). Maintenance of late replication of X-chromosome in hyperdiploid cells of barking deer, *Muntiacus muntjak*. Ind J Exp Biol 22; 229–32.

Reddy, KF, Savage JRK, Papworth OG (1988). Replication kinetics of X-chromosomes in fibroblasts and lymphocytes. Human Genet 79; 44–8.

Redfern CPF (1981). DNA replication in polytene chromosomes: similarity of termination patterns in somatic and germ line derived polytene chromosomes of *Anopheles stephensi* Liston (Diptera-Culicidae). Chromosoma 84; 33–47.

Rieger R, Michaelis A, Green M (1976). Glossary of Genetics and Cytogenetics. Springer-Verlag, Berlin.

Roy JK (1983). Chromosome organization in *Drosophila*. A comparative study in different cell types of *D. nasuta*. Ph.D. Thesis, Banaras Hindu University, Varanasi, India.

Roy S, Lakhotia SC (1979). Replication in *Drosophila* chromosomes. Part II. Unusual replicative behaviour of two puff sites in *D. kikkawai*. Ind J Exp Biol 17; 231–8.

Rudkin GT (1963). The structure and function of heterochromatin. In Genetics Today, vol. 2 (ed. Geerts SJ). Pergamon Press, Oxford; 359–74.

Rudkin GT (1972). Replication in polytene chromosomes. In Results and Problems in Cell Differentiation vol. 4 (ed. Beermann W). Springer-Verlag, Berlin, Heidelberg, New York; 59–85.

Schmidt M, Stolzmann WM, Baranovskaya LF (1982). Replication variants in human inactive X-chromosome. I. Variability within lymphocytes of single individuals. Chromosoma 85; 405–12.

Selig A, Ariel M, Goitein R, Marcus M, Cedar H (1988). Regulation of mouse satellite DNA replication time. EMBO J 7: 419–26.

Sen S, Sharma T (1982). Asynchronous replication of constitutive heterochromatin on X-chromosome: possible influence of facultative heterochromatin of the adjacent constitutive heterochromatin. Chromosoma 85; 119–26.

Shah VC, Lakhotia SC, Rao SRV (1973). Nature of heterochromatin. J Sci Ind Res 32; 467–80.

Sharma T, Dhaliwal MK (1974). Relationship between patterns of late S DNA synthesis and C- and G- banding in muntjac chromosomes. Exp Cell Res 87; 394–7.

Sheehan MA, Mills AD, Sleeman AM, Laskey R, Blow JJ (1988). Steps in the assembly of replication-competent nuclei in a cell-free system from *Xenopus* eggs. J Cell Biol 106; 1–12.

Sheldon S, Nichols WW (1981a). Comparison of the patterns of chromosomal late replication. I. Human renal epithelium and lung fibroblasts *in vitro*. Cytogenet Cell Genet 29; 40–50.

Sheldon S, Nichols WW (1981b). Comparison of the patterns of chromosomal late replication. II. Chick embryo, lung and kidney *in vivo* and *in vitro*. Cytogenet Cell Genet 29; 51–9.

Sinha P, Mishra A, Lakhotia SC (1987). Chromosomal organization of *Drosophila* tumours. I. Polytene chromosome organization and DNA synthesis in ovarian pseudonurse cells in *otu* mutants of *D. melanogaster*. Chromosoma 95; 108–16.

Slesinger SI, Prokofieva-Belgovskaya AA (1968). Replication of human chromosomes in primary cultures of embryonic fibroblasts. II. Inter-chromosomal asynchrony of DNA replication. Cytogenetics 7; 337–46.

Slesinger SE, Lozovskaya ER, Alaayeva DM, Prokofieva-Belgovskaya AA (1974). Comparative study of human chromosome replication in primary cultures of embryonic fibroblasts and in cultures of peripheral blood leukocytes. I. Chromosomal replication at the beginning and termination of the S-period. Chromosoma 49; 173–84.

Sonnenschein C (1970). Somatic cell hybridization: patterns of chromosome replication in viable Chinese hamster X Armenian hamster hybrids. Exp Cell Res 63; 195–9.

Sperling K, Rao PN (1974). Mammalian cell fusion. V Replication behaviour of heterochromatin as observed by premature chromosome condensation. Chromosoma 45; 125–31.

Spierer A, Spierer P (1984). Similar level of polyteny in bands and interbands of Drosophila giant chromosomes. Nature 307; 176–8.

Spradling A, Orr-Weaver T (1987) Regulation of DNA replication during *Drosophila* development. Ann Rev Genet 21; 373–403.

Steinemann M (1980). Chromosomal replication in *Drosophila virilis*. I. Diploid karyotype of brain cells. Chromosoma 78; 211–23.

Steinemann M (1981a). Chromosomal replication in *Drosophila virilis*. II. Organization of active origins in diploid brain cells. Chromosoma 82; 267–88.

Steinemann M (1981b). Chromosomal replication in *Drosophila virilis*. III. Organization of active origins in the highly polytene salivary gland cells. Chromosoma 82; 289–307.

Stubblefield E (1975). Analysis of the replication patterns of Chinese hamster chromosomes using 5-bromodeoxyuridine suppression of Hoechst 33258 fluorescence. Chromosoma 53; 209–21.

Sung WK, Van t Hof J, Jagiello G (1986). DNA synthesis studies in premeiotic mouse oogenesis. Exp Cells Res 163; 370–80.

Swift H (1965). Molecular morphology of the chromosomes. In Vitro 1; 26–49.

Taylor JH (1960). Asynchronous duplication of chromosomes in cultured cells of Chinese hamster. Biophys Biochem Cytol 7; 455–66.

Taylor JH (1978). Control of initiation of DNA replication in mammalian cells. In DNA Synthesis: present and future (eds Molineux I, Kahiyama M.) Plenum Publ Corp, New York, London; 143–59.

Taylor JH (1984). Origins of replication and gene regulation. Mol Cell Biochem 61; 99–109.

Taylor JH. Hozier JC (1976). Evidence for a four micron replication unit in CHO cells. Chromosoma 57; 341–50.

Thireos G, Griffin-Shea R, Kafatos RC (1980). Untranslated mRNA for a chorion protein of *Drosophila melanogaster* accumulates transiently at the onset of specific gene amplification. Proc Nat Acad Sci 77; 5789–93.

Tiwari PK, Lakhotia SC (1984). Replication in *Drosophila* chromosomes. XIII. Comparison of late replicating sites in two polytene cell types in *D. hydei.* Genetica 65; 227–34.

Triboli C, Biamonti G, Giacca M, Colonna M, Riva S, Falaschi A (1987). Characterization of human DNA sequences synthesized at the onset of S-phase. Nucl Acids Res 15; 10211–31.

Tsuji H, Tobari I (1980). Effects of 5-bromodeoxyuridine on metamorphosis and on cell cycle kinetics of ganglion cells in *Drosophila melanogaster.* Exp Cell Res 125; 79–85.

van de Flierdt K (1975). No multistrandedness in mitotic chromosomes of *Drosophila melanogaster.* Chromosoma 50; 431–4.

Van't Hof J (1975), DNA fibre replication in chromosomes of a higher plant (*Pisum sativum*). Exp Cell Res 93; 95–104.

Van't Hof J, Bjerkens CA (1979). Chromosomal DNA replication in higher plants. Bioscience 29; 18–22.

von Kiel K, Hameister H, Somssich IE, Sabine A (1985). Early replication banding reveals a strongly conserved functional pattern in mammalian chromosomes. Chromosoma 93; 69–76.

Wargent M, Hartmann-Goldstein IJ (1976). Replication behaviour and morphology of a rearranged chromosome region in *Drosophila.* Chromosomes Today 5; 109–16.

Wienberg J (1977). Brdu-Giemsa technique for the differentiation of sister chromatids in somatic cells of *Drosophila melanogaster.* Mutat Res 44; 283–6.

Willard HF (1977). Tissue-specific heterogeneity in DNA replication of human X-chromosome. Chromosoma 61; 61–73.

Willard HF, Breg WR (1980). Human X-chromosme: synchrony of DNA replication in diploid and triploid fibroblasts with multiple active or inactive X-chromosomes. Somat Cell Genet 6; 187–98.

Zacharias H (1986). Tissue specific schedule of selective replication in *Drosophila nasutoides.* Roux Arch Dev Biol 195; 378–88.

Zakharov F, Egolina NA (1972). Differential spiralization along mammalian mitotic chromosome. I. BudR-revealed differentiation in chinese hamster chromosomes. 38; 341–65.

Zhimulev IF, Semeshin VF, Kulichkov VA, Belyaeva ES (1982). Intercalary heterochromatin in *Drosophila.* I. Localization and general characteristics. Chromosoma 87; 197–228.

7 Dosage Compensation in *Drosophila* and Mammals: A Review of the Past and Speculation for the Future

A.S. MUKHERJEE

Genetics Research Unit
Department of Zoology
University of Calcutta
35 Ballygunge Circular Road
Calcutta 700 019, India

Introduction

Nearly sixty years ago H.J. Muller recognized that in heterogametic species with XX-XY type sex determination, the possible hyperproduction of gene products caused by two-to-one dosage difference between XX2A and XY2A is balanced by a certain regulatory device. This device that balances the hyperproduction and hypoproduction was termed 'dosage compensation' (Muller, 1928). Shortly after, Stern (1929) showed that compensation results from the genic balance. The significance of the concept of this phenomenon was realized only after the discovery of the sex chromatin in female mammals (Barr and Bertram, 1949), and its identification with the X chromosome (Ohno *et al.*, 1959). Later, Lyon (1961) correlated the sex chromatin with the X that is inactivated in female mammals. This interpretation opened up a new line of thinking, not only on the mechanism of balance between hypoproduction and hyperproduction of X-coded products, but also on that of regulation of eukaryotic genes in general. Within a few years of the discovery of the X-inactivation theory, it was found that in the male *Drosophila,* contrary to X-inactivation, the X chromosome is hyperactive (Mukherjee and Beermann, 1965). Subsequently, this was substantiated both in polytene as well as other diploid tissues, and in a variety of species of *Drosophila* (Lakhotia and Mukherjee, 1969, 1970, 1972; Korge, 1970a,b; Maroni and Plaut, 1973a; Chatterjee and Mukherjee, 1975; Mukherjee and Chatterjee, 1976). The two parallel but different systems of balancing mechanism of the X-coded products rendered the phenomenon of dosage compensation all the more important. Furthermore, although the expression of X-linked genes is independent of the development of the sex types, in both systems, *Drosophila* and mammals, the early determinants of both sex and dosage compensation have common elements in the genetic programme.

Dosage compensation has been observed for many systems other than *Drosophila* and mammals. Each of them seems to have a unique set of intermediate pathways. For example, in Coccids, dosage compensation is achieved by regulating the cell divisions (Berlowitz *et al.*, 1968); in Gryllotalpa, it operates by inactivation of one X in females (with 2 X's) (Rao and Bhattacharya, 1984). There are yet others in which despite the XX (or ZZ) male and XY (or ZW) female, no evidence for a definite dosage compensation has been recorded.

I shall restrict myself in this review to the works reported on *Drosophila* and mammals, and discuss the different mechanisms suggested by different workers in this area. I shall discuss the different discoveries since 1959 when Muller's theory of dosage compensation was rediscovered upon the identification of the sex chromatin with the X chromosome.

As we shall see, dosage compensation has been of interest since 1960, but most of the investigations for molecular explanation of the mechanism of dosage compensation were done during the last decade. I shall emphasize those aspects in this article. For the details of genetic and cytological aspects the readers may refer to the recent reviews (Gartler and Riggs, 1983; Baker and Belote, 1983; Steinmann-Zwicky and Nothiger, 1985; Cline, 1985; Mukherjee, 1986).

Sex Determination and Dosage Compensation

The phenomena of sex determination and dosage compensation, although guided by two independent parallel sequentially-ordered programmes of genetic elements, are linked by a kind of convergent ancestry, and their expressions cannot be separated from each other. Therefore, in order to understand dosage compensation and its molecular explanation, it is necessary to understand the relation between the two phenomena.

In man and other mammals, the Y chromosome has some very important properties for determination of maleness (Ohno, 1967). It is well established that a normal male must have one Y and one X chromosome, besides the normal set of autosomes, and a normal female must carry two X chromosomes besides the autosomes, and no Y. The presence of a Y, regardless of the number of X's in the complement, would result in a male (leaving aside the secondary abnormalities). In other vertebrates (with the exception of birds and certain reptiles), true sex chromosomes are generally not well established. The differentiation of sex is primarily controlled by certain unestablished chromosomal factors. In birds, the mechanism is similar to that in mammals, except that the heterogametic sex is female (ZZ♂/ZW♀). In *Drosophila* and many other insects, females are XX and males are XY (or XO). However, in them, the Y is not strictly required for primary determination of maleness. It is the X to A ratio that is decisive in determining femaleness or maleness.

It is an interesting paradox that while the sex chromosomes evolved for conserving a pattern of distinction between the two sexes, and thereby leading to a dosage difference (Lucchesi, 1978), dosage compensation tended to reduce the difference to a level of equilibrium between hyperproduction and hypoproduction of X-coded gene products (Muller, 1950; Stern, 1960).

Sex determination and dosage compensation in both mammals and *Drosophila* show some interesting parallelism. Both require certain common elements of control at the initial part of the ontogeny. Yet, they do not necessarily depend on each other. For example, in mammals, while the Y chromosome is crucial for maleness, it has no direct role on dosage compensation. In *Drosophila,* the Y has no role in either sex determination or dosage compensation. The autosomes play an important role in sex determination in *Drosophila* and they are important for switching on certain genes required for establishment, finer tuning and maintenance of the level of activity in a given genome. In mammals, on the other hand, the extra X chromosomes over one are silenced, forming the sex chromatins (Barr and Bertram, 1949; Ohno *et al.*, 1959; Lyon, 1961). Thus, in a normal 2X2A genomic complement, only one X chromosome is active. However, in tetraploid (4X4A), two X's are active in a cell; and in triploids (3X3A), either one or two X's may be active in a cell (Weaver *et al.*, 1975; Migeon *et al.*, 1981; Gartler and Riggs, 1983). This implies that autosomes do have a role in mammalian dosage compensation. The single active X concept is therefore to be modified as *single active X per diploid set*. In *Drosophila*, the activity of the X is directly correlated with the X:A ratio, and not with the absolute number of either X or A. Obviously, these facts demand that the programme of operation for the initiation and maintenance of dosage compensation must involve functional attributes from both X chromosome and autosome.

Thus, in order to successfully carry out the information through the pathway leading to dosage compensation, an ordered hierarchy of elements of the genetic programme has to be developed that should involve a sequential array of elements. These are: (a) establishment of the X and A values in the complement; (b) reading and assessment of the X/A ratio as in *Drosophila* or the absolute numbers of the X per diploid set as in mammals; (c) switching on of some genes to produce certain inhibitor (I) and enhancer-like (E) molecules; (d) interaction between the I and E; and (e) finer tuning of the net product through specific genes and maintenance.

Dosage Compensation in Mammals: Facts in Perspective

Genesis of X-inactivation

After the identification of the sex chromatin or Barr body with the X chromosome by Ohno *et al.* (1959) and the X chromosome inactivation hypothesis by Lyon (1961) several important questions emerged. Answers to those questions were extremely significant from the point of view of implications of X-inactivation to regulation and selective value of dosage compensation. It was understood that in soma the inactivation is random and clonal. Either the paternal or maternal X can be inactivated in a random manner. This was formally demonstrated by Mukherjee and Sinha (1964) in the female mule, in which the inactivation of horse and donkey X was distributed in the ratio 1:1.The inactivation process is initiated early in development, but once this differentiative event takes place it is fixed in all subsequent cell lineages. The X-inactivation process is associated with two other features, namely, heterochromatization of the inactive X and generally late replication of the inactive X. In the early embryo, both X chromosomes are euchromatic, and both are active, showing no dosage compensation (Gartler and Riggs, 1983). This should impose a serious effect on the X-coded gene activity during embryogenesis. Yet, it does not. Superimposed on these facts is yet another, that is, the X chromosomes in the germ line are first inactivated during early gametogenesis, and are then reactivated during the time of entry into meiosis. These facts have made the developmental genesis of X-inactivation all the more significant.

The X-linked enzyme activity was found to go through a sequential change from unimodal distribution upto eight-cell stage, followed by a bimodal distribution at a later stage of embryogenesis. The bimodal pattern reflects the activity of the enzyme formed by the embryo. Clearly, at the eight-cell stage, there is no dosage compensation. The bimodal distribution is again transformed to unimodal pattern in the morula to early blastocyst period. This implies that the initiation of X chromosome inactivation takes place somewhere at or after the 16-cell stage.

Although a vast majority of experiments has borne out the validity of Lyon's hypothesis, exceptions of certain kinds have been encountered. It has been demonstrated, for instance, that the onset of X-inactivation during embryogenesis proceeds in steps; the first cells to display inactivation belong to trophectoderm and primitive endoderm followed by the three germinal layers of embryo proper (Monk and Harper, 1979). At least in mouse, and probably also in man, the inactive X in both trophectoderm and primitive endoderm is nonrandomly of paternal origin (Takagi, 1978, 1980; West *et al.,* 1977) as against randomly paternal or maternal inactive X in the three germinal layers. It is important to remember that the cell lineages of trophectoderm and primitive endoderm do not contribute to the formation of the organism, instead to extraembryonic membranes. Interestingly, in primitive mammals (e.g. marsupials) the inactive X is always paternal even in body cells, a situation comparable to that obtained in the extraembryonic cells in eutherians (Vande Berg *et al*, 1983). Similarly, while the X chromosome inactivation is irreversible in somatic cells, occurrence of both active X's in meiotic cells is a consequence of reversal of the inactive X in premeiotic oogonia (Andina, 1978). Exceptions to the last tenet of the Lyon's hypothesis, that the entire X chromosome is rendered inactive, have also been encountered. Genes for the enzyme, steroid sulphatase (sts) and antigens X_g and 12E7 (Mic2) in man and for STS in mouse are known to escape inactivation (Mohandas *et al.*, 1979; Goodfellow *et al.*, 1983), although the level of their activity on the inactive X may be lower than that on the active X. However, several explanations can be suggested to account for this exception. Since all these genes have been shown to have an allele, functional or nonfunctional, also on the Y chromosome (Goodfellow *et al.*, 1983; Keitges *et al.*, 1985) it is possible that the need for their compensation does not arise. Alternatively, these genes may have evolved recently on the X chromosome (STS is non X-linked in marsupials) so that the 'spread' of the inactivation control may not cover this region.

Inactivation centres

It is more or less an accepted view that all inactivated genes are on the inactive X. Cases in which some genes are active on one X while others are active on the homolog, were not found. This fact suggests that the X has a single centre of inactivation. While there is evidence from works on non-human mammals supporting the multiple inactivation centres, strong evidence for the single inactivation centre hypothesis comes from studies on inactivation of the X in the human female. The evidence comes from the studies on long-arm isochromosomes and X-autosome translocation in humans (Therman and Patau, 1974). Gartler and Riggs (1983) in the review have suggested that at least two functional inactivation centres may be essential for inactivation to proceed; if one is nonfunctional, the resulting X-chromosome dosage imbalance (2 active X's) may be lethal.

Methylation and inactivation

Since 1975, Riggs and his co-workers have been proposing the role of methylation in X-chromosome inactivation (Riggs, 1975; Razin and Riggs, 1980; Riggs and Jones, 1983). Two very important results support the role of methylation in X-inactivation: (a) there is an inverse correlation between methylation of 5'-flanking sites and transcriptive activity; and (b) inhibitors of DNA methylation activate genes and cellular phenotypes. The new state of differentiation, then, is maintained even after removal of the inhibitor from the medium. In the context of X chromosome inactivation, it has been demonstrated that 5-azacytidine, an inhibitor of methylation, remarkably reactivates more than one X-linked cellular phenotype, e.g. HPRT− to HPRT+ reversion, human PGK gene. Interestingly, after X-chromosome reactivation by 5-azacytidine, the new phenotype remains stable (Mohandas *et al.*, 1981). This stability has been supported by the DNA-mediated gene transfer following 5-azacytidine reactivation (Venolia *et al.*, 1982, Gartler and Riggs, 1983). However, as pointed out by Gartler and Riggs in the review, most methylation may be unimportant and methylation of only a few critical sites may be of significance.

Notwithstanding the experimental demonstration of the role of methylation in X chromosome inactivation, it seems unlikely that it is the primary cue for the initiation of inactivation. Regulations at gross chromosomal level need to be envisaged to understand the mechanism of inactivation. Gartler *et al*'s (1985) suggestion that heterochromatization of the X may initiate inactivation whose finer tuning is done by the process of methylation, looks interesting. In this context, it is worthwhile recalling that the nonrandomly inactivated X chromosome (paternal) in marsupials and extraembryonic eutherian cells does not show any difference in the state of methylation from their active homologues. Also, experiments using DNA from extra embryonic membranes show transfection capability of the inactive X (Kratzer *et al.*, 1983). Obviously, inactivation in these cell types does not involve DNA modification (e.g. methylation) but possibly only heterochromatization. In view of the fact that in marsupials as well as extraembryonic cells the frequency of spontaneous reactivation is considerable (VandeBerg *et al.*, 1983; Migeon *et al.*, 1986), it appears logical to assume that while X-inactivation is initiated by chromatin modification, its maintenance in subsequent cell generations is regulated by methylation of some of its critical sites (Kaslow and Migeon, 1987).

X-chromosome inactivation and spreading effect

It was shown by Catanach and co-workers (Catanach and Issacson, 1967; Catanach, 1975), that when an autosomal segment is inserted into the X, the autosomal genes in the relocated condition are inactivated. However, the inactivation is stronger at the two ends of the inserted autosome than at the centre. A clear explanation for the weak inactivation at the centre is not available. However, the spreading effect as such suggests that the controlling centre of inactivation must have an important role to play. Gartler and Riggs (1983) have tried to explain this effect through a model called the 'Way station' model.

Many aspects associated with the spreading effect phenomenon remain unexplained and understood. The most important of them are perhaps the relation between the inactivation centre(s) and their control on specific genes, and secondly, the problem of programming of replication and transcription of the X-chromosomal and autosomal genes.

Dosage Compensation in *Drosophila*

Hyperactivation of the X in *Drosophila*

When the relation between sex chromatin and dosage compensation was established in the early sixties, it was taken for granted that dosage compensation operates through favouring the activity of only one X chromosome in any sex. This dogma was found to be wrong. It was soon realized that different animal systems may have adopted different means for achieving dosage compensation.

The first evidence for this prediction was obtained by Mukherjee and Beermann (1965) who showed that *in vivo* chromosomal RNA synthesis of the X chromosome of *Drosophila* male is twice as much as that in the female. The experiments were carried out on the polytene chromosomes of larval salivary glands of *D. melanogaster*, and later extended to other species by the co-workers of Mukherjee (Lakhotia and Mukherjee, 1972; Chatterjee and Mukherjee, 1975; Mukherjee and Chatterjee, 1976). In all species the X chromosome of the male was found to be hyperactive. It was concluded by Mukherjee and Beermann (1965) and Mukherjee (1966), and later by others (Mukherjee and Chatterjee, 1975, 1976; Korge, 1970a, b) that dosage compensation in *Drosophila* operates in the male through hyperactivation of the single X chromosome and is initiated at the level of transcription. This conclusion has recently been corroborated by Ganguly *et al.* (1985) at the level of the G6PD gene-specific transcript.

The question that immediately attracted the Drosophilists was: what happens to the X-coded gene products? The foremost investigations in this regard came from Seecof *et al.* (1969), Maroni and Plaut (1973b) and Schwartz (1973). In all these studies, the amount or activity of X-coded protein/pigment/enzyme was examined. It was concluded from all studies that the primary gene product in male *Drosophila* is indeed elevated to a higher level by a factor of two.

It was argued that although the single X of the male shows a higher transcriptional competence, (than the X of the female), it is not established whether the same end-point could not be realized through a reduced activity of both X's in female. Alternatively, one X may be completely repressed in the female, as in mammals. The first alternative was ruled out by the findings that the X-coded enzyme activity was greater than the per gene activity in the female,whenever the X:A ratio was $\geqslant 0.5$ (Lucchesi, 1973, 1978; Lucchesi *et al.*, 1974). The second alternative was ruled out by two significant findings: (a) In *Drosophila,* females heterozygous for an X-linked gene do not show mosaicism for the expression of the two alleles. (b) Female flies heterozygous for the X-linked gene for 6PGD enzyme variants, fast [6PGD (f)] and slow [6PGD (s)] show three bands when the proteins are run on SDS-PAGE (Kazazian *et al.*, 1968). These bands are a fast, a slow and an intermediate hybrid band. These results reinstated the hypothesis that hyperactivity of the male X in *Drosophila* is responsible for dosage compensation.

Cellular autonomy of hyperactivity of the male X and its significance

Lakhotia and Mukherjee (1969), using unstable ring X strain of *D. melanogaster*, demonstrated that the hyperactivity of male X is cell-autonomous. Later, Prasad-Sinha and Mukherjee (1985) showed that the cellular autonomy of hyperactivity is observed not only in normal XY2A cells but also in males which are hyperploid for small segments of the X. These findings have great significance from two points of view: first, the results ruled out a direct role of sex on the cellular expression of dosage compensation, and second, the manifestation of hyperactivity is decidedly programmed at an early stage of development and long before the establishment of the template activity of the chromatin for a high transcriptional competence. Another important implication of the cellular autonomy is that hyperactivity is decided at every cell level by functions immanent in the pool of cell components and is not dependent on extraneously diffusing factors.

Early completion of replication of the male X and its implications

Berendes (1968) gave the important information that the entire male X of *Drosophila* replicates asynchronously with respect to the autosomes of the genome. Subsequently, Lakhotia and Mukherjee (1970) and Chatterjee and Mukherjee (1977) demonstrated that the male X completes replication earlier than the X's of the female, and that the rate of replication is considerably higher. Furthermore, like the transcription, the early replication of the male X is also cell-autonomous. We suggested that both early replication and higher transcriptional competence of the X in the male are the cytogenetic counterpart of dosage compensation. This was not very readily accepted (see Hägele and Kalisch, 1974). However, the intimate correlation between early replication and transcriptional hyperactivity in various kinds of genomic complements of *Drosophila* and their parallel conservation in evolution (Mutsuddi *et al.*, 1984) supports the contention. Moreover, in the reverse case, where the X is repressed or inactivated as in case of the inactive X of the female mammal, it invariably shows late replication. This has been formally demonstrated by Goldman *et al.* (1984). They concluded that genes that can be transcribed must replicate early and vice versa, and that a gene in order to be. transcriptionally competent must be transformed into early replicating form. Thus, it is not unreasonable to contend that early replication of the male X in *Drosophila* is a predisposed condition for hyperactivity. However, it is not clear what part of the programme of genetic regulation of differentiation dictates the relation.

Entire X vs piecemeal hyperactivation

It has been shown in autoradiographic monitoring of the male X activity as well as conformation of chromatin especially in polytene cells that the hyperactivity of the male X involves the entire X chromosome. Polytenic X with one part male-like and the rest narrow and less active, is never observed in a normal male, except in certain hybrids (Mutsuddi *et al.*, 1984). Muller and Kaplan (1966), however, suggested that hyperactivity may be a piecemeal affair. Stewart and Merriam (1980) suggested that the regulation of dosage compensation in *Drosophila* is a wholly chromosomal affair and that the level of activity is modulated in a stepwise fashion according to the total number of the X chromosomes in the genome. Their proposition was based on the comparison of X-coded enzyme activity in flies with different doses of X chromosome (Stewart and Merriam, 1975). On the other hand, more recently, Jaffe and Laird (1986) have argued that one of the steps of the regulatory system for dosage compensation must be through certain cis-acting sequences located close to each compensated X-linked gene. Their argument was derived from the data on the P-mediated transformation of X-linked sequences to autosomes. Analysis of compensation of three X-linked gene sequences, white, Sgs4 and Hsp82 of *D. pseudoobscura* revealed that the proximity of cis-acting sequences is necessary for compensation of the X-linked genes located elsewhere. However, it is important that in order to assume a full compensatory effect the cis-acting sequences must be in close proximity with the gene and in 'a chromosomal domain suitable for its expression'. Explaining the compensation of the autosomal genes rosy and dopa-decarboxylase, Jaffe and Laird argued that such partial or full compensation of autosomal genes translocated to X chromosomes through P transformation (Spradling and Rubin, 1983; Scholnick *et al*; 1983) might be due to the influence by the cis-acting regulatory elements of the X chromosome. Whether the different cis-acting elements located near the compensated genes have a homology of concensus sequence(s) is currently under study. However, a 7-bp sequence homology between the presumed cis-elements of white and Hsp82 has been found by Jaffe and Laird (1986). They have also reported that in non-compensated alleles of X-linked genes there is a lesion at their 5′ ends.

It appears therefore possible that there may be two distinct levels of regulation: (a) one that might involve a decision at the whole chromosome level, and (b) the other that would read the specific gene frame and tune it to compensate.

Non-histone Chromosomal Proteins and Hyperactivity

All eukaryotic chromosomes contain both histone and non-histone proteins. While the histones are known to participate in forming an integral part of the chromatin, the structural role of non-histones is not quite well established. However, they are invariably found to be associated with the active as well as inactive chromatin (Elgin and Weintraub, 1975). It has been reported in various systems that association and perhaps a certain degree of specificity of non-histone chromosomal proteins may be obligatory to gene expression.

In *Drosophila,* during puff formation, accumulation of non-histone proteins has been demonstrated by Berendes (1966, 1972). Bautz *et al.* (1983) suggested that non-histone protein and RNA polymerase binding takes place at the very beginning of gene activity. It has been shown that the hyperactive X chromosome of male *Drosophila* has a higher binding affinity for non-histone chromosomal proteins (Chatterjee *et al.*, 1981; Ukil *et al.*, 1986). Ukil *et al.* have shown that NHCP from *Drosophila* binds with the male X with greater affinity than with female X's. Furthermore, Chatterjee *et al.* (1981) have shown that when histones and non-histones are removed by high salt treatment and the X is transcribed *in situ* by *E. coli* RNA polymerase (holoenzyme), the hyperactivity of the male X is considerably reduced. Chatterjee *et al.* suggested that the preferential reduction of male X activity is due to immanently existing bound non-histones on the male X resulting in an inflated chromosomal conformation. It was presumed that the puffy conformation of the X is consequent to a modulatory regulation that evokes an organizational change in the structure of the X prior to activation.

Hypomethylation and hyperactivation

Methylation of *Drosophila* DNA has been a generally controversial issue. However, we postulated that it may be possible that certain unique methylation-demethylation cycle might be responsible for moduation of X chromosome organization in *Drosophila.* Existence of methylation at a low level in *Drosophila* has been claimed by Achwal *et al.* (1984). An indirect evidence from our laboratory, showing exogenous methylation with methyl donors, has reinstated the possibility that DNA methylation-demethylation might also operate in *Drosophila.* Here also 5-Azacytidine induced hypomethylation appears to activate the X-linked genes in both sexes but more so in the male X (unpublished data, Chaudhuri and Mukherjee).

Genetic Elements for Dosage Compensation in *Drosophila*

The genetic elements that affect sex determination and/or dosage compensation in *Drosophila* can be grouped into three categories: those which act at the maternal gametic level, those which assess/count the X/A ratio, and those which maintain the specific level of X-linked gene activity. Mutations of these genes can affect either the viability or fertility, or they may upset the sex phenotype and or the net activity level of X-linked genes. I shall discuss the action of these genes in a little detail in the following.

Maternal gametic gene activity

The only gene that acts at the maternal gametic level and affects sex and dosage compensation is the gene daughterless (da) (Cline, 1983). The da$^+$ maternal gene function is necessary to programme the pathway for sex determination and dosage compensation. The function of da$^+$ gene is mediated through the sis-a$^+$ gene action and subsequently through the Sxl$^+$ function at the zygotic level. The Sxl$^+$ gene function plays an important role in determining the turning point for maleness or femaleness, and for hyperactivity vs female level activity. However, the normal function of this gene is dictated by the normal maternal gene function of da$^+$. The organization and functions of the Sxl gene complex has been thoroughly examined by Cline (1983) (see also Maine *et al.*, 1985 a,b). It has been claimed that the Sxl functions only in the female, but its loss of function mutants affect both sex determination and dosage compensation.

Genetic elements for counting and assessment of X/A ratio

Analyses of the levels of activity of X-linked genes in flies with different X/A ratios suggest that the level of X-linked gene activity is set by the X/A ratio and not by the absolute number of either X chromosomes or autosomes. However, the precision with which the level per gene copy would be set seems to depend on certain other elements of the programme. For example, the levels of activity in metafemales (3X2A) and some other hyperploids (our unpublished data) are not consistent with the automatic programming through X/A ratio assessment. The questions such as what genetic elements are responsible for the assessment of the X/A ratio, and (after counting the ratio) what sets the subsequent elements of the programme into operation, are not yet answered. The first question remains open. As regards the second question, much important information has become available mainly through the works of Cline and his collaborators (Cline, 1983, 1984, 1985, 1986; Maine *et al.*, 1985a, b). The most significant finding is about the functions of the sex specific lethal gene complex, Sxl. The effect of the mutant gene is zygotic, and it evokes zygotic lethality in females. When the X/A ratio is >0.5, the Sxl$^+$ is activated by da$^+$, and sis-a$^+$. The Sxl gene complex is located at the cytological region 6F5-7A1 of the polytenic X chromosome and at the genetic map position 1–19.2. The sis-a gene is also located on the X, at the cytological region, 10B4, and at the genetic map site 1–34.3 (Maine *et al.*, 1985 a,b; Cline 1986). Molecular genetic and molecular biological studies on the Sxl loss of function and gain of function mutants have revealed important controlling functions of the Sxl. It appears that Sxl gene functions may have several overlapping features. For example, Jaffe and Laird (1986) have reported that 'the transcription pattern of this locus is surprisingly complex' and that 'there are at least three overlapping pairs of male and female-specific transcripts'. As reported by Maine *et al.* (1985a) there may be some more transcripts of unknown specificity. Furthermore, the sex-specific transcripts are all synthesized from the same strand, thus 'precluding an antisense function for the male transcripts' (Jaffe and Laird, 1986). It is claimed that Sxl has functions not only in dosage compensation and sex determination but also for its own regulation. Therefore, Sxl occupies a central position in the array.

Genetic elements for modulation of X for transcriptional competence

Earlier, from the results on segmental hyperploids for the X, it was contended that there is probably no single gene that would fulfil the requirement for regulating or counting the X/A ratio (Baker and Belote, 1983; Lucchesi, 1983). However, Steinmann-Zwicky and Nothiger (1985) reported that there is a region of the X, 3F3-4B1, that can activate Sxl$^+$, when present in two doses, and that the activation of Sxl$^+$ is necessary for the function of Sxl for sex determination and dosage compensation in females. The results also suggest that Sxl^+ can be activated to different levels of activity depending on the number of Sxl^+, and also that of the 3F3–4B1 segments. Thus, it is evident that in 2X2A the Sxl^+ is activated to produce a product that results in basal low level of X-linked gene activity, and, in flies with 2 X's plus duplications for different segments of the X, may produce intermediate levels of Sxl product and consequent different levels of X activity. Interestingly, Steinmann-Zwicky and Nothiger recognized that there might be some genetic elements in the proximal part of the X that might be necessary for the differentiation of female sex. I suggest by the same reasoning that the proximal part of the X may also contribute some factors that regulate the level of X chromosomal gene activity in both sexes. This is evident from the poor viability of the 1X2A flies with duplication for a small part of the proximal X (Maroni and Lucchesi, 1980; Prasad-Sinha and Mukherjee, 1985). This prediction has been confirmed by us recently (Ghosh and Mukherjee, 1986; Mukherjee and Ghosh, 1986).

All these results suggested that an intermediate step following the assessment of X/A might exist that would programme the transcriptional competence of the X chromosome. Obviously, this would demand 'modulation' of the organization of the DNA-histone-non-histone of the chromatin of the X. As discussed earlier, the modulation of the X is indeed a prerequisite in the programme of hyperactivation of the X in the male. The modulation of the X chromosome takes place first upon converting the DNA of the X to an early replicating form and then to higher template conformation. A gene complex located at 16A7-B2 may have the role of the modulation.

Genes for maintenance and fine tuning

Two sets of genes appear to become active following the function of Sxl^+, and subsequent to the modulation of the X for higher transcriptional competence. These genes are, (a) the autosomal genes, msl 1, msl 2, msl 3 and mle, and (b) certain X-linked cis-acting genes (Jaffe and Laird, 1986). The former sets are required for deciding the final level of activity upon interaction with the latter set of genes. The maintenance of the activity level that is set forth through the interaction and consequent to the counting of X/A ratio, is very likely done by continuous monitoring of the information passed on following the Sxl^+ function and modulation of the X.

I shall leave this episode at this point and turn toward discussion on some of the more practical models which seem to have been validated by at least some experimental data.

Programming of Random Inactivation of X in Mammals

The programming of dosage compensation in mammals requires the consideration of a number of events. The events start with the formation of early blastocyst; before this stage both X's (Xm and Xp) are active. The events are sequentially programmed as follows: initiation →→ counting→→establishment of control centre(s)→modulation of chromatin and spreading→→methylation→→nonrandom inactivation of X→→permanent random inactivation of the X→→maintenance of inactivation in soma→→reactivation of the inactive X in the germ line.

It is rather surprising that, unlike the story in *Drosophila* (discussed above), the involvement of specific genetic elements at the different stages of the programme is not well documented. It may be rewarding to explore the different sex-specific lethal mutants in mammals for their effects on sex differentiation and dosage compensation or X inactivation.

Programming of Hyperactivation of Male X in *Drosophila*

The programming of the genetic events leading to the hyperactivation of the male X in *Drosophila* is similar to that in mammals described above. In this system, specific genetic elements have been identified, mostly due to the work of Tom Cline. There are certain points in which the programme for *Drosophila* differs from that in mammals. Nonetheless, the programme is also sequentially ordered: counting of X/A ratio (sis-a, sis-b and others)→→da-maternal product→→Sxl loci activated→→modulation of chromatin template→→msl loci activated→→interaction of cis-acting elements with msl products→→hyperactivation of the X in male (no hyperactivation in females)→→finer tuning and finally dosage compensation.

The programme delineated above suggests that a certain part of it is activated in the female with X/A = 1.0, while certain others are activated in the male with the ratio = 0.5. It is not yet clear after all these years of work what really reads and establishes the X/A ratio of a genome.

Models for Inactivation of X in Mammals

Several models have been postulated to explain the mechanism of operation of dosage compensation in mammals. The only one that has attempted to give a molecular explanation is by Chandra (1985). However, a number of factors have been assumed by Chandra in his model. Although a practical demonstration of these factors is yet to be made, the model itself takes into account various aspects underlying both sex determination and dosage compensation in mammals. And with a little modification the model also explains the hyperactivation of the X in male *Drosophila*. A simplified version of the model is shown in Figure 1. Essentially, the model proposes that the key element of the regulatory programme is a 'multicopy, non-coding sequence', which is Y-linked and which binds with high affinity to a repressor originating from autosome(s). In the absence of a Y, the repressor would bind to the gene Tdx (testis determining gene) that is located on the X. Transcription of Tdx would occur in XY but not in XX or XO cells. Thus the model proposed by Chandra explains sex determination reasonably well but does not take care of the mechanism or the programming elements leading to X inactivation. Furthermore, it is not clear from this model as to how the phenomena like spreading, initiation and maintenance would take place. In an earlier model, Brown and Chandra (1973) proposed an imprinting model for explaining the random inactivation of one of the two X's in recent mammals, and non-random inactivation in primitive mammals. However, this model does not explain either the reactivation-inactivation cycle or the non-random inactivation of the X in early embryo.

A more rationalized model has been proposed by Gartler and Riggs (1983). In this model, they have suggested an interplay of two kinds of antagonistic factors to decide upon the programmed inactivation of one of the X's in the female. The different elements of the programme are shown schematically in Figure 2. Nothing has, however, been suggested about the molecular nature of the factors, except that one is inhibitory and the other has enhancing activity. The significant aspect of the model is that it visualizes the problem of reactivation, the non-random inactivation in early embryogenesis, the randomness in the permanent inactivation, spreading effect and maintenance through methylation. The model also explains the mechanism of initiation and counting of the X and A. Their analysis further reveals that the number of inactive X does not depend simply on

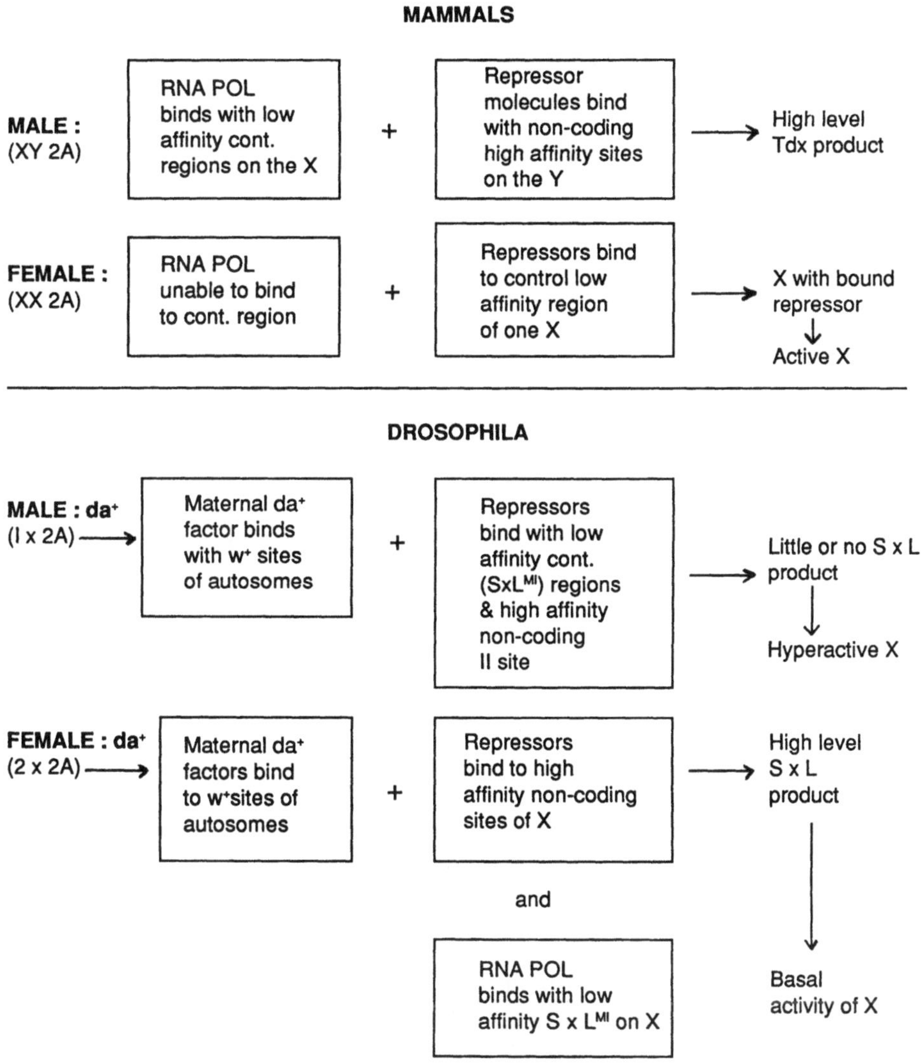

FIGURE 1. Essential feature of Chandra's model based on non-coding sequences. In mammals, the testis determining gene (Tdx) product establishes maleness. When autosomal gene-coded repressor molecules bind with non-coding high affinity sites on the Y, high level Tdx product is formed; when they bind to low affinity control region on the X, the repressor-bound X becomes active in the female. In *Drosophila* males, the repressors bind with low affinity control regions in Sxl (M1) as well as with high affinity non-coding region of the X; in females they bind only to high affinity non-coding regions of the X. In the male little or so Sxl product is formed and the X is hyperactive; in the female high level of Sxl product is formed and only basal activity of the X is maintained.

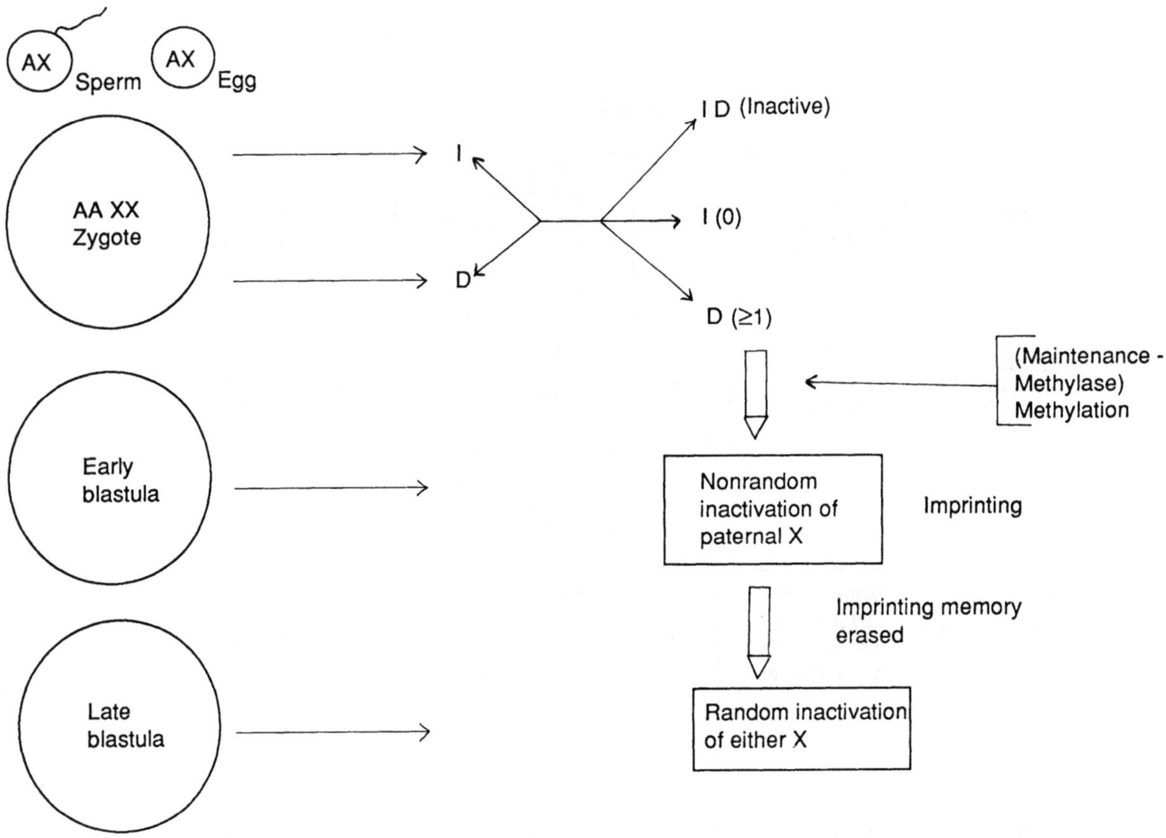

FIGURE 2. The model for the genetic programme for X inactivation in mammals, as proposed by Gartler and Riggs (1983). I = inhibitor; D = differentiator. The I is presumed to be located on the autosome(s), and D on the X chromosome.

the number of the X's in the genome but also on the sets of autosomes. And, therefore, they suggested that the 'the single active X' theory should be rephrased as 'single active X per diploid set of the autosomes'. The model has several testable components, for example, the role of methylation in inactivation of X. It may be worthwhile to test the proposed mechanism of counting by biochemical techniques available now. There are, however, a number of cases of abnormal genome with abnormal X chromosomal complement in which the programme of X inactivation seems to fall through.

Models for Hyperactivation of the Male X in *Drosophila*

Several models have been proposed in the past to explain the mechanism of hyperactivation of the male X in *Drosophila*. Maroni and Lucchesi (1980) suggested that the gene Sxl plays a significant role in determination of sex as well as in dosage compensation, while the switching on and off of the male specific genes, such as tra, dsx on the one hand and msl, mle on the other, decide the establishment of the sex and the activity level of the X. A more refined version of the model was proposed by Baker and Belote (1983). This is presented in Figure 3. This model proposes a three-step programme: (1) a prezygotic determination and counting device, (2) an early embryonic initiation device, and (3) an embryonic maintaining and tuning device. The three steps are under precise control of specific genes which are activated in a sequential order. The model has survived many genetic tests (Lucchesi and Skripsky, 1981; Cline, 1983, 1984, 1985). The decisive genetic element of the programme appears to be the Sxl. On the basis of the results of Cline and Lucchesi and Skripsky, Gadagkar *et al.* (1982) suggested a theoretical model encompassing the possible molecular events that might be operative in deciding the hyperactivity of the male X. They proposed that a *w* factor represses the Sxl function in the male but keeps it active in the female. This results in no Sxl product in the male but high level Sxl product in the female. However, the model does not explain the activity levels in tripliod intersexes, metamales and segmental hyperploids (Maroni and Lucchesi, 1980; Prasad *et al.*, 1981; Prasad-Sinha and Mukherjee, 1985). Furthermore, the discovery of certain X-linked sites regulating the activity of X also stands negatively toward the model (Steinmann-Zwicky and Nothiger, 1985). A more plausible model has come from Jaffe and Laird (1986). This model and the one by Mukherjee and Ghosh (1986) have many aspects in common. The model of

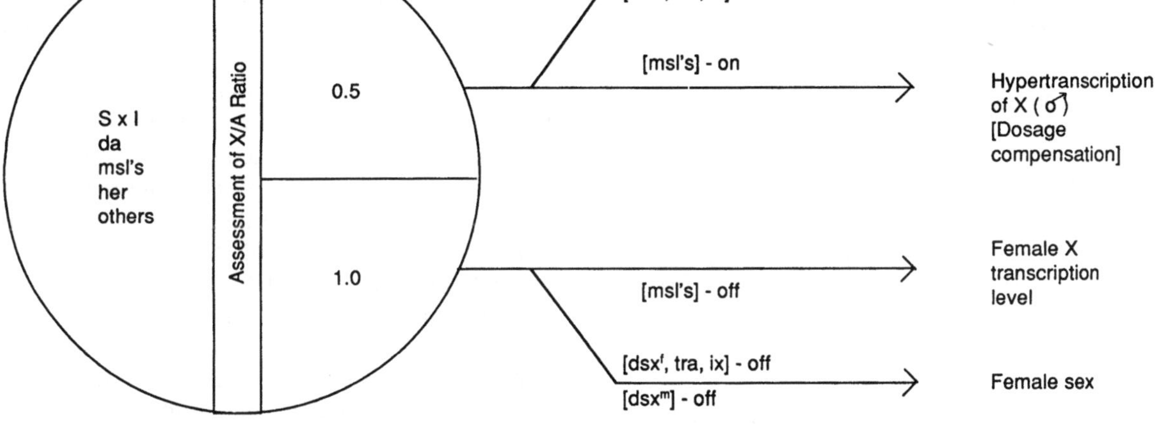

FIGURE 3: Outline of regulatory hierarchy controlling sex and dosage compensation as proposed by Baker and Belote (1983).

Jaffe and Laird (1986) is presented in Figure 4. They suggested that a number of cis-acting regulatory genes become active in the later phase of the programme. These X-linked sequences interact with certain hitherto unknown genes in both sexes and in addition also with the msl genes in the male resulting in a 'basal level of transcription' in the female, and 'hyperactivation and altered chromatin configuration' of the male X. It is clear therefore from the proposed components of the model that Jaffe and Laird have recognized the existence of genetic components which are necessary for modulating the chromatin of the male X. In a recent report, Waring and Pollack (1987) have shown the existence of dispersed A+T rich middle repetitive sequences that hybridize with 20 major and 10 minor sites on the X chromosomes, and these sequences are thought to be involved in the primary establishment of sex and dosage compensation. They suggested that these sequences, termed Xdm, might be involved in assessing the X/A ratio. It appears that the programme of dosage compensation visualized earlier must be revised in the light of the new findings.

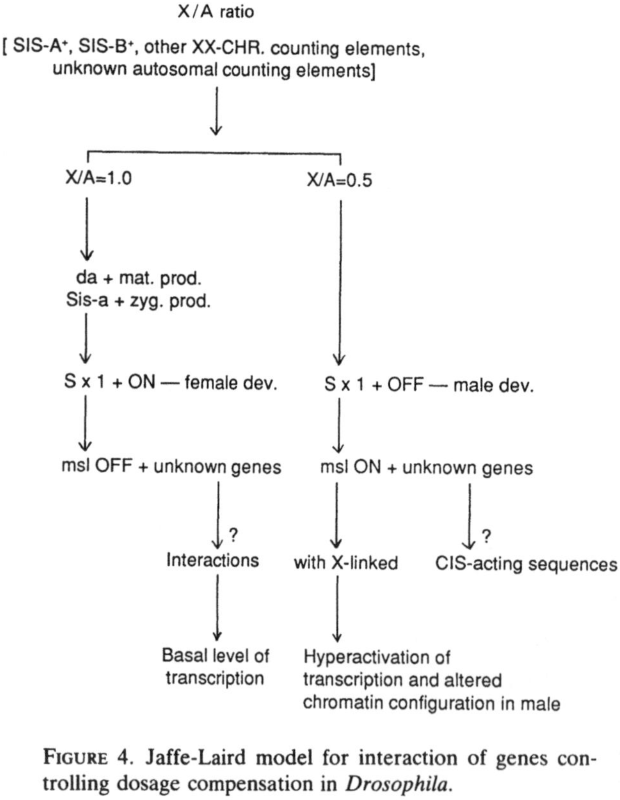

FIGURE 4. Jaffe-Laird model for interaction of genes controlling dosage compensation in *Drosophila*.

Lastly, I shall briefly describe the model proposed by Mukherjee and Ghosh (1986). The model is based on the action of a gene complex (M^+) at 16A7–B2 region of the X chromosome. A hypomorphic mutant (M^m) was discovered in a strain of *Drosophila melanogaster*, called In(1)BM2 (reinverted). In the male larval salivary glands of this strain, the X is hyperinflated and hyper-hyperactive twofold, but it is perfectly normal in the female. A duplication for 16A7–B2 region retrieves the normal level activity. Mukherjee and Ghosh (1986) proposed that this region of the X contains a sequence that might produce an inhibitor signal either by itself or upon switching on certain other dispersed regulatory genes. The products interact with the enhancer signals from the msl loci and a stoichiometric interaction then results in an inactive substance. If the interaction leaves certain *enhancers* unused, hyperactivity would ensue; if some inhibitors remain unused a basal level activity as in the female would follow. It has in addition been suggested that the (M^+) gene also needs to be activated by the Sxl.

Now, there are two aspects in the model that are in disagreement with the known female and male specific activity of Sxl and msl loci, respectively: (a) at least for dosage compensation, Sxl may have effect also in the male, and (b) the msl^+ loci are able to produce the enhancer signals in both sexes. It is not unlikely that while a sex specific gene might be active for one phene in a sex and at a specific stage of development, it might have a slightly different range of specificity at another stage with a different phene. Furthermore, it is implied that both kinds of signals, inhibitor and enhancer, should be produced regardless of sex or chromosomal complement. This prediction has been fulfilled by the finding in our laboratory. We have isolated and fractionated three protein fractions, one of which upon cytological assay shows enhancing activity (peak I), another shows an inhibitory action (peak III) and the third shows no activity (peak II) (Mukherjee and Ghosh, in preparation). The operative mechanism of the model is presented in Figure 5.

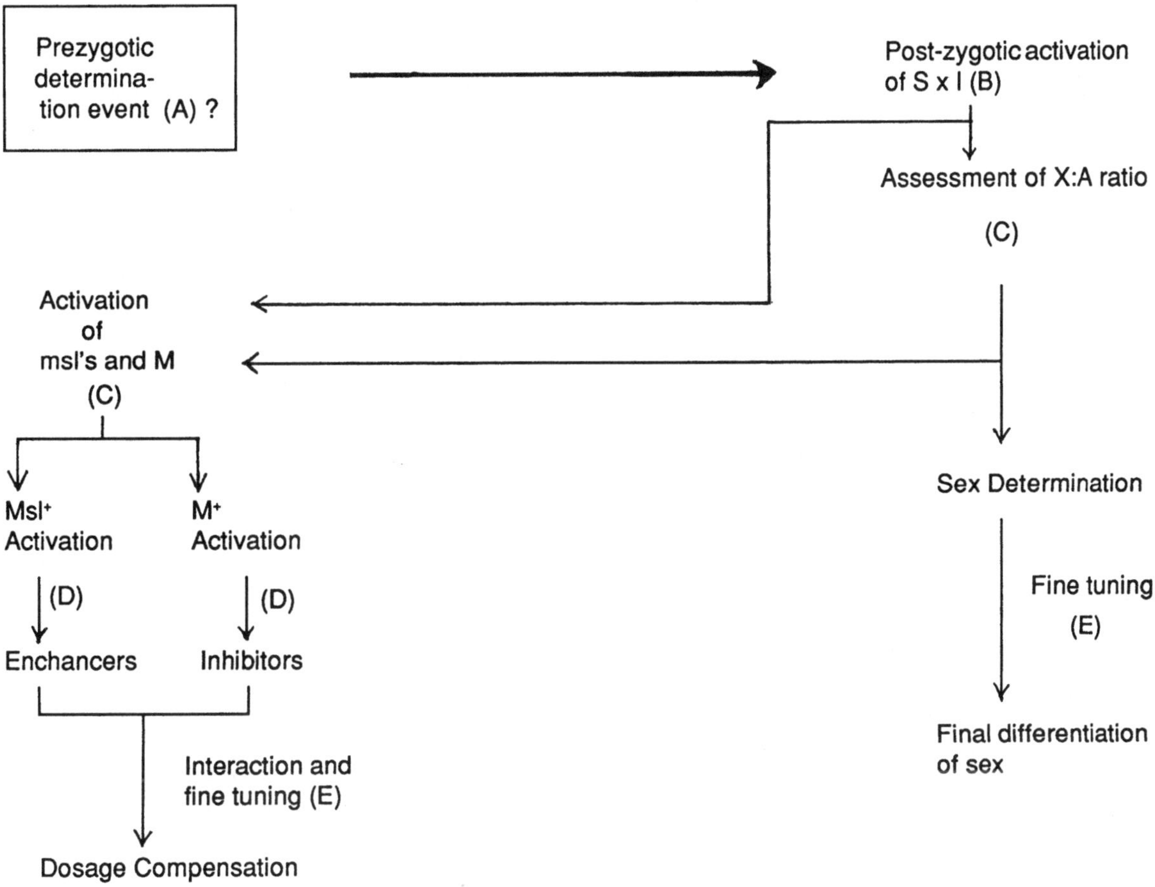

FIGURE 5. The genetic programming of events leading to sex determination and dosage compensation by an interplay of autosome-coded msl-product and X-coded M+ gene products. The autosomal regulatory gene products are presumed to have enhancing activity, while the M+-coded signals are inhibitory (Mukherjee and Ghosh, 1986).

Concluding Remarks

The review presented above reveals only certain aspects of the mechanism and programme of the pathway for dosage compensation in the two systems. Many more have been left out, and many more are yet to be known. Perhaps, a combined investigation of the genetic elements, molecular biological dissection of the sequences of the compensated genes, and biochemical isolation of the regulatory proteins might be useful in unravelling the genetic programme. We believe that the goal is not too far. Let us look forward to the nearing end of a chapter.

Acknowledgements

I acknowledge with gratitude the untiring co-operation of my students and co-workers for the work done in this laboratory. I am also thankful to various sponsoring agencies, like DST, CSIR, DAE and UGC whose continuous support during the last decade has resulted in successful building up of the story on hyperacti vity of the male X of Drosophila.

References

Achwal CW, Ganguly P and Chandra HS (1984). Estimation of the amount of 5-methylcytosine in *Drosophila melanogaster* DNA by amplified ELISA and photo acoustic spectroscopy. The EMBO J. 3; 263-6.

Andina RJ (1978). A study of X chromosome regulation during oogenesis. Exp. Cell Res., 111; 211-3.

Baker BS and Belote JM (1983) Sex determination and dosage compensation in *Drosophila melanogaster*. Ann. Rev. Genet. 17; 345-94.

Barr NL and Bertram EG (1949) A morphological distinction between neurones of the male and female, and the behaviour of the nucleolar satellite during accelarated nucleoprotein synthesis. Nature 163; 676-7.

Bautz EKF, Dangli A, Kabisch R and Kloetzel P (1983). Functional organization of polytene chromosomes. In Proc. Intern. Congr. Genet. (eds. Chopra VL, Joshi BC, Sharma RP and Bansal HC), Oxford & IBH, New Delhi, 1; 99-103.

Berendes HD (1966). Differential replication of male and female X chromosome in *Drosophila*. Chromosoma 20; 32-43.

Berendes HD (1972). The control of puffing in *Drosophila hydei*. In Results and problems in cell differentiation (ed. W. Beermann), Springer-Verlag, Berlin, 4; 181-207.

Berlowitz L, Loewus MW and Pallota D (1968). Heterochromatin and genetic activity in mealy bugs. 1. Compensation for inactive chromatin by increase in cell number. Genetics 60; 93-9.

Brown SW and Chandra HS (1973). Inactivation system of the mammalian X chromosome. Proc. Natl. Acad. Sci. (USA) 70; 195-9.

Catanach BM (1975). Control of chromosome inactivation. Ann. Rev. Genet. 9; 1-18.

Catanach BM and Isaacson JH (1967). Controlling elements of the mouse X-chromosome. Genetics 57; 331-46.

Chandra HS (1985). Sex determination: A hypothesis based on noncoding DNA. Proc. Natl. Acad. Sci., (USA) 82; 1165-9.

Chatterjee RN and Mukherjee AS (1977). Chromosomal basis of dosage compensation in *Drosophila*. IX. Celluar autonomy of the faster replication of the X chromosome of haplo-X cells of *D. melanogaster* and synchronous initiation. J. Cell Biol. 74; 168-80.

Chatterjee RN, Dube DK and Mukherjee AS (1981). In situ transcription analysis of chromatin template activity of the X chromosome of *Drosophila* following high molar NaCl treatment. Chromosoma 82; 515-23.

Chatterjee SN and Mukherjee AS (1975). DNA replication in polytene chromosomes of *Drosophila pseudoobscura*: New facts and their implications. Ind. J. Exptl. Biol. 13; 452-9.

Cline T (1983). The interaction between daughterless and Sex-lethal in triploids: a novel sex-transforming maternal effect linking sex determination and dosage compensation in *Drosophila melanogaster*. Develop. Biol. 95; 260-74.

Cline T (1984). Autoregulatory functioning of a *Drosophila* gene product that establishes and maintains the sexually determined state. Genetics 107; 231-77.

Cline T (1985). Primary events in the determination of sex in *Drosophila melanogaster*. In The Origin and Evolution of Sex (eds. Halvorson H and Monroy A), Alan R. Liss, Inc., New York; 301-27.

Cline TW (1986). A female-specific lethal lesion in an X-linked positive regulator of the Drosophila sex determination gene, Sex lethal. Genetics 113; 641-63.

Das M. Mutsuddi D, Duttagupta AK and Mukherjee AS (1982). Segmental heterogeneity in replication and transcription of X_2 chromosome of *Drosophila miranda* and conservativeness in the evolution of dosage compensation. Chromosoma 87; 373-88.

Elgin SCR and Weintraub H (1975). Chromosomal protein and chromatin structure. Ann. Rev. Biochem. 44; 725–74.

Gadagkar R, Nanjundiah V, Joshi NV and Chandra HS (1982). Dosage compensation and sex determination in *Drosophila*: mechanism of measurement of the X/A ratio. J. Biosci. 4; 377–90.

Ganguly R, Ganguly N and Manning JE (1985). Isolation and characterization of Glucose-6-Phosphate dehydrogenase gene of *Drosophila melanogaster*. Gene 35; 91–101.

Gartler SM and Riggs AD (1983). Mammalian X-chromosome inactivation. Ann. Rev. Genet. 17; 155–90.

Gartler SM, Dyer KA, Graves JAM and Rocchi M (1985). A two-step model for mammalian X chromosome inactivation. In Biochemistry and Biology of DNA methylation (eds. Cantoni GL and Razini A), Allan R. Liss, New York, 225–35.

Ghosh M and Mukherjee AS (1986). Enhanced hyperactivity of the X chromosome in In(1)BM2 (rv) strain of *Drosophila melanogaster* and evidence for the existence of controlling elements in the X. Ind. J. Exptl. Biol. 24; 203–8.

Goldman MA, Holmquist GP, Gray MC, Caston LA and Nag A (1984). Replication timing of genes and middle repetitive sequences. Science 224; 686–91.

Goodfellow PN, Banting G, Shear D, Ropers HH, Caine A, Fergusson Smith MH, Povey S and Voss R (1983). Genetic evidence that a Y-linked gene in man is homologous to a gene on the X chromosome. Nature 302; 346–8.

Hägele K and Kalisch W-S (1974). Initial phases of DNA synthesis in *Drosophila melanogaster*. 1. Differential participation in replication of the X chromosomes in males and females. Chromosoma 47; 403–43.

Jaffe E and Laird C (1986). Dosage compensation in *Drosophila*. Trends in Genetics, 2; 316–21.

Kaslow DC and Migeon BR (1987). DNA methylation stabilizes X chromosome inactivation in eutherians but not in marsupials: Evidence for multistep maintenance of mammalian X dosage compensation Proc. Natl Acad. Sci. USA 84; 6210–4.

Kazazian HH, Young WJ and Childs B (1968). X-linked 6-phosphogluconate dehydrogenase in *Drosophila*: Subunit associations. Science 150; 1601–2.

Keitges E, Rivest M, Siniscalco M and Gartler SM (1985). X-linkage of steroid sulfatase in the mouse is evidence for a functional Y-linked allele. Nature, 315; 226–7.

Korge G (1970a). Dosage compensation and effect for RNA synthesis in chromosome puffs of *Drosophila melanogaster*. Nature 225; 386–8.

Korge G (1970b). Dosis-kompensation und Dosis-effekt für RNA Synthese in Chromosomen-Puffs von *Drosophila melanogaster*. Chromosoma 30; 430–64.

Kratzer PG, Chapman VM, Lambert H, Evans EE and Liskay RM (1983). Differences in the DNA of the inactive X chromosomes of fetal and extraembryonic tissues of mice. Cell 33; 37–42.

Lakhotia SC and Mukherjee AS (1969). Chromosomal basis of dosage compensation in *Drosophila*. I. Cellular autonomy of hyperactivity of male X-chromosome in salivary glands and sex differentiation. Genet. Res. 14; 137–50.

Lakhotia SC and Mukherjee AS (1970). Chromosomal basis of dosage compensation in *Drosophila*. III. Early completion of replication by the polytene X-chromosome in male: Further evidence and its implication. J. Cell Biol. 47; 18–33.

Lokhotia SC and Mukherjee AS (1972). Chromosomal basis of dosage compensation in *Drosophila*. IV a Hyperactivity of polytene X-chromosome in male *D. kikkawai* and *D.bipectinata*. Proc. Zool. Soc. 25; 1–9.

Lucchesi JC (1973). Dosage compensation in *Drosophila*. Ann. Rev. Genet. 7; 225–37.

Lucchesi JC (1978). Gene dosage compensation and evolution of sex chromosomes. Science 202; 711–16.

Lucchesi JC (1983). The relation between gene dosage, gene expression and sex in *Drosophila*. Develop. Genet. 3; 275–82.

Lucchesi JC, Rawls Jr. JM and Maroni G (1974). Gene dosage compensation in meta-females (3X;2A) of *Drosophila*. Nature 248; 564–7.

Lucchesi JC and Skripsky T (1981). The link between dosage compensation and sex differentiation in *Drosophila melanogaster*. Chromosoma 82; 217–27.

Lyon MF (1961). Gene action in the X-chromosome of the mouse (*Mus musculus*, L.). Nature 190; 372–3.

Maine EM, Salz HK, Cline TW and Schedl P (1985a). The sex-lethal gene of Drosophila: DNA alterations associated with sex-specific lethal mutations. Cell 43; 521–9.

Maine EM, Salz HK, Schedl P and Cline TW (1985b). Sex lethal, a link between sex determination and sexual differentiation in *Drosophila melanogaster*. Cold Spring Harb. Symp. Quant. Biol. 50; 595–604.

Maroni G and Lucchesi JC (1980). X chromosome transcription in *Drosophila*. Chromosoma, 77; 253–61.

Maroni G and Plaut W (1973a). Dosage compensation in *Drosophila melanogaster* triploids. I. Autoradiographic study. Chromosoma 40; 361–77.

Maroni G and Plaut W (1973b). Dosage compensation in *Drosophila melanogaster* triploids. II. Glucose-6-phosphate-dehydrogenase activity. Genetics 74; 331–4.

Migeon BR, Sprenkle JA, Do TT (1981). Stability of the two active X' phenotype in triploid somatic cells. Cell 18; 637–41.

Migeon BR, Schmidt M, Axelman J and Cullen CR (1986). Complete reactivation of X chromosomes from human chorionic villi with a switch to early DNA replication. Proc Natl Acad. Sci. USA 83; 2182–6.

Mohandas T, Shapiro LJ, Sparkes RS and Sparkes MC

(1979). Regional assignment of the steriod sulfatase X-linked ichthyosis locus: Implications for a noninactivated region on the short arm of human X chromosome. Proc. Natl Acad. Sci. USA 76; 5779–83.

Mohandas T, Sparkes RS and Shapiro LJ (1981). Reactivation of an inactive human X chromosome: Evidence for X-inactivation by DNA methylation. Science, 221; 393–6.

Monk M and Harper MI (1979). Sequential X chromosome activity in preimplantation mouse embryos. Nature, 281; 311–3.

Mukherjee AS (1966). Dosage compensation in *Drosophila*: An autoradiographic study. The Nucleus 9; 83–96.

Mukherjee AS (1986). Genetics of dosage compensation: An old concept in new perspective. Presidential Address, 73rd Session, Indian Sci. Congr., New Delhi; 1–38.

Mukherjee AS and Beermann W (1965). Synthesis of ribonucleic acid by the X chromosomes of *Drosophila melanogaster* and the problem of dosage compensation. Nature 207; 785–6.

Mukherjee AS and Chatterjee SN (1975). Chromosomal basis of dosage compensation in *Drosophila*. VIII. Faster replication and hyperactivity of both arms of the X chromosome in males of *Drosophila pseudoobscura* and their possible significance. Chromosoma 53; 91–105.

Mukherjee AS and Chatterjee SN (1976). Hyperactivity and faster replication property of the two arms of the male X in *Drosophila pseudoobscura*. J. Microscopy 106; 199–208.

Mukherjee AS and Ghosh M (1986). A different level of X chromosomal transcription in an In(1)BM2 (reinverted) strain and its hyperploid derivatives resolves and X-coded regulatory activity for dosage compensation in *Drosophila*. Genetical Res. 48; 65–75.

Mukherjee BB and Sinha AK (1964). Single active X hypothesis and cytological evidence for random inactivation of X-chromosomes in a female mule complement. Proc. Natl. Acad. Sci. USA 51; 252–9.

Muller HJ (1928). The problem of genic modification. Zeits. induk. Abstam. Vererb. Suppl. I. 234; 160.

Muller HJ (1950). Evidence of the precision of genic adaptation. Harvey Lect. Ser. 43; 165–229.

Muller HJ and Kaplan WD (1966). The dosage compensation in *Drosophila* and mammals as showing the accuracy of the normal type. Genet. Res. 8; 41–59.

Mutsuddi M, Mutsuddi D, Mukherjee AS and Duttagupta AK (1984). Conserved autonomy of replication of the X chromosomes in hybrids of *Drosophila miranda* and *Drosophila persimilis*. Chromosoma 89; 55–62.

Ohno S (1967). Sex chromosomes and sex linked genes. Springer-Verlag, Berlin.

Ohno S, Kaplan WD and Kinosita R (1959). Formation of the sex chromatin by a single X-chromosome in liver cells of *Rattus norvegicus*. Exp. Cell Res. 18; 415–18.

Prasad J, Duttagupta AK and Mukherjee AS (1981). Transcription in X chromosomal segmental aneuploids of *Drosophila melanogaster* and regulation of dosage compensation. Genet. Res. 38; 103–13.

Prasad-Sinha J and Mukherjee AS (1985). Cellular autonomy of hyperactivity in segmental X-chromosomal aneuploids of *Drosophila* and dosage compensation. Genet. Res. 46; 19–20.

Rao SRV and Bhattacharya M (1984). Insect sex chromosomes: IX. X-chromosome linkage of Glucose-6-phosphate dehydrogenase (G-6-PD) locus in *G ryllotalpa fossor* (Orthoptera). Heredity 53; 235–8.

Razin A and Riggs AD (1980). DNA methylation and gene function. Science 210; 604–10.

Riggs AD (1975). X inactivation, differentiation and DNA methylation. Cytogenet. Cell Genet. 14; 9–25.

Riggs AD and Jones PA (1983). 5-methylcytosine, gene regulation and cancer. Adv. Cancer Res. 40; 1–30.

Scholnick SB, Morgan BA and Hirsch J (1983). The cloned Dopa carboxylase gene is developmentally regulated when reintegrated into the Drosophila genome. Cell 34; 37–45.

Schwartz D (1973). The application of the maize-derived gene compensation model to the problem of dosage compensation in *Drosophila*. Genetics 75; 639–41.

Seecof RL, Kaplan WD and Futch DG (1969). Dosage compensation of enzyme activities in *Drosophila melanogaster*. Proc. Natl. Acad. Sci. USA 62; 528–35.

Singh L, Purdom IF and Jones KW(1981). Conserved sex chromosome-associated nucleotide sequences in eukaryotes. Cold Spring Harb. Symp. Quant. Biol. 45; 805–14.

Spradling AC and Rubin GM (1983). The effect of chromosomal position on the expression of the *Drosophila* Xanthine dehydrogenase gene. Cell 34; 47–57.

Stern C (1929). Über die additive Wirkung multiple Allele. Biol. Zentralbl. 49; 261–90.

Stern C (1960). Dosage compensation—Development of a concept and new facts. Canad. J. Genet. Cytol. 2; 105–18.

Steinmann-Zwicky M and Nothiger R (1985). The heirarchial relation between X chromosomes and autosomal sex-determining genes in *Drosophila*. The EMBO J. 4; 163–6.

Stewart B and Merriam J (1975). Regulation of gene activity by dosage compensation at the chromosomal level in *Drosophila*. Genetics 79; 635–47.

Stewart B and Merriam J (1980). Dosage compensation. In The gentics and biology of *Drosophila*. (eds. Ashburner M and Wright TRF). Academic Press, London, 2; 107–40.

Takagi N (1978). Preferential inactivation of the paternally-derived X-chromosome in mice. In Genetic Mosaic and Chimeras in Mammals (ed. Russel LB): 341–360.

Takagi N (1980). Primary and secondary nonrandom X-chromosome inactivation in early female mouse embryos carrying Searle's translocation T(X;16)16H. Chromosoma 81; 439–59.

Therman E and Patau K (1974). Abnormal X chromosomes in man: origin, behaviour and effects. Humangenetik 25; 1–16.

Ukil M, Chatterjee K, Dey A, Ghosh S and Mukherjee AS (1986). Affinity binding of nonhistone chromatin proteins to the X chromosome of *Drosophila* by *in situ* chromatin reconstitution and its significance. J. Cell Sci. 86; 35–45.

VandeBerg JL, Robinson ES, Samollow PB and Johnston PG (1983). X-linked gene expression and X-chromosome inactivation: Marsupials, mouse and man compared. Isozymes: Current topics in biological and medical research.

Venolia L, Gartler SM, Wassman ER, Yen P, Mohandas T and Shapiro LJ (1982). Transformation with DNA from 5-azacytidine reactivated X chromosomes. Proc. Natl Acad. Sci. USA 79; 2352–4.

Waring GL and Pollack JC (1987). Cloning and characterization of a dispersed multicopy, X chromosome sequence in *Drosophila melanogaster*. Proc. Natl. Acad. Sci. USA 84; 2843–7.

Weaver DD, Gartler SM, Boue A and Boue JG (1975). Evidence for two active X chromosomes in a human XXY triploid. Humangenetik 28; 39–42.

West JD, Freis WI, Chapman M and Papaioannou VE (1977). Preferential expression of the maternally-derived X-chromosome in the mouse yolk sac. Cell 12; 873–82.

8 Steroid Sulphatase Inactivation Patterns and X-chromosome Inactivation

DANIEL F. SCHORDERET AND STANLEY M. GARTLER

Department of Medicine and Genetics
and the Center for Inherited Diseases
University of Washington
Seattle, WA 98195, USA

Summary

Dosage compensation is achieved in mammals by random, clonal inactivation, early in embryo development, of one of the female X chromosomes. This process takes place first in trophectoderm, next in extra-embryonic endoderm and last in embryonic endoderm and spreads through the whole X chromosome except for a small region responsible for pairing with the Y chromosome. Several genes, among them steroid sulphatase (*Sts*), have been mapped to or near this pairing region. In humans, *Sts* has been mapped just outside the X-Y pairing region, has a non functional homologous sequence on the Y chromosome with which it does not recombine; and shows only partial inactivation; in the mouse, *Sts* is located inside the X-Y pairing region; has an allele on the Y chromosome with which it recombines; and is not inactivated.

Because of their location and mode of inactivation, the human and mouse *Sts* genes are of great importance in the study of X inactivation. By using 5-azacytidine or ethyl methane sulphonate treatment followed by selection for STS activity, we were able to reactivate the mouse *Sts* gene. Inactivation and reactivation of this gene, which is not normally inactivated, indicate that genes in the pairing region may not be prevented by particular sequences from being inactivated. Further analysis of these 2 genes will help in understanding X-chromosome inactivation and how the X-Y pairing region escapes inactivation.

Introduction

Dosage compensation in mammals is achieved by inactivation of the genes on one of the X chromosomes of the female (Lyon, 1961). This inactivation, however, does not include the whole X chromosome, but appears to be excluded from the X-Y pairing region. The first X-linked gene for which there was quantitative evidence of escape from X inactivation was the human steroid sulphatase (*Sts*) gene which is located just outside the X-Y pairing region (Curry *et al.*, 1984). Later the murine *Sts* gene was shown to be located within the pairing region and to normally completely escape X inactivation. In this chapter we will consider the human and murine *Sts* genes with respect to the overall X chromosome inactivation process.

Developmental and Molecular Biology of X Inactivation

X chromosome inactivation first occurs in the early blastocyst when one of the two X chromosomes in female cells becomes inactivated, bringing about an effective X chromosome to autosome ratio equal to that in the male. At the cytological level, inactivation appears as the formation of sex chromatin (Lyon, 1961). The process follows sequential steps in embryonic differentiation.

It takes place first in trophectoderm, next in extra-embryonic endoderm, and last in the embryonic ectoderm. In the extra-embryonic lineages, X inactivation occurs preferentially in the paternal X chromosome (Takagi and Sasaki, 1975), while in the embryonic lineage the inactivation process is random with respect to paternal and maternal X chromosomes. Once X inactivation is established, this cytologically recognized differentiative event becomes fixed in all subsequent cell generations. The inactive X chromosome is heteropycnotic throughout the cell cycle and replicates asynchronously as do other forms of heterochromatin (Morishima *et al.*, 1962). The one exception to the permanence of X inactivation occurs in the germ line where the inactive X chromosome is reactivated around the time of entry into meiosis (Witschi, 1957).

At the molecular level three sequential, and possibly related, steps must be considered in the X inactivation process: initiation of inactivation at a single site or control centre, spreading of inactivation along the chromosome, and maintenance of the inactivated state throughout subsequent cell generations.

There is evidence that DNA methylation is involved in the maintenance aspect of X inactivation. This evidence includes 5-azacytidine (5AC) induced reactivation of several X-linked genes (Mohandas *et al.*, 1981; Graves, 1982; Lester *et al.*, 1982), transfection experiments with the *HPRT* gene (Venolia *et al.*, 1982), and methylation-sensitive restriction enzyme analyses which have shown differential DNA methylation between active and inactive X-linked genes (Riggs and Jones, 1983). Clusters of 5′ CpG dinucleotides associated with several X-linked constitutive genes are hypomethylated in active genes and hypermethylated in inactive ones. These same regions contain nuclease hypersensitive sites which also differentiate active and inactive alleles (Weintraub *et al.*, 1981; Elgin, 1981). While these methylation and nuclease sensitivity differences are clearly important, there is some evidence that indicates that they are not the primary events leading to X inactivation. For example in extra-embryonic tissues where X inactivation is initiated, transfection experiments indicate that DNA modification is not involved in repression of genes on the inactive X chromosome (Kratzer *et al.*, 1983). Furthermore, it seems that the differential methylation observed in embryonic cells does not appear until after the biological onset of X inactivation (Lock *et al.*, 1987). We have proposed a simple two-step model for X inactivation (Gartler *et al.*, 1985) in which the initiating step is sex chromatin formation, an event common to both extra-embryonic and embryonic cells. Sex chromatin formation does not involve differential DNA modification of individual genes and we assume it is the only factor repressing the inactive X in extra-embryonic cells. When X inactivation occurs in embryonic cells, sex chromatin formation is again the initiating step, but in this case it is followed at an interval by differential methylation as a fail-safe or locking mechanism. Thus in embryonic and adult cells, genes on the inactive X are repressed by both heterochromatinization and hypermethylation. The key to understanding the initiation of X inactivation would appear to be the formation of sex chromatin.

The spreading feature of X inactivation is

another difficult problem. One must consider the spread of inactivation from a single site to more than 1×10^8 bp of DNA, a phenomenon which suggests the need for signal amplification stations along the X chromosome. The consideration of the X-Y pairing region and the murine X- and Y-linked *Sts* gene, which escapes inactivation, may give us more insight into this aspect of the inactivation process.

The Human *Sts* Gene

Steroid sulphatase is a ubiquitous microsomal enzyme responsible for hydrolysis of sulphated compounds like cholesterol-sulphate and sulphated steroids. Its deficiency results in a clinical picture called X-linked ichthyosis, a disease characterized by corneal opacities and scaly skin containing high amounts of cholesterol-sulphate. Affected males are usually born after delayed labour, presumably due to diminished placental estriol production (France and Liggins, 1969).

The human *Sts* gene was first shown to escape inactivation when present on an inactive X chromosome in a mouse-human somatic cell hybrid (Mohandas *et al.*, 1980). This demonstration of expression was possible because the parental mouse strain (C3H/An) from which this hybrid was derived is STS deficient. It was thought at that time that the gene completely escaped inactivation, but later dosage and cloning studies showed that the human *Sts* gene on the inactive X chromosome was partially inactivated, possibly to the extent of 50 per cent (Muller *et al,* 1980; Ropers *et al.*, 1981; Chance and Gartler, 1983). The human gene has been mapped to Xp 22.32 which is apparently just proximal to the pairing region. Genes inside the pairing region, such as *MIC2*, a gene recognized by the antibody 12E7 (Goodfellow, 1983), would have functional alleles on the Y and exhibit a high rate of recombination and this is not the case for *Sts*.

The recent cloning of human *Sts* cDNA by several groups should soon lead to some molecular understanding of why this gene exhibits an anomalous pattern of inactivation (Ballabio *et al.*, 1987; Yen *et al.*, 1987). The molecular studies to date have revealed that the gene is quite large (100 kb) and has a non-functional homologous sequence on the long arm of the Y chromosome, well outside the pairing region. When X-linked ichthyotic patients were examined at the molecular level, more than 85 per cent of them were found to have a deletion spanning the whole gene despite the fact that usually no abnormal karyotype is seen. These deletions are so large that they can even be recognized by flanking DNA markers or flow cytometry (Gillard *et al.*, 1987; Cooke *et al.*, 1988). It is tempting to speculate that the many deletions found among STS deficient patients may be the consequence of unequal crossing over in this region due to the proximity of the *Sts* gene to the pairing region.

The Murine *Sts* Gene

Balazs *et al.* (1982) showed that the C3H/An mouse strain was STS deficient. The enzyme deficiency may not be as severe in the mouse as in human X-linked ichthyosis. Furthermore, there is no gross evidence, at least, of ichthyosis in STS deficient mice. These authors carried our breeding studies in this strain and reported genetic data compatible with autosomal inheritance (Balazs *et al.*, 1982). Two other studies (Erickson *et al.*, 1983; Keinanen *et al.*, 1983) with other murine STS variants seemed to confirm an autosomal pattern of STS inheritance in the mouse. We investigated one of these variants (C3H/An) and found in one cross of a deficient female with a normal male, a single deficient female offspring among otherwise normal progeny. This result was unexpected on the basis of autosomal inheritance, but suggested an unusual form of sex-linked inheritance if the exceptional female was XO. The cytology of this female could not be checked; therefore, we undertook a study involving crosses of STS normal XO females to STS deficient males. This XO mating scheme takes advantage of the fertility of XO female mice which therefore can be crossed to males carrying a recessive marker. If this marker is autosomal all offspring will be normal whereas if it is X-linked, the XO progeny will be mutant and the XX and XY progeny normal. The results of these studies showed unambiguously that the STS murine variant was X-linked. To explain the apparent autosomal inheritance of this STS variant, we postulated that there must be a functional Y-linked allele that undergoes an obligatory meiotic crossover with its X-linked allele (Keitges and Gartler, 1986). Further studies confirmed this hypothesis and demonstrated genetic linkage of *Sts* with the sex-reversing factor (*Sxr*), a duplication-translocation of the murine testis determining factor (Keitges *et al.*, 1987), and MOV15, a retrovirus inserted into the Y chromosome (Soriano *et al.*, 1987). In these same studies we found some apparent multiple crossover events. No double crossovers were found in the human X-Y pairing region, but only a maximum of 2.4 crossovers between DXYS14, DXYS17, and sex phenotype would be expected (Rouyer *et al.*, 1986). Rouyer *et al.* (1986) hypothesized that this region of obligatory recombination is restricted to a single crossover. This assumption of a single crossover appears to be unnecessary at this time. There is no apparent selective advantage to mandating a single crossover event in this region. The absence of double crossovers is more likely due to the limited size of the pairing region. The pairing region in humans has been estimated to be on the order of 5000 kb which is less than 1 per cent of the genome (Rouyer *et al.*, 1986). Considering the total number of chiasmata observed in humans (Hulten, 1974), it is not likely that double crossovers would occur with any frequency in such a restricted region.

On the basis of XO infertility in humans, and fertility in mice, the pairing region has been hypothesized to be smaller in mice (Rouyer *et al.* 1986). The argument is that the region is not dosage compensated and, therefore, a gene in XO individuals is underexpressed as compared to XX or XY offspring (Ferguson-Smith, 1966; Burgoyne, 1982). If this region is important in normal female development, as Ferguson-Smith has hypothesized, then a smaller pairing non-dosage compensated region in the mouse could be the basis of the difference in XO fertility in mice and humans. However, human females with visible X chromosome deletions of the distal part of the short arm may have normal fertility (Wyss *et al.*, 1982) which is contrary to the importance of this region in normal female development. Finally, the only evidence on the relative genetic size of the mouse and human pairing regions comes from studies of the steroid sulphatase locus. In two recent papers (Keitges *et al.*, 1987; Soriano *et al.* 1987) we have presented definitive evidence for the presence of the *Sts* gene well within the murine pairing region. Furthermore, the murine *Sts* gene is not subject to X inactivation as would be expected from its location inside the pairing region and the presence of a functional Y-linked allele (Gartler and Rivest, 1983; Keitges and Gartler, 1986). However, the human *Sts* gene is partially inactivated (Muller *et al.*, 1980; Ropers *et al.*, 1981; Chance and Gartler, 1983) and proximal to the region of recombination. We interpreted these findings to indicate that the human *Sts* gene was formerly in the pairing region, but that evolutionary 'shrinkage' of the homologous region has moved it outside that region. Thus from an evolutionary standpoint, the argument can be made that the murine X-Y pairing region is larger than its human counterpart. It does not appear that a sound argument can be made for explaining the XO murine, human fertility differences on the basis of a difference in the size of the pairing region.

The Mouse *Sts* Gene and X Inactivation

As indicated, it was first thought that the human *Sts* gene on the inactive X chromosome escaped inactivation, but it is now clear that the gene is subject to partial inactivation. Since there are both X and Y *Sts* alleles in the mouse there would appear to be no need for dosage compensation at this locus. However, the mouse sex-reversed mutation, which is closesly linked to *Sts* can be subject to X inactivation (McLaren and Monk, 1982). Therefore we compared STS level in XO, XX, and XY mice and carried out a clonal analysis in fibroblast cell cultures from mice heterozygous for the *Sts* deficiency gene and heterozygous at the X-linked electrophoretic phosphoglycerate kinase locus. The combined results showed that the murine *Sts* locus is not dosage compensated nor subject to X inactivation under normal conditions (Keitges and Gartler, 1986).

However from observations on STS levels in murine cell lines we had to consider the possibility that some form of *Sts* inactivation might take place under cell culture conditions. We have found that all long-term murine cell cultures so far examined were STS deficient (Schorderet *et al.*, 1988). To our knowledge, only one of these cultures, the L cells, is derived from an STS deficient mouse strain (C3H/An). Therefore, STS deficiency in the other cultures must have taken place after their initiation. Two explanations can be considered for this loss of enzyme activity. *Sts* could have been inactivated by a *de novo* mutation or even a deletion, as appears to be the rule in human X-linked ichthyosis (Gillard *et al.*, 1987, Ballabio *et al.*, 1987, Yen *et al.*, 1987). Alternatively, some part or parts of the X-inactivation process could have spread to the X-Y pairing region affecting both the inactive and active X chromosomes. To distinguish between these possibilities we developed a selective system based on steroid sulphatase activity that would permit us to study the induction of STS^+ mutants from STS^- cultures. We reasoned that if loss of activity was due to a classical gene mutation, induction of revertants would be very infrequent or, in the case of a deletion, impossible. On the other hand, if aberrant methylation was the basis of the STS deficiency, the reactivation at high frequencies should be possible with demethylating agents like 5-azacytidine (5AC).

Development of a *STS* Selective System

Mammalian cells are dependent for their growth on cholesterol which can be obtained by two pathways. In the exogenous pathway cholesterol is bound to low density lipoprotein (LDL) and enters the cell through receptor-mediated endocytosis (Goldstein *et al.*, 1983). In the absence of LDL, cholesterol is synthesized from acetyl-CoA. In this system, synthesis of mevalonate catalyzed by hydroxymethyl glutaryl CoA reductase (HMG CoA reductase) is the rate-limiting step (Mosley *et al.*, 1983). HMG CoA reductase can be inhibited by mevinolin, a potent drug derived from a strain of *Aspergillus terreus* (Alberts *et al.*, 1980). The STS selective system is based on the possibility of making cells dependent on an external source of cholesterol by inhibiting HMG CoA reductase with mevinolin and by cultivating these cells in a lipoprotein deficient environment. Only STS positive cells are able to use the added cholesterol-3β-sulphate as a source of cholesterol. It should be possible to recover STS deficient cells by a modification of the system described above. Cells would be cultured in medium supplemented with lipoprotein-deficient serum, cholesterol-3β-sulphate and mevinolin. Bromodeoxyuridine (BrdU) would then be added to the culture and incorporated into dividing cells which should be subsequently killed by exposure to light. STS deficient, non-dividing cells would then be rescued in normal medium.

Induction of Revertants

Four different STS deficient murine cell lines (L, CAK, WEHI and XO) were treated with N-Methyl-N′-nitro-N-nitrosoguanidine (MNNG), Ethyl methane sulphonate (EMS) or 5AC (Schorderet *et al.*, 1988) and selection for STS positive cells was applied. MNNG failed to produce any revertant, regardless of the cell line, and none of the agents produced revertants in the L cell line. However 5AC led to a high rate of revertants in the other cell lines (2×10^{-3}–5×10^{-5}) and EMS which was tested only on CAK also produced a high level of revertants (2×10^{-4}).

5AC is known to be a highly efficient demethylating agent, and our results suggest that hypermethylation may be the basis of STS inactivation in these cell lines. Other workers have shown that EMS treatment induces reactivants in other systems and have suggested that this alkylating agent may bring about demethylation under certain conditions and act in a fashion similiar to 5AC (Harris, 1984). Our negative results with MNNG on these same cell lines indicate that not all alkylating agents can act as effective demethylating agents.

Since L cells are derived from a genetically deficient STS mouse strain, the reversion frequency would be expected to be low even for the case of a point mutation. However, the absence of L-cell revertants may also suggest a deletion in the *Sts* gene as this is the case in a majority of humans with X-linked ichthyosis.

Replication studies with the CAK revertant indicate that both active and inactive X chromosomes are present. This means that since STS is not subject to regular X-chromosome inactivation, both STS alleles must have been inactivated in the STS^- parent. It is however possible that the revertant may only have a single allele reactivated. Since CAK has been in culture for nearly 15 years, it is almost certain that these cells have gone through more than 1000 cell doublings. When the line became STS^- is not known.

To better understand the evolution of STS deficiency in cell cultures, we established two STS^+ cell lines: an XX and an XO. The XX cell line was still STS^+ after more than 200 doublings while the XO cells became STS^- rather suddenly after 80 cell doublings.

This demonstration of inactivation of an X-linked gene that normally escapes inactivation may be another example of the spreading of X inactivation into chromosomal regions not normally subject to X inactivation. From the earliest studies on mammalian X inactivation, it has been known that inactivation can spread into an adjoining translocated autosomal region (Cattanach *et al.*, 1971; Cattanach, 1974). More recently it has been shown that a Y-linked gene (*Sxr*) translocated to the X chromosome can occasionally become inactivated (McLaren and Monk, 1982), and now we have demonstrated that an X-linked gene that normally escapes inactivation can be inactivated in cell culture. Our results imply that aberrant methylation may be the basis of *Sts* inactivation, and we suggest that aberrant methylation may also be invoked in the spread of X inactivation into other ectopic locations, as in the cases mentioned above.

There is now a considerable amount of evidence that the initial event in X-chromosome inactivation does not involve DNA modification. DNA methylation is most likely a secondary event serving as a fail-safe or locking mechanism of repression, possibly helping to keep genes turned off during DNA synthesis (Kratzer *et al.*, 1983; Gartler *et al.*, 1985; Lock *et al.*, 1987). However, differential methylation of the active and inactive X chromosomes may play a key role in the spreading of X inactivation along the chromosome and into adjoining ectopic regions. Since a gene turned off by methylation in an ectopic region might not be subject to the initial event of X inactivation, such genes might be more easily reactivatable than X-linked genes normally subject to X inactivation. Because methylation patterns must be maintained in every cell division, the possibility for aberrant methylation in continually dividing cell culture systems could be high. This would not only apply to the *Sts* gene but to autosomal genes as well, since methylation appears to be a ubiquitous mechanism of repression in mammalian cells. In fact, Harris and others have recently shown that several autosomal loci in cell culture can be inactivated by methylation and reactivated by 5AC treatment (Harris, 1982; Steglich *et al.*, 1985). It seems likely that aberrant methylation can occur occasionally in any dividing cell, but that, unless there is a selective advantage for the cell carrying the inactivated gene, the inactivation event will go undetected. In this respect it is surprising that the STS deficiency characteristic of murine cell cultures is not found in either HeLa or CHO, two long-term permanent cell lines of human and hamster origin, respectively, or in canine cells. It is possible that the selective difference found between STS^- and STS^+ murine cells is not manifested in human, hamster, or canine cultures, or it could be that the somewhat unique location of the murine *Sts* gene may make it more susceptible to aberrant methylation and inactivation.

Acknowledgements

D.F.S. is supported by the Swiss National Science Foundation and the Swiss Foundation for Medical and Biological grants; S.M.G. is the recipient of a National Institute of Health Career award.

References

Alberts AW, Chen J, Kuron G, Hunt V, Huff J, Hoffman C, Rothrock J, Lopez M, Joshua H, Harris E, Patchett A, Monaghan R, Currie S, Stapley E, Albers-schonberg G, Hensens O, Hirshfield J, Hoogsteen K, Liesch J and Springer J. (1980). Mevinolin: a highly potent competitive inhibitor of hydroxymethylglutaryl-coenzyme A reductase and cholesterol-lowering agent. Proc Natl Acad Sci USA 77; 3957–61.

Balazs I, Purello M, Rocchi M, Rinaldi A and Siniscalco M (1982). Is the gene for steroid sulfatase X-linked in man. An appraisal of data from humans, mice and their hybrids. Cytogenet Cell Genet 32; 251–2.

Ballabio A, Parenti G, Carrozzo R, Sebastio B, Andria G, Buckle V, Fraser N, Craig I, Rocchi M, Romeo G, Jobsis AC and Persico MG (1987). Isolation and characterization of a steroid sulfatase cDNA clone: genomic deletions in patients with X-chromosome-linked ichthyosis. Proc Natl Acad Sci USA 84; 4519–23.

Burgoyne PS (1982). Genetic homology and crossing over in the X and Y chromosomes of mammals. Hum Genet 61; 85–90.

Cattanach BM, Pollard CE and Hawkes SG (1971). Sex-reversed mice: XX and XO males. Cytogenetics 10; 318–37.

Cattanach BM (1974). Position effect variegation in the mouse. Genet Res 23; 291–306.

Chance PF and Gartler SM (1983). Evidence for a dosage effect at the X-linked steroid sulfatase locus in human tissues. Am J Hum Genet 35; 234–40.

Cooke A,. Gillard EF, Yates JRW, Mitchell MJ, Aitken DA, Weir DM, Affara NA and Ferguson-Smith MA (1988). X chromosome deletions detectable by flow cytometry in some patients with steroid sulfatase deficiency (X-linked ichthyosis). Hum Genet 79; 49–52.

Curry CJR, Magenis RE, Brown M, Lanman JT, Tsai J, O'Lague P, Goodfellow P, Mohandas T, Bergner EA and Shapiro LJ (1984). Inherited chondrodysplasia punctata due to a deletion of the terminal short arm of an X chromosome. N Engl J Med 311; 1010–15.

Elgin SCR (1981). DNase I-hypersensitive sites of chromatin. Cell 27; 413–5.

Erickson RP, Harper K and Kramer JM (1983). Identification of an autosomal locus affecting steroid sulfatase activity among inbred strains of mice. Genetics 105; 181–9.

Ferguson-Smith MA (1966). X-Y chromosomal interchange in the etiology of true hermaphroditism and of XX Klinefelter's syndrome. Lancet ii; 475–6.
France JT and Liggins GC (1969). Placental Sulfatase deficiency. J Clin Endocrinol Metab 29; 138–41.
Gartler SM and Rivest M (1983). Evidence for X-linkage of steroid sulfatase in the mouse: steroid sulfatase levels in oocytes of XX and XO mice. Genetics 103; 137–41.
Gartler SM, Dyer KA, Graves JAM and Rocchi M (1985). A two step model for mammalian X-chromosome inactivation. In Biochemistry and Biology of DNA Methylation (eds. Cantoni GL and Razin A). Alan R. Liss, New York; 223–35.
Gillard EF, Affara NA, Yates JRW, Goudie DR, Lambert J, Aitken DA and Ferguson-Smith MA (1987). Deletion of a DNA sequence in eight of nine families with X-linked ichthyosis (steroid sulfatase deficiency). Nucleic Acids Res 15; 3977–85.
Goldstein JL, Basu SK and Brown MS (1983). Receptor-mediated endocytosis of low-density lipoprotein in cultured cells. Methods Enzymol 98; 241–60.
Goodfellow P (1983). Expression of the 12E7 antigen is controlled independently by genes on the huamn X and Y chromosome. Differentiation 23; 35–9.
Graves JAM (1982). 5-azacytidine-induced re-expression of alleles on the inactive X chromosome in a hybrid mouse cell line. Exp Cell Res 141; 99–105.
Harris M (1982). Induction of thymidine kinase in enzyme-deficient Chinese hamster cells. Cell 29; 483–92.
Harris M (1984). High-frequency induction by 5-azacytidine of proline independence in CHO-K1 cells. Somat Cell Mol Genet 10; 615–24.
Hulten M (1974). Chiasmata distribution in diakinesis in the normal human male. Hereditas 76; 55–78.
Keinanen BM, Nelson K, Daniel WL and Roque JM (1983). Genetic analysis of murine arylsulfatase C and steroid sulfatase. Genetics 105; 191–206.
Keitges E and Gartler SM (1986). Dosage of the *Sts* gene in the mouse. Am J Hum Genet 39; 470–6.
Keitges EA, Schorderet DF and Gartler SM (1987). Linkage of the steroid sulfatase gene to the *Sex-reversed* mutation in the mouse. Genetics 116; 465–8.
Kratzer PG, Chapman VM, Lambert H, Evans RE and Liskay RM (1983). Differences in the DNA of the inactive X chromosomes of fetal and extraembryonic tissues of mice. Cell 33; 37–42.
Lester SC, Korn NJ, DeMars R (1982). Depression of genes on the human X chromosome: evidence for differences in locus-specific rates of derepression and rates of transfer of active and inactive genes after DNA-mediated transformation. Somat Cell Genet 8; 265–84.
Lock LF, Takagi N and Martin GR (1987). Methylation of the *Hprt* gene on the inactive X occurs after chromosome inactivation. Cell 48; 39–46.
Lyon MF (1961). Gene action in the X-chromosome of the mouse (*Mus musculus L.*) Nature 190; 372–3.
McLaren A and Monk M (1982). Fertile females produced by inactivation of an X chromosome of 'sex-reversed' mice. Nature 300; 446–8.
Mohandas T, Sparkes RS, Hellkuhl B, Grzeschik KH and Shapiro LJ (1980). Expression of an X-linked gene from an inactive human X chromosome in mouse-human hybrid cells; further evidence for the noninactivation of the steroid sulfatase locus in man. Proc Natl Acad Sci USA 77; 6759–63.
Mohandas T, Sparkes RS and Shapiro LJ (1981). Reactivation of an inactive human X chromosome: evidence for X inactivation by DNA methylation. Science 211; 393–6.
Morishima A, Grumbach MM and Taylor JH (1962). Asynchronous duplication of human chromosomes and the origin of sex chromatin. Proc Natl Acad Sci USA 48; 756–63.
Mosley ST, Brown MS, Anderson RGW and Goldstein JL (1983). Mutant clone of Chinese hamster ovary cells lacking 3-hydroxy-3-methylglutaryl coenzyme A reductase. J Biol Chem 258; 13875–81.
Muller CR, Migl B, Traupe H and Ropers HH (1980). X-linked steroid sulfatase: evidence for different gene dosage in males and females. Hum Genet 54; 197–9.
Riggs AD and Jones PA (1983). 5-methylcytosine, gene regulation, and cancer. Adv Cancer Res 40; 1–30.
Ropers HH, Migl B, Zimmer J, Fraccaro M, Marschio PP and Westerveld A (1981). Activity of steroid sulfatase in fibroblasts with numerical and structural X chromosome aberrations. Hum Genet 57; 354–6.
Rouyer F, Simmler M-C, Johnsson C, Vergnaud G, Cooke HJ and Weissenbach J (1986). A gradient of sex linkage in the pseudoautosomal region of the human sex chromosomes. Nature 319; 291–5.
Schorderet DF, Keitges EA, Dubois PM and Gartler SM (1988). Inactivation and reactivation of Sex-linked steroid sulfatase gene in murine cell culture. Somat Cell Mol Genet 14; 113–21.
Soriano P, Keitges EA, Schorderet DF, Harbers K, Gartler SM and Jaenisch R (1987). High rate of recombination and double crossovers in the mouse pseudoautosomal region during male meiosis. Proc Natl Acad Sci USA 84; 7218–20.
Steglich C, Grens A and Scheffler IE (1985). Chinese hamster cells deficient in ornithine decarboxylase activity: reversion by gene amplification and by azacytidine treatment. Somat Cell Mol Genet 11; 11–23.
Takagi N and Sasaki M (1975). Preferential inactivation of the paternally derived X chromosome in the extraembryonic membranes of the mouse. Nature 256; 640–2.
Venolia L, Gartler SM, Wassman ER, Yen P, Mohandas T

and Shapiro LJ (1982). Transformation with DNA from 5-azacytidine-reactivated X chromosomes. Proc Natl Acad Sci USA 79; 2352–4.

Weintraub H, Larsen A and Groudine M (1981). Alpha-globin-gene switching during the development of chicken embryos: expression and chromosome structure. Cell 24; 333–44.

Witschi E (1957). Sex chromatin and sex differentiation in human embryos. Science 126; 1288–90.

Wyss D, DeLozier CD, Daniell J and Engel E (1982). Structural anomalies of the X chromosome: personal observation and review of non-mosaic cases. Clin Genet 21; 145–59.

Yen PH, Allen E, Marsh B, Mohandas T, Wang N, Taggart RT and Shapiro LJ (1987). Cloning and expression of steroid sulfatase cDNA and the frequent occurrence of deletions in STS deficiency: implications for X-Y interchange. Cell 49; 443–54.

9 Chromosomal Aberrations from Radiation-induced DNA Lesions

A.T. NATARAJAN

Dept. of Radiation Genetics & Chemical Mutagenesis
State University of Leiden
Sylvius Laboratory, Wassenaarseweg 72, 2333 AL Leiden
and
J.A. Cohen Institute for Radiopathology and Radiation Protection
Leiden, The Netherlands

Introduction

One of the important biological consequences of exposure to ionizing radiations in eukaryotic cells is the formation of chromosomal aberrations. Classical basic radiobiological concepts such as target theory have been based on the quantitative data on radiation-induced chromosomal aberrations in *Tradescantia* microspores (Lea, 1955). Mechanistic aspects of chromosomal aberration formation following treatment with ionizing radiations were considered by Sax (1940) who proposed the breakage first hypothesis and Revell (1955) who proposed the contact first hypothesis. These hypotheses were proposed at a time when the knowledge of the structure of the chromosome and DNA was not well understood. Bender *et al.* (1974) proposed a unified model for the formation of different types of chromosomal aberrations following treatment of cells with short-wave UV, ionizing radiations and chemical mutagens. In this model DNA double strand break has been considered as the most important lesion leading to radiation-induced chromosomal aberrations.

Ionizing radiations induce mainly four different classes of DNA lesions, namely single strand breaks (SSBs), double strand breaks (DBSs), various classes of base damage (BD) and DNA-protein cross links. The relative proportion of these lesions varies with the quality of radiation employed. For example high LET (linear energy transfer) radiations, such as fast neutrons, alpha particles etc., induce more DSBs and less SSBs for a given dose in comparison to low LET radiations like X- and gamma rays. Ionizing radiations induce chromosome type of aberrations in G1 cells and chromatid type of aberrations in S and G2 cells. Short-wave UV and most of the chemical mutagens induce mainly chromatid type of aberrations in all stages of cell cycle and the cells have to go through an S-phase before the aberrations can be visualized. The latter group of agents—the so-called S-dependent agents—are potent inducers of sister chromatid exchanges, i.e. exchanges between sister chromatids at homologous loci. In contrast, ionizing radiations (S-independent agents) are poor inducer of SCEs. Thus, from the pattern of aberrations formed one can try to deduce the type of DNA lesion responsible for the formation of aberrations.

DNA Lesions Leading to Radiation-induced Chromosome Aberrations

Chromosomal aberration has been considered to be the major factor leading to cell death (Joshi *et al.*, 1982). If one estimates the relative frequencies of different types of lesions induced by high LET radiation (fast neutrons) and low LET radiation (gamma rays) at D37 dose, one gets approximately 40 DSBs, 1500 SSBs and 3000 BDs for gamma rays, and 40 DSBs, 800 SSBs and 800 BDs for fast neutrons. This would indicate that DSBs are the most important lesions responsible for cell killing and chromosomal aberrations if these two end-points are interrelated. It is also known that high LET radiations are more efficient in inducing chromosomal aberrations for a given dose compared to low LET radiations. All this indirect evidence points to the involvement of DNA DSBs in the formation of radiation-induced chromosomal aberrations.

Both DSBs, and BDs have been considered as possible candidates for the formation of aberrations (Natarajan, 1983; Preston, 1980). Most of the radiation-induced lesions are repaired and the others remain unrepaired or misrepaired. Unrepaired and misrepaired lesions lead to biological effects including aberrations.

1. Experiments favouring the involvement of DSBs

Most of the SSBs are repaired by direct ligation and others go through an excision repair process during which small single strand gaps are formed (Regan and Setlow, 1974, Henner *et al.*, 1982). If one can convert the single strand gaps into DSBs, one would expect an increase in the frequency of X-ray induced aberrations, provided DSBs lead to chromosomal aberrations. When Chinese hamster ovary cells (CHO) are X-irradiated in the G2 or G1 phase of the cell cycle, made permeable with inactivated Sendai virus and Neurospora endonuclease (NE—an enzyme which recognizes single strand DNA stretches and digests them, thus introducing DSBs) the frequencies of all types of aberrations increase by a factor of 2 (Natarajan and Obe, 1978). When CHO cells are irradiated with 0.5 MeV neutrons and post-treated with NE, no potentiation in the yield of aberrations is observed, indicating that no SSBs or single strand gaps are available for the action of NE (Obe *et al.*, 1982). Post-treatment with NE of X-irradiated cells also increases the frequency of DSBs by a factor of 2 as determined by neutral sucrose gradient centrifugation analysis (Natarajan *et al.*, 1980). These data can be interpreted to mean that X-ray induced DSBs are directly involved in the formation of chromosomal aberrations, as concomitant with the increase in the frequency of DSBs by NE, there is an increase in the frequency of aberrations. However, these correlations give only indirect evidence for the involvement of DSBs in the aberration formation.

An ideal way to demonstrate unequivocally the involvement of DSBs in the formation of aberrations would be to induce only one class of DNA lesion, namely DSBs, and ask the question whether these lesions lead to chromosomal aberrations and, if so, whether the pattern of the formation of these aberrations is similar to that observed following irradiation. Restriction endonucleases (REs) which exclusively induce DNA DSBs are ideal candidates for this purpose. When REs are introduced in CHO cells permeabilized by inactivated Sendai virus, a very high frequency of chromosomal aberrations is observed (Bryant, 1984; Natarajan and Obe, 1984). REs induce predominantly chromosome-type aberrations in cells treated in G1, and chromatid-type aberrations in G2, a pattern which is similar to that observed following X-irradiation (Natarajan and Obe, 1984; Obe and Winkel, 1985). Generally, REs which induce 'blunt' DNA DSBs are more efficient in inducing chromosomal aberrations than those which induce cohesive DSBs (Bryant, 1984; Natarajan and Obe, 1984). The high efficiency of REs in inducing chromosomal aberrations, especially exchanges, is due to the non-random induction of DSBs in identical DNA sequences, unlike the radiations which induce DSBs in a random manner. For this reason REs, unlike X-rays, are very efficient in inducing SCEs when cells are treated in S-phase of the cell cycle (Natarajan *et al.*, 1985). The studies with REs clearly show that DSBs lead to chromosomal aberrations.

2. Experiments favouring involvement of X-ray induced base damage in the formation of aberrations

Circulating human peripheral blood lymphocytes are in a dormant G0 stage of the cell cycle and yield chromosome type of aberrations (dicentrics, rings, etc.,) following irradiation. If X-irradiated cells are post-treated with cytosine arabinoside (ara C), an inhibitor of DNA polymerase alpha, the frequency of aberrations increases with the duration of the post-treatment (Preston, 1980; Natarajan *et al.*, 1986; Holmberg and Gumauskus, 1986). Since most of the DNA strand breaks are repaired very fast, the prolonged influence of ara C on the yield of aberrations has been interpreted to be due to its effect on the slowly repairing DNA base and the fact that X-ray induced aberrations arise from DNA base damage rather than DNA strand breaks (Preston, 1980). This is based on the earlier observation that SSBs accumulate if UV irradiated cells are post-incubated with ara C, due to inhibition of the polymerization step or ligation step during the excision repair of pyrimidine dimers (Hiss and Preston, 1977). An analysis of the data on X-ray induced chromosomal aberrations in human peripheral blood lymphocytes in studies involving ara C shows that most of the increase in the frequency of aberrations occurs during the first hour following irradiation, indicating that most of the repair of the lesions which can be inhibited by ara C occurs during the first hour following irradiation (Preston, 1980; Natarajan *et al.*, 1986). The influence of ara C treatment for shorter durations (15 to 60 min.) on the yield of X-ray induced chromosome aberrations in human lymphocytes has been studied (Natarajan *et al.*, 1986; Holmberg and Gumauskus, 1986). The maximum increase was found to be in the first 15 min. In a study involving blood samples from 4 donors, it was found that the increase in the frequency of aberrations due to ara C post-treatment saturates by 30 min. (Holmberg and Gumauskus, 1986). These results indicate that ara C influences the repair of lesions which occurs immediately after irradiation rather than that which occurs more slowly. When X-irradiated lymphocytes are allowed to repair for one hour and then challenged with ara C, no potentiation in the frequency of aberrations occurs, which indicates that the majority of lesions whose repair is inhibitable by ara C are repaired within one hour following irradiation.

When blood lymphocytes are processed for metaphase analysis there is a possibility that the cells do not reach mitosis in a random manner, mitosis being delayed in those cells which are more affected. In order to overcome this, we employed the technique of premature chromosome condensation (PCC). Isolated lymphocytes were irradiated with 2 Gy and fused immediately to mitotic CHO cells, or allowed to repair for 30 min. in the presence or absence of ara C and then fused (Natarajan *et al.*, 1986). The results show that ara C completely inhibits the repair of chromosomal breaks as visualized as PCC. In 30 min. 60 per cent of the breaks were repaired, but in the presence of ara C there was no repair at all (Natarajan *et al.*, 1986). To get a biochemical insight into this problem, the frequencies of DNA strand breaks induced by X-rays in lymphocytes and their repair in the presence or absence of ara C were evaluated (Natarajan *et al.*, 1986). Since it is not possible to label the DNA of lymphocytes with radioactive precursors to determine the strand breaks and their repair, we employed the nucleoid sedimentation technique using ethidium bromide as DNA stain (Mullenders *et al.*, 1983; Natarajan *et al.*, 1986). In lymphocytes irradiated with 10 Gy of X-rays, the majority of the induced DNA strand breaks were repaired in the one hour recovery time. When the irradiated cells were post-treated the repair of strand breaks was completely inhibited (Natarajan *et al.*, 1986). If the lymphocytes were allowed to recover for one hour and then challenged with ara C, there was no further increase in the frequency of breaks, indicating that most of the strand breaks, the repair of which could be inhibited by ara C, were already repaired (Natarajan *et al.*, 1986). Neutral conditions are employed in the nucleoid sedimentation technique, and thus the strand breaks detected do not represent alkaline labile sites such as base damage. The results obtained from analysis of metaphase chromosomes, prematurely condensed chromosomes and measurement of DNA strand breaks nucleoid sedimentation technique, indicate that the majority of the aberrations induced by X-rays are formed by lesions which are repaired very fast; that these lesions are mainly strand breaks and their repair can be effectively inhibited by ara C. This conclusion is also supported by studies on induction and repair of DNA strand breaks in lymphocytes in the presence and absence of ara C using the DNA unwinding technique (Holmberg and Gumauskus, 1986).

Experimental evidence exists for the ability of ara C to inhibit repair of radiation-induced DNA DSBs (Iliakis and Bryant, 1983). It was of interest, therefore, to see if the rejoining of DNA DSBs induced by restriction endonucleases is inhibited by ara C, thus potentiating restriction endonuclease induced chromosomal aberrations. Ara C was found to increase the frequency of all types of aberrations induced by Pvu II in Chinese hamster ovary cells (CHO) and mouse fibroblasts—PG 19 (Natarajan and Obe 1984). Further, experiments have demonstrated that a dose-dependent increase by Ara C on Alu I induced aberrations in CHO cells (Obe and Natarajan, 1985). From these observations, it appears that the potentiation by Ara C of the frequency of X-ray induced aberrations need not be due to its influence on slowly repairing base damage as suggested by Preston (1980), but rather on fast repairing DNA double strand breaks. In fact, when the influence of Ara C on the yield of dicentric chromosomes in lymphocytes over a wide range of doses is evaluated, it appears that its influence is predominant at low doses (less than 2 Gy) indicating an influence on the linear part of the dose response curve which is assumed to be the contribution of DSBs arising from single ionization track (Holmberg and Gumauskus, 1986).

3. Studies with X-ray sensitive mutants

The human hereditary disease ataxia-telangiectasia (A-T) is sensitive to ionizing radiations. Cells derived from patients yield higher frequencies of chromosomal aberrations following irradiation in comparison to the cells from healthy individuals (Taylor, 1978; Natarajan and Meijers, 1979). It has not been possible to detect any deficiency in DNA strand break repair in these cells using existing biochemical techniques. The increased frequency of induced chromosomal aberrations in A-T cells may arise due to a few unrepaired or misrepaired residual DNA breaks which cannot be detected biochemically (Taylor, 1978). Using recombinant DNA plasmids, it has been shown that in A-T cells, though there may be efficient repair of DSBs, the fidelity of this repair is not efficient (Cox *et al.*, 1984).

There are several rodent mutant cell lines sensitive to ionizing radiations which have become available recently (Jeggo and Kemp, 1983). These cell lines are characterized by different levels of deficiency of DSB repair. There is a very good correlation between the extent of DSB repair and the frequencies of X-ray induced chromosomal aberrations in CHO mutant lines (Kemp and Jeggo, 1986; Darroudi and Natarajan, 1987). As with A-T cells, when X-ray sensitive mutant CHO cells (xrs5 and xrs6) are irradiated in the G1 stage of the cell cycle, both chromosome and chromatid-type aberrations are induced (Darroudi and Natarajan, 1987). This implies that apart from the defect in the repair of DSBs, these cells might have other repair deficiencies, allowing the lesions to persist till the S-phase giving rise to chromatid-type aberrations. These cell lines are known to have no defect in DNA SSB repair (Kemp *et al.*, 1984). One possible candidate is DNA base damage. In anology to chemically-induced base damage, which leads to increase in the frequency of SCEs, one might expect also an increase in the frequency of SCEs in these cell lines following X-irradiation. It was indeed found that in xrs6 mutants, there is a dose-dependent increase in the frequency of SCEs following X-irradiation (Darroudi and Natarajan, 1987) suggesting an additional defect, namely repair of base damage.

Earlier studies from our laboratory employing CHO cells with chromosomes in G1 unifiliarly substituted with 5-bromodeoxyuridine and exposed to UVA (to induce SSBs in one strand of the DNA) demonstrated that induced SSBs mainly led to chromatid-type aberrations (Natarajan *et al.*, 1980). These results rule out the possibility of X-ray induced SSBs in G1 leading to chromosome-type aberrations, unless one assumes a qualitative difference between the SSBs induced by X-rays in unsubstituted DNA and UVA on BrdUrd substituted DNA.

Conclusion

The experimental results reviewed above support the conclusion that ionizing radiation-induced DSB is the most important lesion responsible for the origin of chromosomal aberrations in wild-type repair competent cells.

Acknowledgements

These investigations were partly supported by EURATOM-Leiden University contract No. B16–166 NL.

References

Bender MA, Griggs HG, Bedford JS (1974). Mechanisms of chromosomal aberration production III: Chemicals and ionizing radiations. Mutation Res. 23; 197–212.

Bryant PE (1984). Enzymatic restriction of mammalian DNA using Pvu II and Bam H1: Evidence for double strand break origin of chromosomal aberrations. Int. J. Radiat. Biol. 46; 57–65.

Cox R, Masson WK, Debenham PG, Webb MBT (1984). The use of recombinant DNA plasmids for the determination of DNA repair and recombination in cultured mammalian cells. Brit. J. Cancer 49 (suppl.6); 67–72.

Darroudi F, Natarajan AT (1987). Cytological characterization of Chinese hamster ovary x-ray sensitive mutant cells xrs 5 and xrs 6.I. Induction of chromosomal aberrations by X-irradiation and its modulation with 3 amino benzamide and caffeine. Mutation Res. 177; 133–48.

Hanner WD, Grunberg SN, Haseltine WA (1982). Sites and structure of radiation induced DNA strand breaks. J.Biol. Chem. 257; 11750–4.

Hiss EA, Preston RJ (1977). The effect of cytosine arabinoside on the frequency of single strand breaks in mammalian cells following irradiation or chemical treatment. Biochim. Biophys. Acta 478; 1–8.

Holmberg M, Gumauskus E (1986). The role of short lived DNA lesions in the production of chromosome exchange aberrations. Mutation Res. 160; 221–9.

Iliakis G. Bryant PE (1983). Effects of nucleoside analogue alpha-araA, beta-araA, and beta-araC on cell growth and repair of potential lethal damage and DNA double strand breaks in mammalian cell cultures. Anti-Cancer Res. 3; 143–50.

Jeggo PA, Kemp LM (1983). X-ray sensitive mutants of Chinese hamster ovary cell lines. Isolation and cross sensitivity to other DNA damaging agents. Mutation Res. 112; 313–27.

Joshi GP, Nelson WJ, Revell SH, Shaw CA (1982). X-ray induced chromosome damage in live mammalian cells, and improved measurement of the effect on the colony forming ability. Int. J. Rad. Biol. 41; 161–81.

Kemp LM, Jeggo PA (1986). Radiation-induced chromosome damage in X-ray sensitive mutants (xrs) of the Chinese hamster cell line. Mutation Res. 166; 225–63.

Kemp LM, Sedwick SG, Jeggo PA (1982). X-ray sensitive mutants of Chinese hamster ovary cells defective in double strand break rejoining. Mutation Res. 132; 189–96.

Lea DW (1955). Actions of Radiations on Living Cells. University Press, Cambridge.

Mullenders LHF, van Zeeland AA, Natarajan AT (1983). Comparison of DNA loop size and super coiled domain size in human cells. Mutation Res. 112; 245–52.

Natarajan AT (1983). Mutagenesis and chromosome changes—Review. In Radiation Research. Proc. 7th Int. Cong. Rad. Res. (eds. Broerse JJ *et al.*); 239–47.

Natarajan AT, Darroudi F, Mullenders LHF (1986). The nature and repair of DNA lesions that lead to chromosome aberrations induced by ionizing radiations. Mutation Res. 160; 231–6.

Natarajan AT, Kihlman BA, Obe G (1980), Use of the 5-bromodeoxyuridine-labelling technique for exploring mechanisms involved in the formation of chromosomal aberrations. II. G1 experiments with Chinese hamster ovary cells. Mutation Res. 73; 307–17.

Natarajan AT, Meijers M (1979). Chromosomal radiosensitivity of ataxia telangiectasia cells at different cell cycle stages. Hum. Genet. 52; 127–32.

Natarajan AT, Mullenders LHF, Meijers M, Mukherjee U (1985). Induction of sister chromatid exchanges by restriction endonucleases. Mutation Res. 144; 33–9.

Natarajan AT, Obe G (1978). Molecular mechanisms involved in the production of chromosome aberrations. I. Utilization of Neurospora endonuclease for the study of aberration production in G2 stage of the cell cycle. Mutation Res. 52; 137–49.

Natarajan AT, Obe G (1984). Molecular mechanisms involved in the production of chromosomal aberrations. III. Restriction endonucleases. Chromosoma (Berl.) 90; 120–7.

Natarajan AT, Obe G, van Zeeland AA, Palitti F, Meijers M, Verdegaal-Immerzeel EAM (1980). Molecular mechanisms involved in the production of chromosome aberrations. II Utilization of Neurospora endonuclease for the study of aberration production by x-ray in G1 and G2 stages of the cell cycle. Mutaton Res. 69; 293–305.

Obe G, Natarajan AT, Palitti F (1982). Role of DNA double strand breaks in the formation of radiation induced chromosomal aberrations. In DNA Repair, Chromosomal Alterations and Chromatin Structure (eds. Natarajan AT, Obe G. Altman H), Elsevier Biomedical Press; 1–9.

Obe G, Natarajan AT (1985). Chromosomal aberrations induced by restriction endonucleases Alu I in Chinese hamster ovary cells: Influence of duration of treatment and potentiation by cytosine arabinoside. Mutation Res. 152; 205–10.

Obe G, Winkel EU (1985). The chromosome breaking activity of restriction endonuclease Alu I in CHO cells is independent of the S phase of the cell cycle. Mutation Res. 152; 25–9.

Preston RJ (1980). The effect of cytosine arabinoside on the frequency of X-ray induced chromosome aberrations in normal leukocytes. Mutation Res. 69; 71–9.

Regan JD, Setlow RB (1974). Two forms of repair in the DNA of human cells damaged by chemical carcinogens and mutagens. Cancer Res. 34; 3318–25.

Revell SH (1955). A new hypothesis for chromatid changes. In Proc. Radiol. Symp. Liege (eds. Bacq ZM, Alexander P); 243–53.

Sax K (1940). An analysis of X-ray induced chromosomal aberrations in Tradescantia, Genetics 25; 41–68.

Taylor AMR (1978). Unrepaired DNA strand breaks in irradiated ataxia telangiectasia lymphocytes suggested by cytogenetic observations. Mutation Res. 50; 225–36.

10 Current and Future Trends in Human Meiotic Research

ANN C. CHANDLEY

MRC Clinical and Population Cytogenetics Unit
Western General Hospital
Edinburgh EH4 2XU
Scotland

1 Introduction

In accepting this invitation to write in honour of Prof. S.P. Ray Chaudhuri, I am conscious of the great contribution made by him in the field of animal genetics over the years and of his great skill and enthusiasm imparted through his teachings. If my chapter can also contribute in a small way to the enlightenment of present and future generations of students, it will have served its intended purpose.

To write on the general topic of meiosis would require not just a chapter but a whole book. One is forced therefore to limit the scope in some way. As my own involvement has been in the field of mammalian meiosis with particular reference to man, it is principally with that species that I will deal.

Over many years now, our laboratory has been well-placed to exploit a situation which sees human testicular material arriving on a regular basis from the infertility clinics of the Edinburgh hospitals. Currently, we have the excellent support and co-operation of Mr T.B. Hargreave, Consultant Urologist, Department of Surgery, Western General Hospital, who provides us with material. Our aim throughout has been to investigate the manner in which, and extent to which, genetic and chromosomal factors influence gametogenic development and contribute to infertility in the human male. The results of ten years (1968–78) of systematic study into the somatic chromosome constitution, seminal characteristics, testis histology and meiosis in a large series of infertile men have already been published (Chandley *et al.*, 1975; Chandley *et al.*, 1976; Chandley, 1979). Now, selected patients, mainly within the low sperm count categories are analysed, and with new developments in technique, these are yielding much valuable information on the meiotic behaviour of male-sterilizing anomalies. Structural rearrangements in the Y chromosome, which we have found in some of our azoospermic patients, give us the opportunity to study the consequence of loss of the gene for spermatogenesis, situated on the Y long arm (Chandley *et al.*, 1986a).

Meiotic development in oocytes, and in particular in females carrying chromosome anomalies, is also among our interests. The prophase stages in the female take place during foetal life, and in order to study meiosis during this period, it is necessary to prepare oocytes from the gonads of mid-term embryos. In humans, this is possible whenever an aborted foetus is available and the ovaries are sufficiently fresh to allow satisfactory processing. Termination is sometimes carried out following the antenatal diagnosis of a chromosome anomaly in a human foetus, and meiotic study of such anomalies is thus made possible.

The chapter will outline some of our findings made over recent years in normal and aberrant meiotic development in males and females. A final section will look to the future and will provide pointers to the way in which research on meiosis may be going. The molecular age of probes, clones and DNA digests is upon us and tools to investigate the germ line, more powerful than anything we have had available to us before, may soon become a routine part of the armoury. The future holds great possibilities for the development of this area of research.

2 Current Techniques in Meiotic Chromsome Study

A significant advance in human meiotic chromosome preparation has been made approximately once per decade over the past thirty years. The original squash method of preparation (Ford and Hamerton, 1956) was superceded ten years later by the introduction of air-drying of spermatocytes in suspension (Evans *et al.*, 1964). This gave much better detail of the meiotic bivalents at metaphase I and enabled chiasma counting to be carried out with precision. The application of special staining techniques like C-banding and the fluorochromes quinacrine mustard, quinacrine dihydrochloride and distamycin A/DAPI (Schweizer *et al.*, 1978) allow for identification of individual bivalents such as the No. 1, 9, 16 and the Y chromosome, within the cell (Chandley, 1977).

In 1973, a new era dawned when Counce and Meyer introduced their simple and rapid surface microspreading technique for the visualization of the synaptonemal complex at meiotic prophase using insect spermatocytes. Application of the technique to human spermatocytes (Moses *et al.*, 1975; Solari, 1980) soon followed, and the method has been adapted to oocytes in our own laboratory by Speed (1982, 1985, 1988). The spreading technique in one or other of its forms has revolutionized our understanding of basic meiotic chromosome behaviour during prophase and is proving to be a powerful tool in the analysis of anomalous chromosome behaviour.

Spread preparations of synaptonemal complexes (SC) are obtained through surface tension on an aqueous solution. The SC, to which chromatin fibrils are attached, is a proteinaceous structure which appears prior to or at synapsis, and disappears at or following desynapsis. Pairing events in the SC follow exactly the behaviour of chromatin in the meiotic bivalents throughout prophase and detail of structural anomaly such as translocation or inversion can readily be visualized.

Basically there are two principle techniques and two levels of observation, light microscope (LM) and electron microscope (EM). One technique involves the picking up of spread cells from the surface of a saline or sucrose solution onto a slide and is generally used for spermatocyte spreading when an abundance of cells is available in suspension. The second involves the settling of cells onto a slide surface through a sucrose or saline solution and has been used principally by us for oocyte spreading, where numbers of germ cells may be limited. Full details of both methods can be found in Chandley (1987), together with the methods used for sequential LM and EM investigation. Staining of the SC can be carried out using a variety of methods including silver nitrate or phosphotungstic acid (PTA), the latter giving better detail of centromere positioning and the central element of the SC than the former.

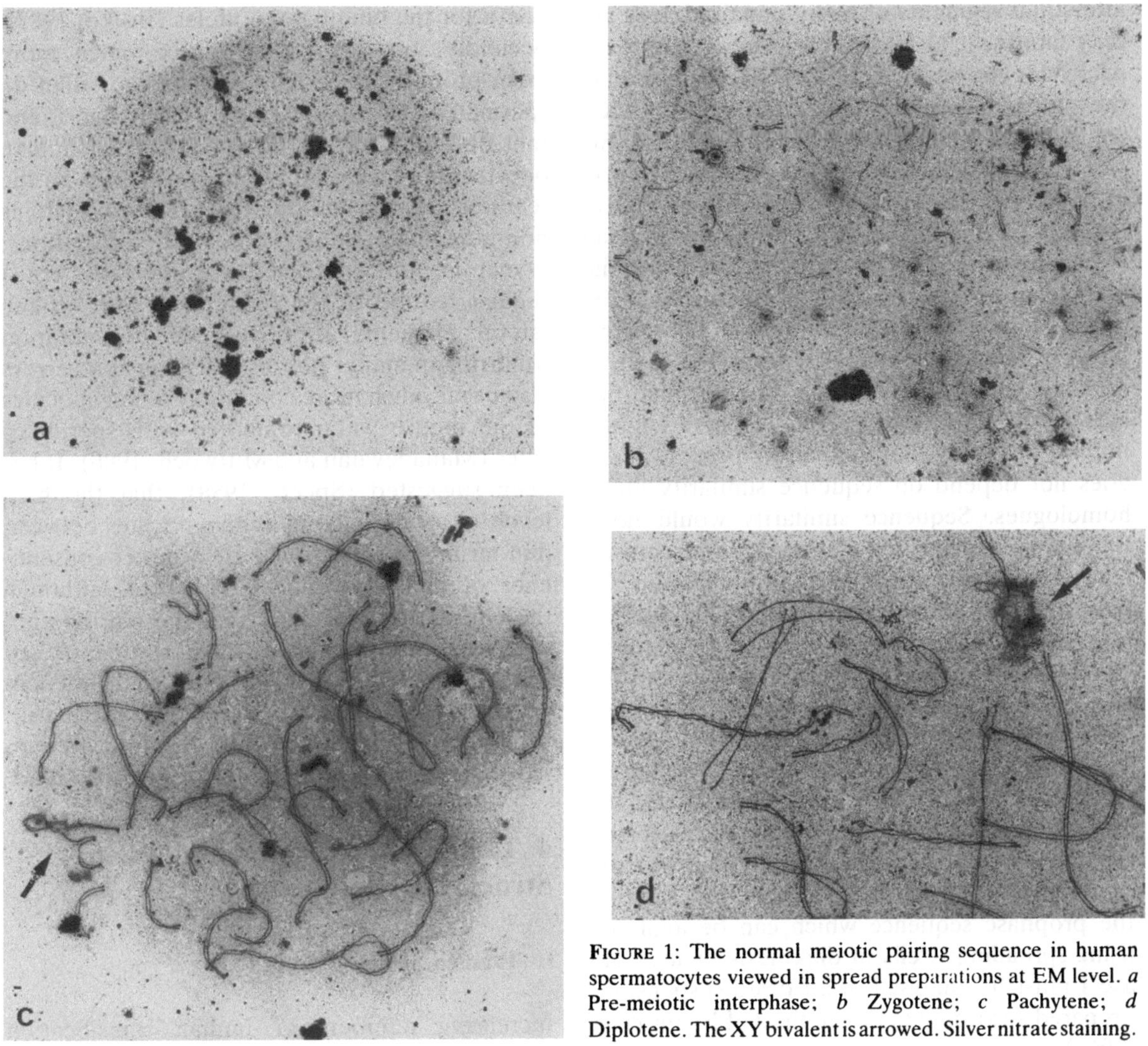

FIGURE 1: The normal meiotic pairing sequence in human spermatocytes viewed in spread preparations at EM level. *a* Pre-meiotic interphase; *b* Zygotene; *c* Pachytene; *d* Diplotene. The XY bivalent is arrowed. Silver nitrate staining.

3 The Normal Meiotic Pairing Sequence at Prophase

Figure 1 illustrates four stages in human meiotic prophase in spermatocytes prepared by spreading, stained with silver nitrate, and viewed at EM level. Cells in pre-meiotic interphase (Figure 1a), show a granular appearance, background chromatin being interspersed with silver staining bodies of variable shape and size, which probably represent nucleolar material. By the zygotene stage, pairing initiation can be observed and particularly in terminal segments (Figure 1b). At this early stage of synapsis, the interstitial segments of many unpaired axes are only faintly, if at all, visible. Once cells reach pachytene, however, full pairing is established in the autosomal SCs (Figure 1c), and the XY complement shows a variable length of SC forming between the short arm-short arm (Xp;Yp) pairing segments. Studies which have focused on the XY pairing sequence in man (Chandley *et al.*, 1984) show that up to half of the total length of the Y chromosome axis can be involved in SC formation with the X short arm. DNA sequence homology within the pairing region is confined however, to the terminal tips (see Goodfellow *et al.*, 1985 for review), the implication being that the pairing process *per se* does not depend on sequence similarity along homologues. Sequence similarity would none the less be expected as a necessary prerequisite for reciprocal recombination. In the human XY pair, only the extreme distal tips of Xp and Yp have shared homology and these sequences, which are highly polymorphic, cross over with a high frequency, the rate of exchange increasing with proximity to the telomere (Rouyer *et al.*, 1986). When spermatocytes reach diplotene (Figure 1d), desynapsis is observed in the autosomes while the XY complement takes on a tangled appearance as the axes break down and undergo splitting. Diplotene is the last stage in the prophase sequence which can be analysed using spreading: Cells in diakinesis or metaphase I must be analysed in air-dried suspensions prepared according to the method of Evans *et al.* (1964).

The prophase sequence in the human female resembles that in the male although there are some important differences to be noted. Firstly, the overall synaptonemal complex length of the oocyte is about twice that found for the spermatocyte given a nucleus of similar dimensions although crossover events, inferred from the occurrence of recombination nodules and bars, are of similar proportions, indicating that synaptonemal complex length *per se* is not a limiting factor for the binding of nodules (Bojko, 1983). Secondly, oocytes appear to be much more error-prone in their pairing behaviour than do spermatocytes. Speed (1985, 1988) has found that only one-half to three-quarters of human oocytes show what he would consider to be the features of 'normal' pairing. Errors such as non-homologus associations, triple associations, asynapsis, univalent axes and interlocks are a regular feature of the remainder, with broken lateral elements suggestive of degeneration occurring in many. Interestingly, synaptic errors also occur much more commonly among oocytes of the mouse when compared with spermatocytes (Mahadevaiah and Mittwoch, 1986). It has been suggested (Speed, 1988), that the high levels of atresia among human oocytes between mid-term and birth could be a direct consequence of those synaptic disturbances, although the alternative possibility that the pairing anomalies themselves are a manifestation of cell death attributable to some third and independent factor, cannot at present be ruled out. Whether pairing disturbance is cause or consequence of cell death has not yet been decided.

4 The Synaptonemal Complex in Structural Heterozygotes

i Translocations

Increasing numbers of human translocation cases are now being analysed by spreading. These are showing a number of interesting facts regarding the pairing behaviour within multivalents, and the interactions that can arise between translocation configurations and the XY pair in male-sterile cases. A human reciprocal translocation between chromosomes No. 9 and 20 was investigated recently in our own laboratory (Chandley *et al.*, 1986b). The translocation was ascertained in an oligospermic man (mean sperm count 9.2×10^6/ml), and the full

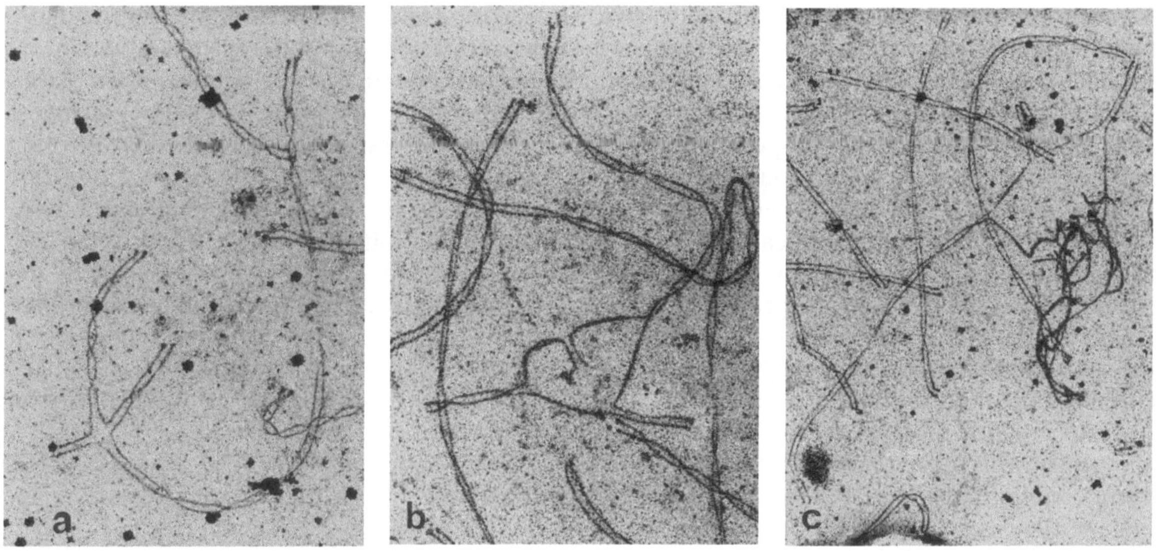

FIGURE 2: Translocation quadrivalents seen at EM level in a t(9;20) translocation heterozygote. *a* Full pairing; *b* Asynapsis around the points of interchange; *c* Association with the tangled XY pair.
(From Chandley *et al.*, 1986b. With permission.)

karyotype was found from the somatic chromosome analysis to be 46,XY,t(9;20)(q34;q11). A spreading analysis showed a clear quadrivalent in most spermatocytes at the pachytene stage (Figure 2a), interpretation of the cross configuration being possible on the basis of the known breakpoint positions in the somatic complement. The most common type of quadrivalent was one in which pairing had occurred along the four arms, although asynapsis remained around the points of interchange (Figure 2b). In other cells, asynapsis was also seen in the short (9q) arm of the quadrivalent, and in some of these cells, contacts between this arm and the XY pair were established (Figure 2c).

Synaptic disturbance at pachytene seen in this and many other cases of chromosome anomaly, has led Burgoyne and Baker (1984) to suggest that germ cell death may ensue as the result of pairing failure which leaves a variable number of sites along chromosome arms 'unsaturated'. Based on the earlier hypothesis of Miklos (1974), they believe that unsaturated pairing sites could become 'activated' during meiosis, setting in motion a process of disturbed metabolism that would eventually lead to the destruction of the nucleus. In the 9;20 translocation heterozygote, however, germ cell destruction appeared to be starting much earlier than the pachytene stage in about one half of all cells examined, and synaptic failure alone could not therefore offer a full explanation for the sterility. The mere presence of the chromosome anomaly in the genome seemed to be of primary importance in triggering the process of cell destruction.

As an alternative to the Miklos (1974) hypothesis, others have postulated that sex chromosome-autosome contacts such as that shown in Figure 2c might be responsible for germ-cell death (Forejt, 1974). In some human male sterile translocations, contacts between the

XY pair and the translocation multivalent have been noted at pachytene with a frequency reaching 60–70 per cent of all cells. The higher the frequency of contact the more severe appears to be the disruption of spermatogenesis and the lowering of the sperm count. The effect is most marked when an acrocentric chromosome is involved in the translocation, whether this be Robertsonian (Luciani *et al.*, 1984; Rosenmann *et al.*, 1985) or reciprocal (Gabriel-Robez *et al.*, 1986a; Johannisson *et al.*, 1987), the short arms of the acrocentric (NOR-bearing) chromosome often forming the association with the XY pair. Translocations involving an acrocentric chromosome are thus found to be more common among male-sterile than among male-fertile cases (Gabriel-Robez *et al.*, 1986a), and because of failed chiasma formation in the short arms of the acrocentric, they also show a tendency to produce chain rather than ring configurations at meiotic metaphase I (Chandley, 1981).

The idea originally put forward by Forejt (1974) to explain sterility in male mice arising out of X-autosome contact was that reactivation of the X chromosome might occur under the spreading influence from the autosome in question. The X and Y are normally transcriptionally inactive during meiotic prophase in the male (Monesi, 1967; Tres, 1975). Alternatively, inactivation of a segment of autosome might equally well take place under the spreading influence of the X chromosome with which it had made contact. The issue remains to be decided, but the observations, both in mice and men, suggest that when all ends of chromosome arms are fully paired, contacts do not arise and spermatogenesis proceeds normally. A remaining possibility is, of course, that contacts are simply a secondary consequence of a primary defect which causes both spermatogenesis to be impaired and pairing to be disrupted.

The whole question of gametogenic impairment remains very complex and is still very little understood. Some translocations, for example, behave differently in different males within the same family or (in the case of mice) when placed on different genetic backgrounds, making it seem that the translocation itself is not the only and consistent cause of infertility. Moreover, many anomalies that produce serious consequences for fertility in males produce far less severe consequences for female carriers and although oogenic disturbances may arise, fertility is usually not impaired to the same extent. Egg losses may however result in a reduced ovarian size or a shortened reproductive span (Mittwoch *et al.*, 1984).

ii Inversions

Translocations are, however, not the only type of rearrangement to be associated with spermatogenic disturbance. Pericentric inversion in chromosome 1 in man appears to place the male heterozygote at great risk for infertility owing to germ-cell maturation impairment. A recent case of inversion 1, with breakpoints at p31 and q43, ascertained through infertility in an oligospermic man attending our clinic (Chandley *et al.*, 1987), caused us to search the literature for other cases of its kind. Since 1970, 20 cases of pericentric inversion in chromosome 1 have appeared in the literature, of which nine presented in adult males. Of these, eight were ascertained through sterility with associated oligo- or azoospermia. Ascertainment of pericentric inversions in man in general is not made through male infertility but by the birth of a child with congenital malformations or spontaneous abortion resulting from a duplication-deficiency product from the carrier parent. It would seem likely in the case of chromosome 1, that asynapsis across the inverted segment and elsewhere in the genome, seen in our own patient in the majority of prophase cells (Figure 3a), might be the responsible factor leading to germ-cell failure and the sterility of the carrier.

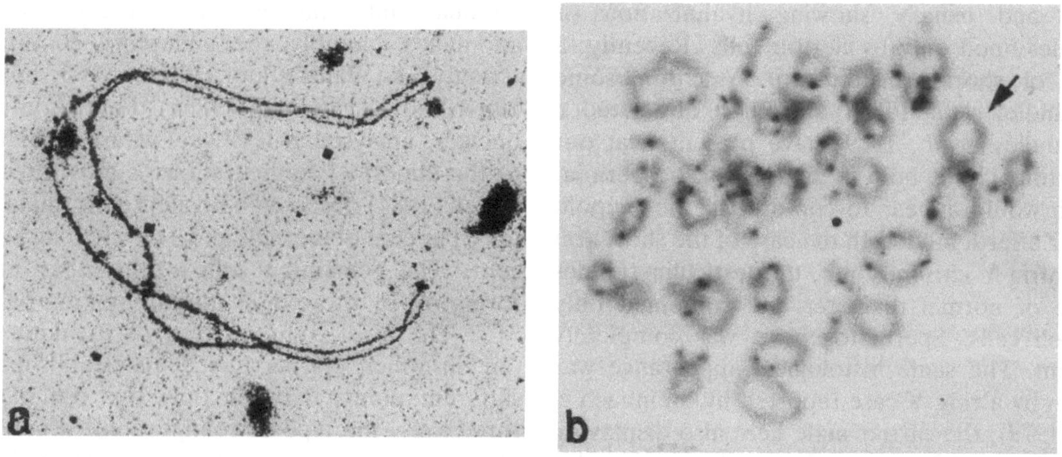

FIGURE 3: Inversion (1) in a human male. *a* EM spread showing asynapsis across the inverted region at pachytene; *b* Metaphase I (C-banded) with No 1 bivalent arrowed. Note the asymmetrical positions of the centromeres and lack of chiasma formation in the inverted regions. (From Chandley *et al.*, 1987. With permission.)

The pairing problem in the No. 1 bivalent could arise initially out of the large heterochromatic blocks on proximal 1q, such blocks tending to delay or inhibit synapsis even in normal individuals (Solari, 1980).

The classic expectation of loop formation in this inversion was not found, most cells showing the asynapsis referred to above, or the early establishment of heterologous pairing across the inverted segment, such cells showing 23 normal-looking SCs. Predictably, few cells at metaphase I showed evidence of chiasma formation within the inverted region (Figure 3b), thus lowering the risk for recombinant offspring, although of course in this particular inversion, this is largely a matter of academic interest since the male is sterilized by the anomaly. The study of other human inversions at meiosis could, however, provide valuable data for genetic counselling of carriers in terms of their risks for recombinant offspring. Only a few other human cases have so far been analysed by spreading (Gabriel-Robez *et al.*, 1986b; Saadallah and Hulten, 1986) and these too indicate that full loop formation is a relatively rare phenomenon, the early initiation of heterologous pairing across the inverted region being seen in many cells. In a number of other animal species in which pericentric inversions have been studied at meiotic prophase by microspreading, the phenomena of asynapsis and/or heterologous pairing across the rearranged segment have also been noted, classical loops seeming here also be relatively rare (see Hale, 1986 for review).

5 Structural Rearrangement in the Y Chromosome

Phenotype-genotype correlations in man show that deletion of the long arm heterochromatin (Yqh) and a small segment of distal non-fluorescent euchromatin on the Y chromosome results in aspermatogenesis (Tiepolo and Zuffardi, 1976). These authors have suggested that a gene controlling spermatogenesis is located at the euchromatin/heterochromatin interface at Yq11. Loss of this gene by deletion or structural rearrangement produces a testis lacking germ cells and usually showing hyalinization or tubules lined only by Sertoli cells. Recently, a case of short arm dicentric Y chromosome (Chandley *et al.*, 1986a) and a case of deleted Y (Chandley *et al.*, 1989) have presented at our infertility clinic, both in men with azoospermia. Each would appear to conform to the Tiepolo and Zuffardi model. In the case of the short arm dicentric Y chromosome, the testicular tubules were of normal diameter and contained only Sertoli cells. Spermatogenesis was completely absent. The same histological appearance was given by a ring Y case reported by Steinbach *et al.* (1979), the carrier male here also displaying a total lack of fluorescent heterochromatin in his somatic Y chromosome. For the deleted Y chromosome mentioned above (Chandley *et al.* 1989), only a Y-DNA probe analysis enabled an unequivocal identification of the tiny metacentric marker chromosome seen in the somatic karyotype. This indeed showed that the abnormal chromosome was the Y chromosome deleted in the long arm. The testicular histology showed marked hyalinization with only minimal germ cell presence in the tubules. A few dividing spermatogonia were present and occasional prophase spermatocytes, but development beyond prophase was arrested completely. A similar case of Yq deletion with an almost identical spermatogenic picture has recently been published by Hartung *et al.* (1988). The action of the gene for spermatogenesis on Yq11 is not known but it may affect proliferation or migration of primordial germ cells in the early embryo, or perhaps impair spermatogonial proliferation in the adult male.

6 Chromosome Anomaly in Oocytes

The application of spreading techniques to oocytes in man (Speed 1985, 1988) and the mouse (Speed 1982; Speed and Chandley 1983), as described earlier, has also enabled studies to be made into the behaviour of chromosome anomalies, notably the aneuploid conditions trisomy 21 (Wallace and Hulten, 1983; Speed, 1984) and XO (Speed, 1986a). For trisomy 21 oocytes, analysed at EM level where resolution of the abnormal configurations is most precise, Speed (1984) observed a bivalent plus univalent in 60 per cent of the cells and a trivalent in 34 per cent. The remaining cells were either uninformative or showed 23 normal-looking paired SCs. The most common type of trivalent was one in which pairing had been established at different points between only two out of the three axes, the type of configuration which has always been suggested as the classic mode of pairing for primary trisomics. In the remaining trivalents however, triple association was observed, in rare cells this being complete along the entire length of the three No. 21 axes. More often, however, triple association occurred along only a portion of the axes, sometimes terminal at other times interstitial. Similar observations have been made by Wallace and Hulten (1983).

For XO foetal ovaries, Speed (1986a) has shown that at least in man, for three cases examined, severe depletion of most germ cells occurred soon after preleptotene. Germ cell failure could thus not be accounted for on the Burgoyne and Baker model by the asynaptic state of the single X axis at pachytene. In the mouse, where fertility for the adult female is known, germ cell development was complete for many although not all oocytes, and an interesting observation was made by Speed (1986a). As gestation progressed, the single X axis in almost two-thirds of oocytes succeeding in pairing non-homologously on itself, or with a pair of autosomal lateral elements making a triradial structure (Figures 4a and b). The survival of oocytes in the XO mouse, Speed (1986a) suggested, might perhaps depend on this ability for intrachromosomal heterologous pairing, since

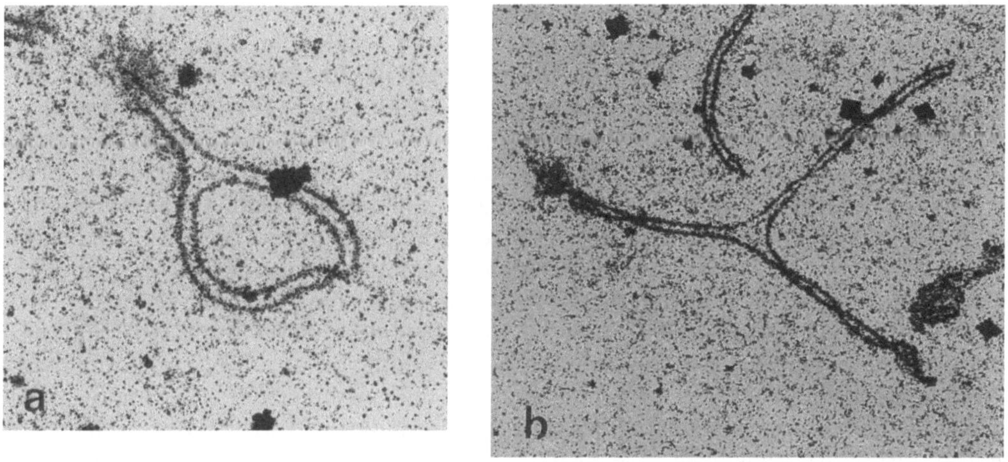

FIGURE 4: The X axis in an XO mouse oocyte. *a* Self-paired; *b* Forming a triradial with an autosome (From Speed, 1986a. With permission.)

oocytes displaying a free-lying X axis decreased in frequency towards the late gestational days, the X axis becoming progressively thickened and dark staining with silver. These oocytes, it was suggested, might become atretic and be lost from the oocyte pool.

The evidence from the human XO foetuses, the t(9;20) translocation in the male described earlier in Section 4(i), and from another human foetus showing an iso-18q karyotype also studied by Speed (1986b), all suggest germ cell destruction at the very earliest stage and long before cytologically recognizable pairing has been established. Such cases therefore test the hypothesis of Burgoyne and Baker (1984), and indicate that asynapsis and the failure to saturate pairing sites at pachytene are, alone, not sufficient to explain germ cell death. If there is a defect in synapsis responsible for death, it would have to be operating prior to prophase and even during the gonial stages. It is of interest that severe reductions in pre-meiotic germ cell numbers have also been recorded in the testes of human foetuses with trisomy 13 and 18 (Coerdt *et al.*, 1985) with lesser reductions being seen in trisomy 21, XXY and XYY foetuses. In all of these aneuploid situations, there would seem to be a disturbance in the germ line starting from the very beginning and not just with the onset of meiotic maturation. Coerdt *et al.* (1985) have suggested aberrant germ cell migration, impaired colonization of the gonadal ridges, low mitotic activity and/or increased pre-meiotic cell losses as possible factors responsible for the paucity of germ cells.

Much therefore seems to remain to be resolved with regard to chromosome anomaly and germ cell death and future research into meiosis and infertility will undoubtedly focus on this as an area of some interest.

7 Meiotic Segregation and the Error Leading to Aneuploidy in the Female

Much painstaking work over the years has revealed that the nondisjunctional error leading to Down syndrome (trisomy 21) and other human aneuploid conditions usually occurs at the first meiotic division in the oocyte (Hassold and Jacobs, 1984), the production of trisomic conceptuses rising markedly with maternal age (Hassold and Chiu 1985). The basis of the maternal age effect for aneuploidy remains obscure and constitutes an outstanding unsolved problem in human meiotic research. Over the years, many factors have been suggested to underlie the phenomenon, but few experimental studies have been attempted, principally because of the complexities involved. As a model system, the CBA mouse—which shows a certain degree of increasing aneuploidy with age, as well as a short reproductive span and an ovary which, as in women, is devoid of oocytes at the end of fertile life—has been investigated recently in our laboratory (Eichenlaub-Ritter *et al.*, 1986, 1988). Ageing oocytes in MI and MII have been examined for changes in spindle structure and function, changes in chromosome alignment on the spindle, and for any other abnormal features which may be of significance in the origin of aneuploidy.

Deterioration in the spindle or its components has frequently been put forward as a possible factor underlying the increasing levels of nondisjunction in ageing women, but only recently has an attempt been made to study the oocyte spindle throughout a reproductive lifespan. Immunofluorescence techniques were used to examine whole-mount spindles of MI and MII CBA mouse oocytes for evidence of change during natural ageing of the female, and after ageing in the fallopian tube after ovulation. The MII spindles of old females (9 mths) were found to be conspicuously shorter than those of young females (<4 mths) and chromosome alignment in old oocytes was much more disordered. Ageing of MII eggs in the fallopian tubes seemed to bring about similar changes. For MI, the division at which most segregational anomalies arise in humans, few changes in spindle structure were detectable with ageing of the female, but quite striking changes in the dynamics of meiotic maturation were found. Breakdown in the germinal vesicle stage of the arrested diplotene oocyte was significantly delayed in old females and this was followed by unusually rapid progression through the first meiotic division with precocious transition to anaphase I. Oocytes of old females did not show any signs of precocious separation of parental chromosomes at prophase, or significant lagging of chromosomes during anaphase and telophase. Disturbances in the timing of the developmental programme thus seem the more likely cause of nondisjunction at the first meiotic division, the irregular arrangement of chromosomes on the MII spindles of old females perhaps reflecting the occurrence of a prior nondisjunctional event (Eichenlaub-Ritter *et al.*, 1988).

8 New Applications of Molecular Technique to Meiotic Chromosomes

Although somatic metaphases provide a ready source of chromosomal material for study using molecular techniques, there are nevertheless examples in the literature of the use of meiotic chromosomes for investigation at the DNA level. It has, for example, been found that pachytene bivalents, suitably prepared to give fully extended arms (Jhanwar *et al.*, 1982), are good material for *in situ* hybridization studies, the homologous pairing at this stage enhancing the efficiency of hybridization over that obtained in the equivalent somatic chromosomes (Chaganti *et al.*, 1985; Jhanwar *et al.*, 1982). Individual bivalents can be identified in pachytene spreads by the patterns of chromomeres along arms, the patterning corresponding to G-band patterning of somatic cells. Using this approach, Chaganti *et al.* (1985) have determined the germ-line positions of the parathyroid hormone, -globin and insulin genes on the short arm of chromosome 11, while Jhanwar *et al.* (1982) have determined the positions of the oncogenes ABL, SIS and FES.

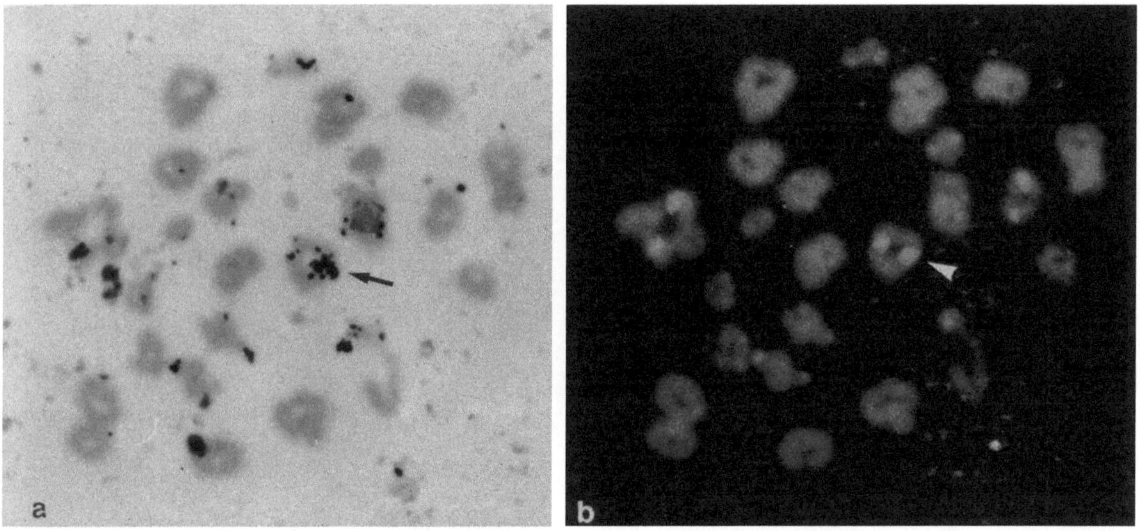

FIGURE 5: *a* Human male MI probed with the sequence Xb1 which binds principally to the secondary constriction region of bivalent No 9 (arrow); *b* here, the same cell shows the use of DA/DAPI to identify the No 9 bivalent (arrow.)

In another set of experiments, Mitchell *et al.* (1986) used human male meiotic chromosomes to order two sequences on chromosome 9. The special characteristic of the No. 9 chromosome at metaphase I and II in displaying a pale staining gap at the paracentromeric secondary constriction in the long arm, was exploited. Attempts to assign the sequences p82H and Xb1 to the centromere and/or secondary constriction region had failed using somatic metaphases, but at meiosis, the pale secondary constriction is clearly distinct from the centromere itself. The use of pachytene cells in which the secondary constriction of chromosome 9 displays a cluster of highly refractile bodies known as 'parameres' (Page, 1973) gave additional value to the use of meiotic cells in the analysis.

The heavy signal produced by probe Xb1 to the secondary constriction of bivalent No 9 at metaphase I, identified by pre-staining with the fluorescent stain combination of distamycin A/DAPI, is shown in Figure 5. The result of the investigation was to demonstrate quite unequivocally the close association of Xbl to the secondary constriction and of p82H to the centromere.

Another type of investigation which has been undertaken using meiotic bivalents is that employing the use of DNase I nick translation to demonstrate sites of active or potentially active genes. The chromatin organization of such sites in interphase nuclei is of a conformation that renders them sensitive to digestion by the nicking enzyme (see Weisbrod, 1982 for review) and mammalian mitotic metaphase chromosomes have been found to retain this characteristic. Sensitive regions can thus be visualized by applying nick translation and autoradiography to fixed chromosome spreads (Gazit *et al.* 1982). DNase I sensitive sites show a close correspondence with light G-bands (Kerem *et al.*, 1984) which tend also to be early replicating (Goldman *et al.*, 1984).

The application of nick translation techniques

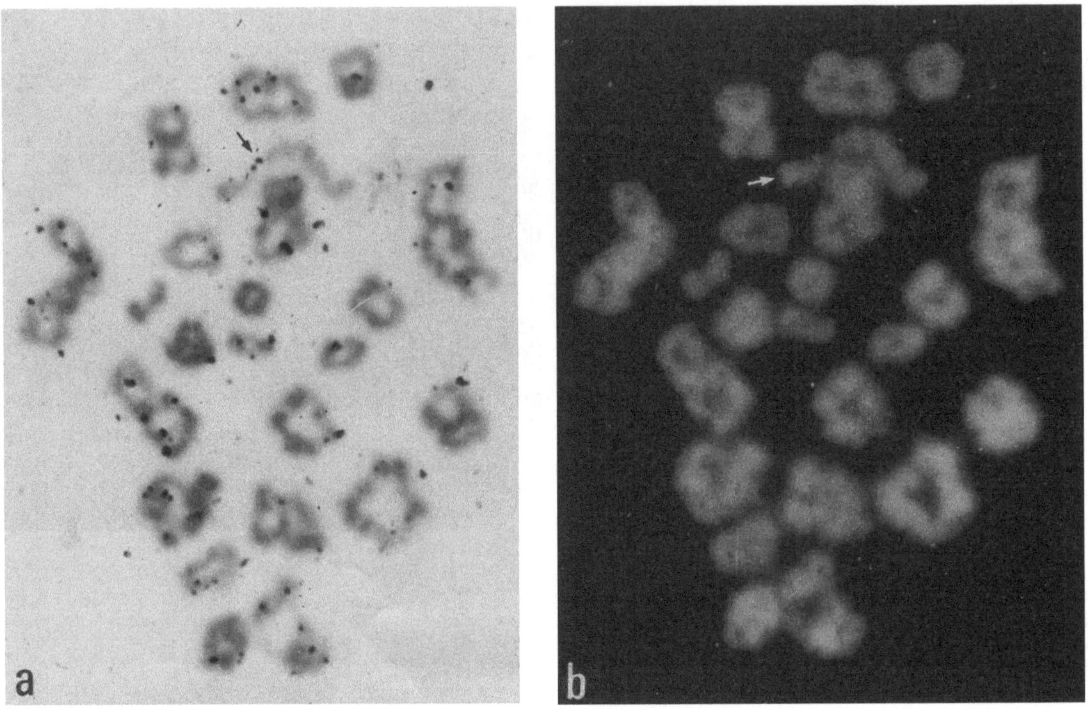

FIGURE 6: *a* Nick translation of the human XY pair showing a hot-spot of label at the pairing tips (arrow); *b* DA/DAPI identifies the Y chromosome long arm heterochromatin (arrow). (From Chandley and McBeath, 1987. With permission.)

to meiotic bivalents of the mouse (Sperling and Marcus, 1984; Rajcan Separovic and Chandley, 1987) and man (Chandley and McBeath, 1987) shows a general overall labelling which is lower for the XY bivalent than for the autosomal bivalents. This is because the sex chromosomes at meiosis in the male are condensed and transcriptionally inactive (Monesi, 1967; Tres, 1975) and of a DNA conformation which in general is restrictive in allowing action by the nicking enzyme. Nevertheless, in both species, hot-spots of DNase I sensitivity are detectable at meiosis. In the human XY pair, Chandley and McBeath (1987) found terminal/telomeric sites to be prone to nicking, including the short arm–short arm pairing region (Figure 6). A second site of sensitivity occurred just below the centromere in the long arm of the X chromosome, a site which it was argued, could correspond to the X-inactivation centre in the female. The principal nicked sites were all known from separate studies to be sites of early replication in the somatic chromosomes (Schempp and Meer, 1983).

For the mouse XY pair, the data are not quite so clear-cut and between laboratories are somewhat conflicting. The claim by Richler *et al.* (1987) that the XY pairing tips are DNase I sensitive at pachytene and MI could not be confirmed in our own laboratory (Rajcan Separovic and Chandley, 1987), at least for the MI cells investigated. The finding that a site in the distal third of the X chromosome, in the approximate D-band region of the somatic karyotype, did show sensitivity to DNase I, was however agreed by the two laboratories. This is an early replicating region housing the mouse X-inactivation centre (Xce). Failure to detect DNase I sensitivity in the pairing region of the mouse XY bivalent argues against the idea that sex chromosome pairing tips in mammals generally might assume an open conformation at meiosis as a prerequisite to obligatory crossing-over (Raman and Nanda, 1986; Chandley and McBeath, 1987). Crossing-over between DNA sequences at the telomeres of Xp and Yp in man (Rouyer *et al.*, 1986) and between Xq and Yq in the mouse (Keitges *et al.*, 1985; Avner *et al.*, 1986) have been demonstrated, and yet only in man do these pairing regions appear, from the nick translation experiments, to be sensitive to DNase I.

For autosomal cross-over sites along meiotic bivalents there was a suggestion in the study of Chandley and McBeath (1987) that points of chiasma formation at MI might also show label clustering. The data were not conclusive on this point however, and further more detailed investigation will be required to test whether cross-overs are localized to specific sites within DNA. A meiotic pairing model proposed by Chandley (1986) puts crossing-over specifically in segments of DNA showing an open conformation, i.e. pale G-band equivalent regions, where it is argued, effective pairing could preferentially be established. Pale bands along the chromosomes also tend to house genes which are transcriptionally active (Goldman *et al.*, 1984) and are early replicating. On this model, pairing initiation would depend on the crude matching of homologous pale bands rather than on strict DNA sequence similarity, although sequence similarity would be a subsequent requirement for reciprocal recombination. It is known that hotspots for recombination occur in the human genome and ends of arms where a pale G-band is almost always located, seem to be particularly prone to chiasma formation (Hulten, 1974).

The future seems bright for meiotic research. The use of probes and other tools of molecular biology has only just begun and, given the good cytogenetic database which has now been established, the field could be transformed over the next few decades. Already it is known that gene expression can vary both in mice and humans, according to whether transmission is through the male or female germ line, the basis for this perhaps lying in the different patterns of methylation parentally imprinted at spermatogenesis and oogenesis (Sapienza *et al.*, 1987; Swain *et al.*, 1987). The need for good and accurate analysis by molecular biologists will require the co-operation of classical cytogeneticists and cell biologists with an understanding of the gametogenic process and with expertise in the identification and handling of meiotic cells. The small band of mammalian meiotic experts could find themselves increasingly in demand over the next few years!

Acknowledgements

The author wishes to thank her colleagues Mr R.M. Speed and Mrs S. McBeath, for their continuing help and loyalty and without whose assistance the work described in this chapter would not have been possible.

References

Avner PH, Cambrou J, Hatat D, Bishop C (1986). X-Y common molecular probes mapping to the distal region of the mouse X chromosome detect recombination events between the X and Y chromosomes. Mouse News Letter 76; 42.

Bojko M (1983). Human meiosis. VIII. Chromosome pairing and formation of the synaptonemal complex in oocytes. Carlsberg Res Commun 48; 457–83.

Burgoyne PS, Baker T (1984). Meiotic pairing and gametogenic failure. In Controlling Events in Meiosis (eds

Evans CW, Dickinson HG). Co of Biologists, Cambridge; 349–62.

Chaganti RSK, Jhanwar SC, Antonarakis SE and Hayward WS (1985). Germ-line chromosomal localization of genes in chromosome 11p linkage: Parathyroid hormone, -globin, c-Ha-ras-l and insulin. Somat Cell Molec Genet 11; 197–202.

Chandley AC (1977). Karyotyping of infertile men. In Techniques of Human Andrology (ed. Hafez, ESE). Elsevier/North Holland Biomedical Press; 321–36.

Chandley AC (1979). Chromosomal basis of human infertility. Brit Med Bull 35; 181–6.

Chandley AC (1981). The origin of chromosomal aberrations in man and their potential for survival and reproduction in the adult human population. Ann Genet 24; 5–11.

Chandley AC (1986). A model for effective pairing and recombination at meiosis based on early replicating sites (R-bands) along chromosomes. Hum Genet 72; 50–7.

Chandley AC (1987). Meiotic analysis in germ cells of man and the mouse. In Mammalian Development: A practical approach (ed. Monk M)., IRL Press, Oxford, Washington DC; 71–91.

Chandley AC and McBeath S (1987). DNAse I hypersensitive sites along the XY bivalent at meiosis in man include the XpYp pairing region. Cytogenet Cell Genet 44; 22–31.

Chandley AC, Edmond P, Christie S, Gowans L., Fletcher J, Frackiewicz A, Newton M (1975). Cytogenetics and infertility in man. I. Karyotype and seminal analysis. Ann Hum Genetics 39; 231–54.

Chandley AC, Edmond P, Maclean N, Fletcher J, Watson GS (1976). Cytogenetics and infertility in man: Results of a five-year survey of men attending a subfertility clinic. Part II. Testicular histology and meiosis. Ann Hum Genet 40; 165–76.

Chandley AC, Goetz P, Hargreave TB, Joseph AM, Speed RM (1984). On the nature and extent of XY pairing at meiotic prophase in man. Cytogenet Cell Genet 38; 241–7.

Chandley AC, Gosden JR, Hargreave TB, Spowart G, Speed RM, McBeath S (1989). Deleted Yq in the sterile son of a man with a satellited Y chromosome (Yqs). J. Med. Genet. 27; (in press).

Chandley AC, Ambros P, McBeath S, Hargreave TB, Kilanowski F and Spowart G (1986a). Short arm dicentric Y chromosome with associated statural defects in a sterile man. Hum Genet 73; 350–3.

Chandley AC, Speed RM, McBeath S, Hargreave TB (1986b). A human 9;20 reciprocal translocation associated with male infertility analyzed at prophase and metaphase I of meiosis. Cytogenet Cell Genet 41; 145–53.

Chandley AC, McBeath S, Speed RM, Yorston L and Hargreave TB (1987). Pericentric inversion in human chromosome 1 and the risk for male sterility. J. Med Genet. 24: 325–34.

Coerdt W, Rehder H, Gausmann I, Johannisson R, Gropp A (1985). Quantitative histology of human fetal testes in chromosomal disease. Ped Pathol 3; 245–59.

Counce SJ, Meyer GF (1973). Differentiation of the synaptonemal complex and the kinetochore in *Locusta* spermatocytes studied by whole mount electron microscopy. Chromosoma 44; 231–53.

Eichenlaub-Ritter U, Chandley AC, Gosden RG (1986). Alterations of the microtubular cytoskeleton and increased disorder of chromosome alignment in spontaneously ovulated mouse oocytes aged *in vivo*: an immunofluorescent study. Chromosoma 94; 337–45.

Eichenlaub-Ritter U, Chandley AC, Gosden RG (1988). The CBA mouse as a model for age-related aneuploidy in man: studies of oocyte maturation, spindle formation and chromosome alignment during meiosis. Chromosoma 96; 220–6.

Evans EP, Breckon G, Ford CE (1964). An air-drying method for meiotic preparations from mammalian testes. Cytogenetics 3; 289–94.

Ford CE, Hamerton JL (1956). The chromosomes of man. Nature 178; 1020–3.

Forejt J (1974) Non-random association between a specific autosome and the X chromosome in meiosis of the male mouse: possible consequence of homologous centromeres separation. Cytogenet Cell Genet 13; 369–83.

Gabriel-Robez O, Ratomponirina C, Dutrillaux B, Carre-Pigeon F and Rumpler Y (1986a). Meiotic association between the XY chromosomes and the autosomal quadrivalent of a reciprocal translocation in two infertile men, 46,XY,t(19;22) and 46,XY,t(17;21). Cytogenet Cell Genet 43; 154–60.

Gabriel-Robez O, Ratomponirina C, Rumpler Y, Marec B Le, Luciani JM, Guichaoua MR (1986b) Synapsis and synaptic adjustment in an infertile human male heterozygous for a pericentric inversion in chromosome 1. Hum Genet 72; 148–52.

Gazit B, Cedar H, Lerer I, Voss R (1982) Active genes are sensitive to deoxyribonuclease I during metaphase. Science 217; 648–50.

Goldman MA, Holmquist GP, Gray MC, Caston LA, Nag A (1984). Replication timing of genes and middle repetitive sequences. Science 224; 686–92.

Goodfellow P, Darling S, Wolfe J (1985). The human Y chromosome. J Med Genet 22; 329–44.

Hale DW (1986). Heterosynapsis and suppression of chiasmata within heterozygous pericentric inversions of the Sitka deer mouse. Chromosoma 94; 425–32.

Hartung M, Devictor M, Codaccioni JL, Stahl A (1988). Yq deletion and failure of spermatogenesis. Ann Génét 31; 21–6.

Hassold TJ, Jacobs PA (1984). Trisomy in man. Ann Rev Genet 18; 69–97.
Hassold TJ, Chiu D (1985). Maternal age specific rates of numerical chromosome abnormalities with special reference to trisomy. Hum Genet 70; 11–17.
Hulten MA (1974). Chiasma distribution at diakinesis in the normal human male. Hereditas 76; 55–78.
Jhanwar SC, Burns JP, Alonso ML, Hew W, Chaganti RSK (1982). Mid-pachytene chromomere maps of human autosomes. Cytogenet Cell Genet 33; 240–8.
Jhanwar SC, Neel BG, Hayward WS, Chaganti RSK (1984). Localization of the cellular oncogenes ABL, SIS and FES on human germ-line chromosomes. Cytogenet Cell Genet 38; 73–5.
Johannisson R, Lohrs U, Wolff HH, Schwinger E (1987). Two different XY-quadrivalent associations and impairment of fertility in men. Cytogenet Cell Genet 45; 222–30.
Keitges E, Rivest M, Siniscalco M, Gartler SM (1985). X linkage of steroid sulphatase in the mouse is evidence for a function and Y-linked allele. Nature 315; 226–7.
Kerem B, Goitein R, Diamond G, Cedar H, Marcus M (1984). Mapping of DNAse I sensitive regions on mitotic chromosomes. Cell 38; 493–9.
Luciani JM, Guichaoua MR, Mattei A, Morazzani MR (1984). Pachytene analysis of a man with a 13q; 14q translocation and infertility. Behaviour of the trivalent and non-randon association with the sex vesicle. Cytogenet Cell Genet 38; 14–22.
Mahadevaiah S, Mittwoch U (1986). Synaptonemal complex analysis in spermatocytes and oocytes of tertiary trisomic Ts (5^{12})31H mice with male sterility. Cytogenet Cell Genet 41; 169–76.
Miklos GLG (1974). Sex chromosome pairing and male fertility. Cytogenet Cell Genet 13; 558–77.
Mitchell AR, Ambros P, McBeath S, Chandley AC (1986). Molecular hybridization to meiotic chromosomes in man reveals sequence arrangement on the No 9 chromosome and provides clues to the nature of 'parameres'. Cytogenet Cell Genet 41; 89–95.
Mittwoch U, Mahadevaiah S, Setterfield LA (1984). Chromosomal anomalies that cause male sterility in the mouse also reduce ovary size. Genet Res Camb 44; 219–24.
Monesi V (1967). Ribonucleic acid and protein synthesis during differentiation of male germ cells in the mouse. Arch Anat Microscop Morphol Experimentale 56; 61–74.
Moses MJ, Counce SJ, Paulson DF (1975). Synaptonemal complex complement of man in spreads of spermatocytes with details of the sex chromosome pair. Science 187; 363–5.
Page BM (1973). Identification of chromosome 9 in human male meiosis. Cytogenet Cell Genet 12; 254–63.
Racjan Separovic E, Chandley AC (1987). Lack of evidence that the XqYq pairing tips at meiosis in the mouse show hypersensitivity to DNAse I. Chromosoma 95; 290–4.
Raman R, Nanda I (1986). Mammalian sex chromosomes. I. Cytological changes in the chiasmatic sex chromosomes of the male musk shrew *Suncus murinus*. Chromosoma 93; 367–74.
Richler C, Uliel E, Kerem B-S, Wahrman J (1987). Regions of active chromatin conformation in 'inactive' male meiotic sex chromosomes of the mouse. Chromosoma 95; 167–70.
Rosenmann A, Wahrman J, Richler C, Voss R, Persitz A, Goldman B (1985). Meiotic association between the XY chromosomes and unpaired autosomal elements as a cause of human male sterility. Cytogenet Cell Genet 39; 19–29.
Rouyer F, Simmler M-C, Johnsson C, Vergnaud G, Cooke HJ, Weissenbach J (1986). A gradient of sex linkage in the pseudoautosomal region of the human sex chromosomes. Nature 319; 291–5.
Saadallah N, Hulten M (1986). EM investigations of surface spread synaptonemal complexes in a human male carrier of a pericentric inversion inv(13)(p12q14): the role of heterosynapsis for spermatocyte survival. Ann Hum Genet 50; 369–83.
Sapienza C, Petersen AC, Rossant J, Balling R (1987). Degree of methylation of transgenes is dependent on gamete of origin. Nature 328; 251–4.
Schempp W, Meer B (1983). Cytologic evidence for three human X-chromosomal segments escaping inactivation. Hum Genet 63; 171–4.
Schweizer D, Ambros P, Andrle M (1978). Modification of DAPI banding on human chromosomes by prestaining with a DNA binding oligopeptide antibiotic, distamycin A. Exptl Cell Res 111; 327–32.
Solari AJ (1980). Synaptonemal complexes and associated structures in microspread human spermatocytes. Chromosoma 81; 315–37.
Speed RM (1982). Meiosis in the foetal mouse ovary. 1. An analysis at the light microscope level using surface spreading. Chromosoma 85; 427–37.
Speed RM (1984). Meiotic configurations in female trisomy 21 foetuses. Hum Genet 66; 176–80.
Speed RM (1985). The prophase stages in human foetal oocytes studied by light and electron microscopy. Hum Genet 69; 69–75.
Speed RM (1986a). Oocyte development in XO foetuses of man and mouse: The possible role of heterologous X-chromosome pairing in germ cell survival. Chromosoma 94; 115–24.
Speed RM (1986b). Prophase pairing in a mosaic 18p ;iso 18q human female foetus studied by surface spreading. Hum Genet 72; 256–9.
Speed RM (1988). The possible role of meiotic pairing

anomalies in the atresia of human foetal oocytes. Hum Genet 78; 260–6.

Speed RM, Chandley AC (1983). Meiosis in the foetal mouse ovary. II. Oocyte development and age-related aneuploidy. Does a production line exist? Chromosoma 88; 184–9.

Sperling K, Marcus M (1984). Mapping of genetic activity on mammalian chromosomes. In Chromosomes Today (eds Bennett MD, Gropp A and Wolf U). George Allen and Unwin, Vol 8; 169–78.

Steinbach P, Fabry H, Scholz W (1979). Unstable ring Y chromosome in an aspermic male. Hum. Genet. 47; 227–31.

Swain JL, Stewart TA, Leder P (1987). Parental legacy determines methylation and expression of an autosomal transgene: A molecular mechanism for parental imprinting. Cell 50; 719–27.

Tiepolo L, Zuffardi O (1976). Localization of factors controlling spermatogenesis in the non-fluorescent portion of the human Y chromosome long arm. Hum Genet 34; 119–24.

Tres LL (1975). Nucleolar RNA synthesis of meiotic prophase spermatocytes in the human testis. Chromosoma 53; 141–51.

Wallace BMN, Hulten MA (1983). Triple chromosome synapsis in oocytes from a human foetus with trisomy 21. Ann Hum Genet 47; 271–6.

Weisbrod S (1982). Active chromatin. Nature 297; 289–95.

11 Meiotic Development Features Relevant to Prophase Specific Chromosomal Characteristics in *Saccharomyces cerevisiae*

SOUMITRA K. SEN, SAMPA DAS, AMIT BANERJEE, NALINI JASANI-GUPTA, MINAL SHAH AND SHYAMALI GHOSH

Plant Molecular and Cellular Genetics
Bose Institute, Centenary Building
P1/12, C.I.T. Scheme VII-M
Calcutta 700054, India

I. Introduction

The fact appears to be remarkably striking that although the meiotic machinery negotiates with varying orders of magnitude of nuclear DNA content in various forms of eukaryotes (Gaddipati and Sen, 1978), the number of genetic crossing over β remains fairly constant. The possibility that the basic mechanism by which the genetic material is transacted may be universal in all the different life forms of the eukaryotic world has earned support from the fact that certain similarities in the crucial aspects of meiosis have been recorded. This has led to the speculation that the functional aspects of the meiotic division and some of the properties of the components of the biological machinery which apparently mediate the function of meiotic division have been conserved in all eukaryotic systems.

Studies of molecular events that are believed to be relevant to a resultant reduction division and crossing over have been initiated in a variety of organisms. As the meiocyte is a highly specialized cell system which produces haploid progeny, the problems of identifying events relating strictly to meiotic special functions are often formidable. From varied systems it has been possible to identify a number of landmarks based upon which speculative schemes have been proposed in the past (Callan, 1967; King, 1970; von Wettstein, 1971; Sen, 1971; Stern and Hotta, 1978; Pukkila, 1977; Sen *et al.*, 1981; von Wettstein *et al.*, 1984) in search of answers to questions relating to meiotic development specialities.

II. The Rationale

The premeiotic DNA replication resulting in the doubling of the nuclear DNA is carried out in meiosis at a much slower pace (Kofman-Alfaro and Chandley, 1970; Benett *et al.*, 1971; Callan, 1972; Stern and Hotta, 1973; Lu and Jeng, 1975; Iyengar *et al.*, 1977) than the same replication in mitosis. The meiotic process thus turns out to be lengthy in comparison to mitosis. It has also been realized, chiefly from studies on lily meiocytes, that the nuclear DNA duplication is carried out on a differential management programme characterized by a small yet distinct fraction of genomic DNA (Zyg-DNA) being replicated and also repaired (P-DNA synthesis) through distinct mechanisms of management. Likewise, at the chromosomal organizational front a different and distinct picture emerges at more or less the same developmental phase. The 200–300 Å chromatin fibres organize into discrete loops (Nelson *et al.*, 1986) arranged in a linear manner. The chromatin loops aggregate into discrete foci and the association of adjacent foci results in the definition of the longitudinal axis of the meiotic chromosome and is coordinated with the appearance of the axial elements of the synaptonemal complex (Hamkalo *et al.*, 1981). The looped organization is maintained through homologous pairing (Sen, 1970a,b) and the maturation of the synaptonemal complexes. The mechanism by which the chromatin fibres (200–300 Å) organize into a brush-like formation at all early meiotic prophase cells is dependent (Sen, 1969a) upon elements which are likely to be present in those cells. The chemical nature of such elements modulating the nucleic acid is likely to be proteinaceous. Moreover, since the brush-like organization follows a specific, characteristic mechanism, uniform for a species—a highly specific order of such mechanism must be encoded in the nucleotide sequences of the genetic material to be organized and in the properties of the presumptive proteinaceous elements. Additionally, the organizational machinery of the chromatin is likely to be temporally activated and it very likely forms the integral parts of meiotic development (Sen, 1971) hand in hand with a score of elements including the formation of synaptonemal complexes.

There are enough indications that chromatins are functionally relevant to the central piece of meiotic function causing reduction division (Sen and Kundu, 1977) and genetic crossing over (Sen, 1969b). The former process is the prerequisite for alternation of generations in the sexual cycle of the eukaryotic world and the latter forms the source of genetic variation upon which selection operates in the biological world making evolution a dynamic process. These as well as many other developmental features are parts of the meiotic process. They may not be interrelated in a strictly cause and effect manner, yet in a holistic view they are all interdependent in ultimately deciding the outcome of the essentialities of meiotic division in the genetical sense. This rationality could be developed from the insight that has evolved from many experimental systems. However our insight into the mechanism of this important biological process has started to gain life more particularly through revelations of the intricacies of meiosis in lily meiocytes (Hotta and Stern, 1978; Hotta *et al.*, 1984, 1985; Stern and Hotta, 1973, 1978; Sen *et al.*, 1977; Bouchard and Stern, 1980). The tenacious exercises chiefly made by Herbert Stern and Yasuo Hotta of the Department of Biology, University of California, San Diego have met with rich dividends. However, all their results are relevant to the molecular events in meiotic recombination where it is pre-supposed that the meiotic chromosomes are organized in a manner to suit their scheme of operations. The aspect of the mechanism by which the chromosomal organization is carried out has often been left out of focus. Yet this seems to be vital and primarily the foundation (Sen *et al.*, 1981) upon which the rationale is built. We in our laboratory over the past decade or so have carried out some observations with regard to meiotic developmental features of baker's yeast, *Saccharomyces cerevisiae* which we believe are relevant to this main theme. The results so far achieved concern three main facets of meiotic development. They are, timing of meiotic DNA synthesis in the calendar of meiotic development, genetic control of meiosis and chromosomal proteins in association with superstructures of meiotic chromatin.

III. The Meiotic Course of Sporulating Yeast Cells

The mitotic cell cycle of yeasts has an initiation point in Gl phase, termed 'start' (Shilo *et al.*, 1978). Stationary phase cells freeze prior to start and may stay viable in this situation for long periods of time. This is the natural resting stage. When a diploid cell reaches start point, it has two alternatives—to remain in the stationary phase or to undergo mitosis or meiosis. Both the divisional processes are initiated at the same point. Initiation of meiosis is closely related to the parallel event of mitosis. There is indication that the initiation of meiosis occurs in Gl at or prior to the mitotic start (Shilo *et al.*, 1978). Under special circumstances cells that have initiated mitosis may continue as meiotic cells utilizing the early functions of mitosis like 'start'. Conversely, meiotic cells can revert to mitosis on transferring the cells to a rich medium before commitment to meiosis has occurred. There is little evidence as to the period over which commitment to meiotic development becomes irreversible. Patterns of meiosis in yeast can be affected by events occurring before the premeiotic DNA synthetic phase (Simchen *et al.*, 1972). Meiotic readiness is distinct from meiotic commitment. *Readiness* is a prolonged phase in which the cells are able to sporulate in water upon transfer from sporulation medium, but in a vegetative medium they revert to mitosis. A short stage follows this approximately at the time of meiotic prophase and is characterized by the inability of the cells to grow vegetatively or to sporulate when transferred to vegetative medium. Next comes *commitment* which is characterized by the ability of the cells to sporulate upon transfer to vegetative medium. Throughout the premeiotic DNA synthesis period, the cells retain their ability to revert to mitosis. Premeiotic DNA synthesis is characteristic and necessary for meiotic cells but not sufficient to commit the cells to meiosis (Simchen *et al.*, 1972). The meiotic cycle includes DNA replication (Croes, 1966), genetic recombination, first and second nuclear division and spore-wall formation (Moens and Rapport, 1971). Premeiotic DNA synthesis and commitment to intragenic recombination (gene conversion) are initiated before cells become committed to meiosis (Sherman and Roman, 1963; Simchen *et al.*, 1972). Commitment to intragenic and intergenic recombination occurs before commitment to meiotic chromosome disjunction. Separate cellular changes contribute to readiness and mark an irrevocable commitment to meiosis. In yeast, only after the initiation of recombination events do the cells get committed to meiotic division (Esposito and Esposito, 1974a).

Although sporulating yeast cells represent one of the simplest forms of eukaryotic cells undergoing meiosis, until recently (Williamson *et al.*, 1983) not much insight into the time course of events in the meiotic growth in yeast during sporulation could be gained. This was mainly due to the fact that the yeast cells enter into meiosis asynchronously as the cells do not respond at the same time to the meiotic initiation stimulus. Marking the time of appearance of a particular metabolite or initiation of a process is possible but its completion can never be marked because of the differential response of the cells making up the population. Besides asynchrony, not all cells actually undergo meiosis, many cells remain non-responsive to the sporulation stimulus. In fact, it has not been possible to induce synchronization in yeast meiotic development at an explicit level during sporulation. Nevertheless, advantages of the particular behaviour of certain strains during sporulation have been made use of by workers in the past (Simchen *et al.*, 1972; Esposito and

Esposito, 1974b; Williamson *et al.*, 1983) and useful information with regard to nuclear DNA metabolism in meiosis has been gained. Late premeiotic S-phase to early meiotic prophase which lasts between 2 and 3 h is the subject of this critical search for characteristics of meiotic development. As a matter of fact it has been seen in the most synchronous strain including that of Williamson *et al.* (1983) that sporulation is built gradually over a period of not less than 12 h between cells which respond first and those that respond finally to enter into meiotic regime. Understandably such a situation complicates the experimental requirement. Our experimental strategy in such a situation has been to use the first batch of cells undergoing meiosis as the experimental probe. The strategic use of the first batch of asci completing meiosis while studying the meiotic growth of yeast cells was first realized by Simchen *et al.* (1972). We have only extended the same strategy. Additionally we were able to recover the nuclei in a pure form from the ascospores of the first batch of asci which remain mixed with other yeast cells in the culture (Chatterjee *et al.*, 1982). The strain NJS-2-1 exclusively used in our study, behaved similarly to that of SK1 (Williamson *et al.*, 1983), although only 65 per cent of the cells sporulate (Figure 1) against 95 per cent in the case of SK1. This seems to be of no particular disadvantage. Accumulation of cells within the first half hour forming asci remained 5 per cent in both the strains. It was furthermore observed that when the growth conditions of sporulation regime of yeast cells are unchanged between experiments, the kinetics of meiotic growth remained constant. Additionally the size of yeast cells responding first to the sporulation stimulus remains fixed and consistent in the remaining cells that follow them in course of time to undergo meiosis. Hence the first batch of cells forming ascus does represent the rest of the cells which undergo meiosis. The timing of meiotic DNA synthesis and the analysis of genetic control of the meiotic developmental pathway have been monitored through the use of the first batch of ascospores. However the same experimental strategy could not be employed while searching for chromosomal proteins at various stages of meiotic growth. In this case, the growth was arrested at particular stages of development and chromatin analysed. Appearance of any protein (s) from the chromatin was taken to be the characteristics of the particular stage of growth under study.

Diploid sporulating strain NJS-2-1 constructed in our laboratory (Jasani, 1981) formed the main experimental material. It has the following genotype.

$$\frac{\text{a}}{\alpha}\quad \frac{\text{ade 1}}{\text{ade 1}}\quad \frac{\text{lys 2 +}}{\text{+ tyr 1}}$$

Additionally, sporulating temperature sensitive cell division defective (cdc) diploid strains, viz. cdc 7 known to have defect in DNA synthesis initiation (Hartwell, 1973); cdc 8 known to have defect in chain elongation during DNA replication (Hartwell, 1971); cdc 15, a late nuclear division defective (Culotti and Hartwell, 1971); and cdc 40 containing a lesion affecting essential functions in DNA synthesis and recombination function, possibly the repair synthesis (Kassir and Simchen, 1978), were utilized in meiotic developmental studies.

The following diploid strains containing double cdc mutants were also utilized in this study. The strains were constructed in our laboratory (Shah, 1981; Jasani, 1981).

$$\text{NJD S:}\quad \frac{\text{cdc 40}}{\text{cdc 40}}\quad \frac{\text{cdc 7}}{\text{cdc 7}}\quad \frac{\text{lys 2 +}}{\text{+ tyr 1}}\quad \frac{\text{a}}{\alpha}$$

$$\text{MSD-3:}\quad \frac{\text{cdc 40}}{\text{cdc 40}}\quad \frac{\text{cdc 15}}{\text{cdc 15}}\quad \frac{\text{lys 2 +}}{\text{+ tyr 1}}\quad \frac{\text{a}}{\alpha}$$

$$\text{NMD-1:}\quad \frac{\text{cdc 7}}{\text{cdc 7}}\quad \frac{\text{cdc 15}}{\text{cdc 15}}\quad \frac{\text{lys 2 +}}{\text{+ tyr 1}}\quad \frac{\text{a}}{\alpha}$$

$$\text{JSD-1:}\quad \frac{\text{cdc 8}}{\text{cdc 8}}\quad \frac{\text{cdc 15}}{\text{cdc 15}}\quad \frac{\text{lys 2 +}}{\text{+ tyr 1}}\quad \frac{\text{a}}{\alpha}$$

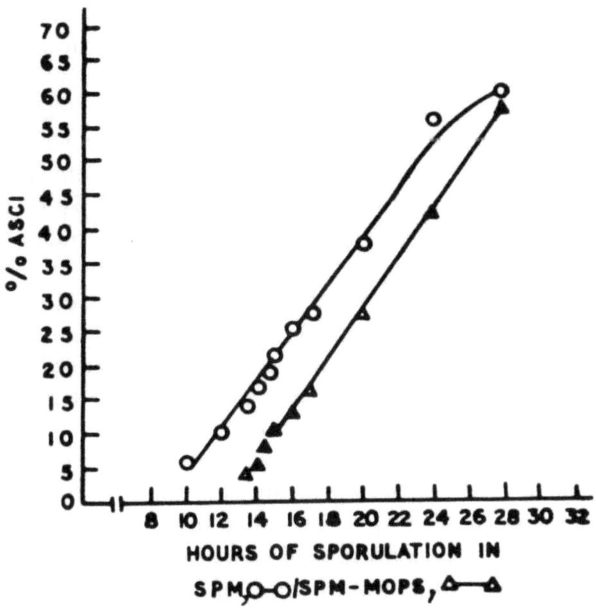

FIGURE 1: Sporulation kinetics of yeasts, strain NJS-2-1 showing increase in asci formation as the function of time in SPM and SPM-MOPS.

A diploid strain SNM-75 developed in this laboratory from cdc 4–1 and X2928–3D was also utilized. This is a temperature sensitive mutant which showed filamentous growth at restrictive temperature (30°C) and produced multispored asci.

The mitotic cultures were grown with constant shaking in YEPD (1% yeast extract; 2% peptone; 2% dextrose), and the pre-sporulation medium (YEPA) contained (1% yeast extract; 2% peptone and 1% potassium acetate). The media were supplemented with 40 mg/l adenine. The sporulation medium contained either 2 per cent potassium acetate alone (SPM) or SPM-MOPS consisting of 2 per cent potassium acetate buffered with 0.2M morpholinopropane sulphonic acid (MOPS) at pH 7.0. Growth and sporulations were carried out at 30°C (unless specified otherwise) following basically the conditions mentioned by Johnston *et al.* (1982).

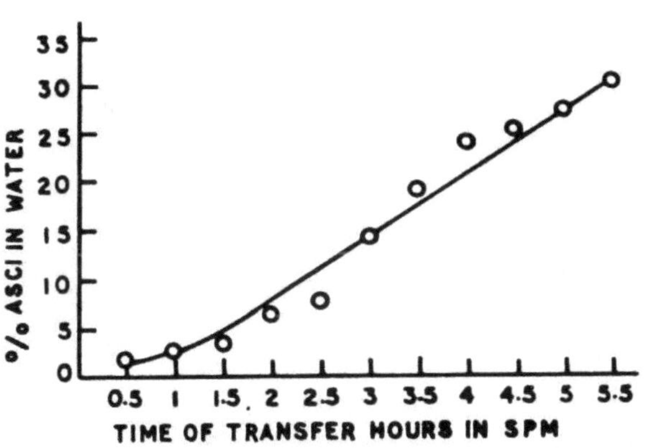

FIGURE 2: Readiness to meiotic growth. Transfers were made every 30 min. Cell density and frequency of asci were scored 24 h after the initiation of SPM growth. For more explanation see text.

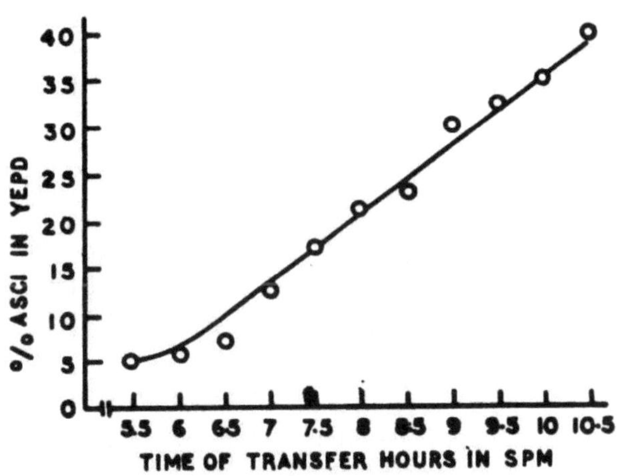

FIGURE 3: Commitment to meiosis. Transfers were made every 30 min. Cell density and frequency of asci were scored 24 h after the SPM growth was initiated. For more explanation, see text.

IV. Time of Meiotic DNA Synthesis in the Calendar of Meiotic Development

In yeast, monitoring the time of DNA synthesis during meiosis becomes difficult because of asynchrony. Additionally the other major constraint in studying DNA synthesis in yeast has been the fact that it does not incorporate thymidine (Grivell and Jackson, 1968) because of lack of the corresponding kinase enzyme. Moreover there is a rise in the pH of the medium as the sporulation progresses in a culture, which in turn reduces uptake from the external precursor pool (Mills, 1972). We overcame this difficulty to a reasonable extent by buffering the sporulation cultures with morpholinopropane sulphonic acid (MOPS) which probably is not metabolized by the cells (McCusker and Haber, 1977).

The sporulation kinetics in SPM and SPM-MOPS of the strain NJS-2-1 were found to follow a pattern as indicated in Figure 1. The asci formation could first be noticed at 9.5 h in SPM and at 12.5 h in case of SPM-MOPS. Within half an hour, i.e., at 10 h in SPM and at 13 h in SPM-MOPS, 5 per cent of the sporulating yeast cells formed asci. These cells follow essentially the same pattern of growth as the cells which succeed them (Figures 2 and 3). This has been evident from their state of readiness and commitment to meiosis in time lapse growth studies.

The first batch of asci were separated from the rest of the yeast cells (Chatterjee *et al.*, 1982). The ascospores of these asci (Figure 4) were further subjected to cell wall degrading zymolyase treatment (1 mg/g of cell) at 37°C for 2.5 h for ultimately isolating clean (Figure 5) nuclei (Banerjee, 1986).

Prelabelling of yeast cells was carried out by growing the cells in YEPA containing 1 μ Ci/ mmole ^{3}H adenine for 14 h before the cells attained 10^{7} cells/ml level prior to their exposure to sporulation medium.

Pulse labelling was carried out at every 30 min of growth interval during sporulation for a period of 30 min by exposure to 10 μ Ci/ml of ^{32}P-orthophosphoric acid (carrier free high sp. act.) in sporulation medium of 100 ml cultures. Pulses were terminated by adding 200-fold cold orthophosphoric acid and returned to radioactive label-free sporulation medium for continuing growth till 5 per cent of the cells formed asci.

The characteristics of DNA synthetic activity was revealed through ^{32}P uptake into the newly synthesizing DNA segments. From the results (Figures 6 and 7) it was evident that the distinct

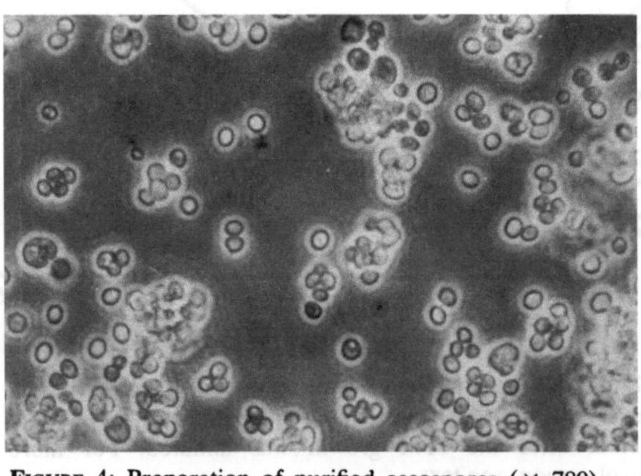

FIGURE 4: Preparation of purified ascospores (× 780).

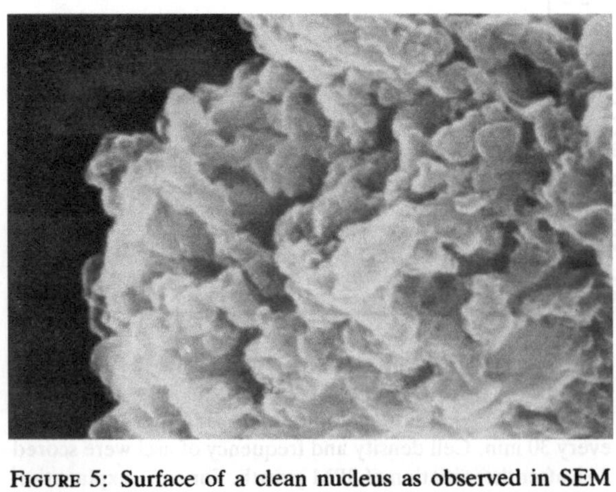

FIGURE 5: Surface of a clean nucleus as observed in SEM (× 22500).

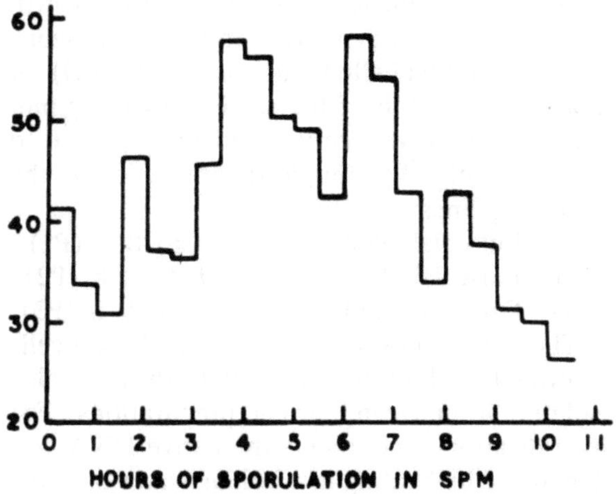

FIGURE 6: Rates of incorporation of ^{32}P-orthophosphoric acid into the nuclear DNA for the first batch of asci appearing in the sporulation culture. Values plotted along ordinate are ^{32}P, ^{3}H cpm ratios, showing incorporation of ^{32}P labelled phosphate into the newly synthesized DNA at each half-hour of pulse labelling in SPM.

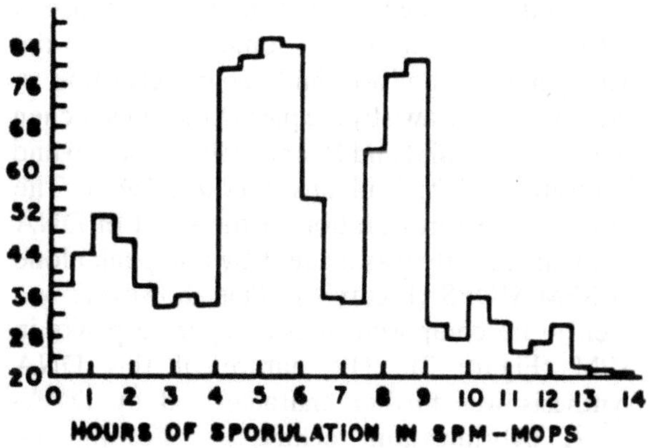

FIGURE 7: Rates of incorporation of ^{32}p-orthophosphoric acid into the nuclear DNA, as in Figure 8. Values are the ^{32}P, ^{3}H cpm ratios into the newly synthesized DNA at each half hour of pulse labelling in SPM-MOPS.

phases of active period of DNA synthesis took place during the growth regime. There were two distinct phases of DNA synthetic activity: one occurring at the early stages of the regime and the other somewhere at the mid-stage. The early stages of activity were confined to between 3 and 5.5 h in case of SPM (Figure 6) and 4 to 6.5 h in case of SPM-MOPS (Figure 7). The lengthening of the meiotic cycle in SPM-MOPS from the one found in SPM was related to the 1 h delay in initiation of the DNA synthetic phase. The timing of this phase of DNA synthesis coincided well with our microfluorometric estimation carried out separately. It was furthermore observed that the termination of the premeiotic S-phase occurred before the cells became committed to meiosis.

The timing of DNA replication in the present case had been assessed by pulse labelling the cells with ^{32}P. It is known that yeast cells have large stores of polyphosphates. There is likelihood that they will influence the pulse labelling of the cells and thus erroneously affect the chase-period with unlabelled phosphate. While we cannot suggest an answer for the possibility, it turned out that our estimation of premeiotic DNA synthesis data agree very well with all alternative schemes of procedures adopted by earlier workers (Simchen *et al.*, 1972; Esposito and Esposito, 1974b; Williamson *et al.*, 1983). The radioactivity that we witnessed in the later phase of the meiotic growth figures as the cornerstone of our observation. Induction of repair synthesis as the consequence of post radiation with γ-rays of meiotically growing yeast cells served as the positive control against our method of estimation. It was possible to assess differential labelling in the case of irradiated and nonirradiated yeast cells by the technique employed. Additionally, cdc 40, a temperature sensitive repair synthesis defective strain also indicated differential radioactive labelling through our method of estimation during meiotic growth at permissive as well as restrictive temperatures. The timing of this second phase of activity appeared to be sometime when the actual nuclear division had set in according to the calendars drawn by the past workers (Simchen *et al.*, 1972; Schild and Byers, 1978; Esposito and Esposito, 1974b; Williamson *et al.*, 1983). The distinct yet short duration of the spurt of DNA synthetic activity was evident by one peak alone in SPM-MOPS (Figure 8). This, however, was seen to be composed of two separate peaks in SPM (Figure 7). The nature of this DNA synthesis was further characterized by DNA-DNA hybridization technique following the basic rationale of Hotta and Stern (1971) for similar purpose. Nuclear DNA of the first batch of yeast cells undergoing meiosis in SPM-MOPS, pulse labelled at three phases of growth between 7.5 and 9 h, were individually hybridized with unlabelled nuclear DNA (PO) of yeast cells for 6 h of the same growth regime. Pulse labelling was carried out by exposure to 5 μCi/ml of ^{32}P-orthophosphoric acid on the following lines:

(i) Pulse labelled from 7.5 h to 8 h (P1)
(ii) Pulse labelled from 8.5 h to 9 h (P2)
(iii) Pulse labelled from 7.5 h to 9 h (P3)

The nuclear DNA of the unlabelled yeast cells growing till 6 h of meiotic growth (PO) was also isolated from clean nuclear preparations. For hybridization experiments the native DNA was denatured by heating in a boiling water bath for 10 min in 4×SSC followed by quick cooling.

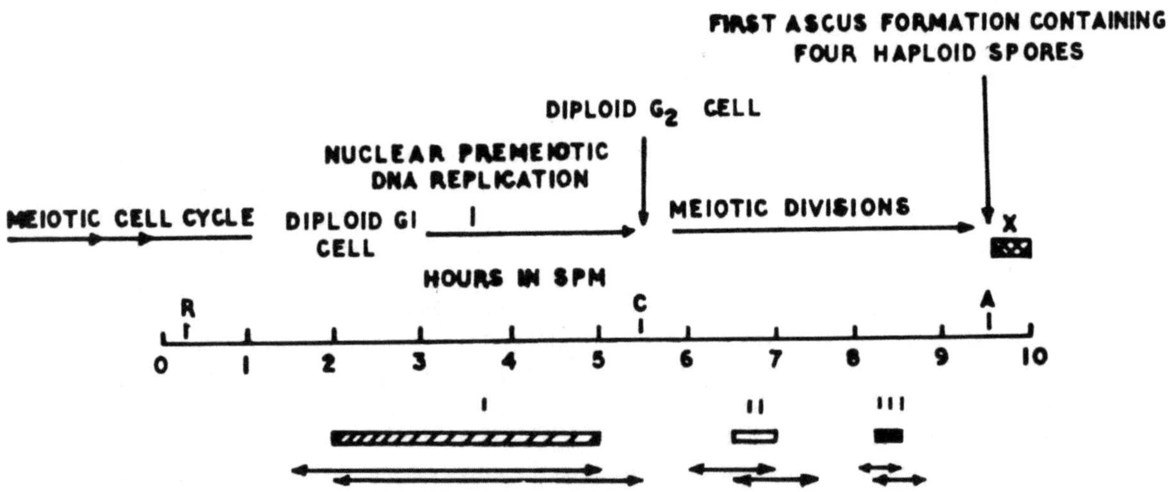

FIGURE 8: The schematic diagram indicates the calendar of development of the meiotic process in individual yeast cells of NJS-2-1. The process builds up on a developmental scale as the yeast cells are exposed to SPM and ends by the time ascus is formed. The block with striation within 2 to 5 h period indicates the period when premeiotic DNA synthesis takes place; the hollow block within 6 to 7 h indicates the time when delayed DNA synthesis for the duplication of the genome is completed; and the solid block within 8 to 9 h shows when the repair DNA synthesis takes place. These inferences are based on the findings of the present study through radioactive labelling. The arrows indicate that the time sequential radioactive pulse labelling of the asynchronous yeast cells allowed 30 min of additional incubation time for the generation of 5% ascus formation. Thus an allowance of 30 min was permitted in the experimental procedure both at the beginning and termination while determining the time specific stages when label was incorporated in the developing yeast cells. The possibility of such marginal influence was eliminated while charting the calendar through blocks, as explained above.

DNA solution was passed through 25 mm nitrocellulose filters in amounts of 100 μg (Denhardt, 1966). Hybridization was carried out with appropriately sheared DNA probes following the method adopted by Hotta and Stern (1971). The extent of hybridizations carried out on membrane filters was monitored in a liquid scintillation counter. It was expected that in the case of segments of DNA replicated during P1 and P2, hybridizability with P0 will show no difference when they are replicated in the same manner. On the other hand, DNA segments which may form delayed DNA synthesis of the genome replicated semiconservatively, will have a smaller possibility of complimentation than that of DNA segments replicated through repair replication. It was found that P1 had lower hybridizability with P0 than P2 in the comparable situation. The lower hybridizability of P1 was additionally evidenced by lower hybridizability of P3 compared to that of P2. This may quite possibly indicate that during P1 growth, DNA synthesis is predominantly semiconservative in nature. On the other hand P2 may quite likely be the stage when mainly repair replication took place. Thus the occurrence of semiconservative replication at the meiotic prophase forms the evidence for meiotic DNA synthesis.

Understandably in the experimental condition only a small fraction of the DNA contained in P0 qualitatively fulfilled the requirement of the experimental rationality of Hotta and Stern (1971) in DNA-DNA hybridization experiments. Nevertheless the experimental results determined through statistical analysis seem to indicate the scheme outlined above. Additionally it was also realized that the amount of DNA during the various phases of meiotic growth could not be directly compared with the total genomic DNA because of problems relating to quantification of ^{32}P labelling.

The calendar of development of the meiotic process can thus be drawn schematically (Figure 8).

V. Genetic Control of Meiotic Developmental Pathway

There exist two possible mechanisms for ordering a fixed sequence of cell cycle events relative to one another (Mitchison, 1974). First there may be a direct causal connection between one event and the next in which case it would be necessary for the earlier event to be completed before the later event could occur. In this case the 'product' of the earlier event provides the substrate for the later event or the completion of the earlier event might activate the other. This is the 'dependent pathway model'. A second possibility is that the events are ordered by the signals from some master timing mechanism. In this the completion of the earlier event is not necessary for the later event to occur, although the events would normally occur in the proper order due to the activity of the timer. This is the 'independent pathway model'. It was quite possible that the yeast meiotic cycle is controlled by a combination of the two mechanisms. We took advantage of the temperature sensitive cell division cycle (cdc) mutants already available in yeast to understand the genetic control of the meiotic events.

Cell division cycle mutants have been isolated in a number of organisms including *Escherichia coli* (Hirota *et al.*, 1968), *Chlamydomonus reinhardii* (Howell, 1974) and Chinese hamster cells (Roscoe *et al.*, 1973). Hartwell and others have isolated a number of temperature sensitive mutants in yeast which help in establishing the dependency relationship between various events of the cell cycle. These temperature sensitive cdc mutants have genes which specify a function which is required at a particular time to mediate one specific step alone in the cell cycle. The cdc mutants represent functions that are indispensible for the mitotic cell cycle. Interestingly, 13 out of 20 cdc genes that have been tested (Simchen, 1974) were found to be essential for meiosis and sporulation. These genes include mutants which show defects in plaque duplication, plaque separation, initiation of DNA synthesis, DNA synthesis, and medial and late nuclear division of the mitotic cell cycle. The stages of the gene function of these mutants in meiosis are, however, not known. Nevertheless it was realized that these are gene products which mediate functions both in mitosis and meiosis keeping in view the fact that there are other gene products besides these cdc mutants which function exclusively in meiosis (Baker *et al.*, 1976).

Techniques for sporulation kinetics, estimation for the time of readiness for sporulation and commitment to sporulation were essentially similar to Simchen *et al.*, (1972). Estimation of the execution points in meiosis of the temperature sensitive cdc mutants was carried out according to the methods of Esposito *et al.* (1970) and Simchen and Hirschberg (1977).

In order to define the various steps of meiotic development in relation to cdc genes, we offered exposure to restrictive temperature for a short period in a pulse manner. One hour exposure of such pulse treatment of restrictive temperature was given during the meiotic cycle and then chased till the completion of ascus formation. Samples from cultures growing in SPM at 23°C for cdc mutants were shifted at every hour of their sporulation regime to SPM maintained at a restrictive temperature, i.e., 34°C for cdc 40 and 36°C for cdc 7, 8 and 15. After one hour of the pulse treatment they were brought back to SPM at 23°C.

Additionally, pulse treatment of hydroxyurea (0.075 M) during the sporulation regime was carried out at 23°C for 1 h in the same manner as the temperature treatment. Schindler and Davies (1975) indicated that at 0.075 M concentration level of hydroxyurea (HU) the yeast nuclear DNA synthesis is depressed while allowing RNA and protein synthesis to continue at the normal rate. The effect of HU is reversible, so that on removal of the inhibitor, DNA synthesis resumes without lag (Slater, 1973; Schindler and Davies, 1975).

Meiotic intergenic recombination amongst the ascospores of the first batch of asci was carried out. The frequency of recombination was determined in the heterozygous genetic loci of lysine 2 and tyrosine 1 located on chromosome II of the linkage group. Random spore analysis was carried out for the estimation of recombination where the normality test was adopted for measuring the statistical significance of our observations.

The gene cdc 7 is known to play a role in the initiation of DNA synthesis in the mitotic course. It was revealed that its gene product is required twice during meiosis. First it is required halfway through the cell cycle and then later during the commitment period. It was also found to influence recombination. Any perturbation during the earlier phase of the cdc 7 gene function altered recombination frequency.

The DNA chain elongation gene cdc 8 is found to be active for a very long stretch of time in meiosis, extending to the commitment period. Perturbation during any stage of cdc 8 gene function alters the recombination frequency by increasing it. According to Stern and Hotta (1973) the formation of synaptonemal complex at meiotic prophase depends upon completion of the delayed replication. Since cdc 8 shows temperature sensitivity even up to commitment to meiosis, it is suspected that the cdc 8 gene product which is required for premeiotic DNA replication is also necessary for delayed replication in meiosis which is essential for synaptonemal complex formation. Hence, directly or indirectly, both cdc 7 and cdc 8 are required for meiotic DNA replication and recombination.

The cdc 40 gene product was also found to be required at the later phase of S-phase stretching to the period of commitment. It had been earlier realized that cdc 40 gene product is involved some how in repair replication (Kassir and

Simchen, 1978). Information that meiotic DNA synthesis, as that in lily and mouse spermatocytes, also takes place in yeast meiocytes was recorded (Das, 1980). Incidentally the timing of sensitivity of cdc 40 gene product in the meiotic course coincided with the time for meiotic DNA synthesis. The HU sensitivity stage for cdc 40 gene product was also found to be at the later part of S-phase. Temperature-shift experiments with cdc 40 and the effect of HU resulted in alterations in recombination frequency. Repair replication defective cdc 40 gene product was indicated to have a role in the process of recombination. It also provided evidence that recombination takes place at the stage of commitment to meiosis.

The function of gene cdc 15 commences after readiness and terminates prior to commitment to meiosis. Although the gene is active in the first half of the cycle, its effect is manifested during the late nuclear division stage. The present study indicates that the functioning of cdc 15 during the G1 stage is necessary for commitment of cells to meiosis. The crucial control events for the regulation of growth seems to reside in G1. Evidence has accumulated for the existence of a point in mid-to-late G1 at which time a cell decides whether to initiate DNA synthesis and undergo division or to cease growth. Similarly the functioning of cdc 15 during G1 indicates that the decision for meiotic commitment also lies in G1. Since cdc 15 plays some role in nuclear separation, we enquired whether it played any role in the timely separation of the synaptonemal complex elements of the meiotic chromosomes. We argued that if it did, then any disturbance in the cdc 15 gene function during the meiotic course would alter the recombination frequency. The recombination data reveal that this gene does not play any role in the process of crossing over at the chromosomal level.

The dependence of late G1 or early S-phase development on cdc 15, cdc 7 and cdc 40 gene products to make finally the cell committed to meiosis is interesting. The behaviour of a double mutant carrying cdc 40 and cdc 7 mutated gene products was judged in order to find interaction between them. It was observed that such a double mutant behaved varyingly at restrictive temperature and in the presence of HU. Gene products of DNA initiation defective cdc 7 and repair replication defective cdc 40 are expected to affect DNA replication, as they are both necessary in the genomic replication process. Their interference during premeiotic DNA synthesis and commitment periods showed no influence on recombination. Reasons for such an effect cannot readily be provided. However, it does imply that the gene products are functionally dependent on each other. On the other hand, a double mutant carrying cdc 40 and cdc 15 showed lack of interdependence at the functional level. That the total development programming of the meiotic build up is gradual and in co-ordination with many gene products, was evident from the stage specific functional role of the cdc gene products. It was interesting to note that cdc 15 gene product is synthesized sometime at the late phase of G1 to influence commitment. This was further made interesting through the fact of its dependence on cdc 7 gene product at S-phase (Shah, 1981). On the other hand, when cdc 15 was checked for its dependence on cdc 40 gene product, it was found they are independent of each other.

The cytokinesis and cell separation defective strain SNM-75 (Jasani, 1981) did not affect the meiotic programming. The multispored condition of the filamentous cells (Figures 9a,b) was derived as the result of the multinuclei entering into meiosis. It was found out that cell separation was not a prerequisite for meiotic nuclear division and ascospore formation.

Thus, it appears that, although mitotic gene function is necessary for meiotic development, the genetic control of such development is modulated in a different manner. The stage

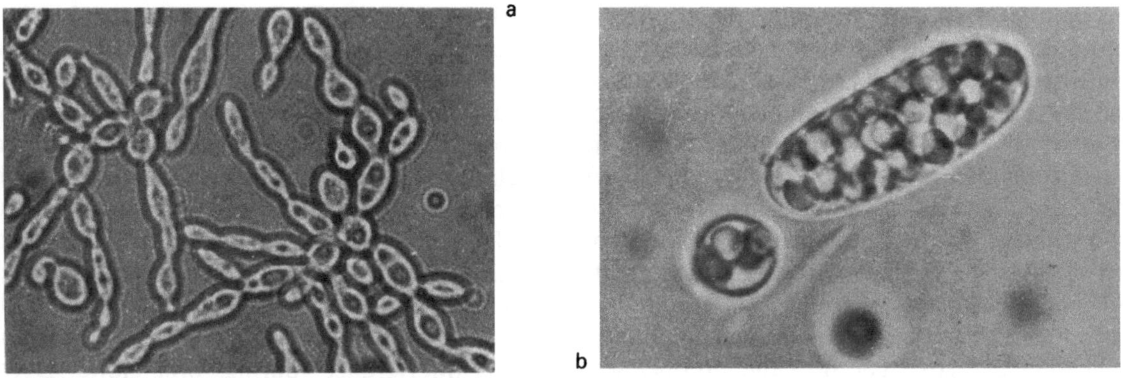

FIGURE 9: *a* Growth at 30°C of strain SNM-75.
b. Multispored ascus produced by a single cell of SNM-75 and a typically normal 4-spored ascus.

specificity of the functions (Figure 10) of such gene products essentially contributes towards the differential control of meiotic development.

VI. Meiotic Prophase-specific Chromatin Protein

The chromatin at pachytene contains histone variants in addition to the somatic histones (Kumaroo *et al.*, 1975; Shires *et al.*, 1976; Meistrich *et al.*, 1978; Scyedin and Kistler, 1980; Trostle-Weige, 1982). The functional significance of these histone variants is not clear but it is suspected that they somehow facilitate the meiotic events. It has been indicated (Rao and Rao, 1987) that histone-DNA interaction with special reference to H2B in the pachytene core particles is weaker than its somatic counterpart. Increased release of subnucleosomes to mononucleosomes upon micrococcal nuclease digestion of pachytene chromatin could also be the result of weak interaction because of variant TH 2 B. Additionally two major components of lateral elements of SC having 30 and 33 kd mol. wt. have been identified (Heyting *et al.*, 1987) from mammalian pachytene cells. Likewise a 110 kd phosphoprotein with DNA binding property has been located in the pachytene specific nuclear matrix (Behal *et al.*, 1987). The protein is speculated to be a putative component of SC (Ieradi *et al.*, 1983). This protein is likely to be conserved in mammals, chickens, insects and plants (Behal *et al.*, 1987). Meiotic stage specific appearances of proteins (Sobel, 1972) such as endonucleases, polynucleotide kinase, DNA ligase, R and U proteins and finally Rec A like m-Rec have already been documented (Stern and Hotta, 1980; Mather and Hotta, 1977; Hotta *et.al.*, 1985). In fact they have been found in plants and in animals, signifying their conserved nature. However some meiotic proteins seem to be species specific (Moens *et al.*, pers. comm.) and not conserved as the anti SC antibodies (Dresser, 1987).

Based on the speculative scheme of meiotic chromatin organization (Hamkalo *et al.*, 1981) we suspect, as earlier mentioned, that there are functional components which mediate the loop-like organization of the 200–300 Å fibres. They are quite likely non-histone chromosomal (NHC) proteins and their presence is specific for meiotic prophase chromatin (Banerjee *et al.*, 1983). We also believe that these elements appear as a normal developmental feature of

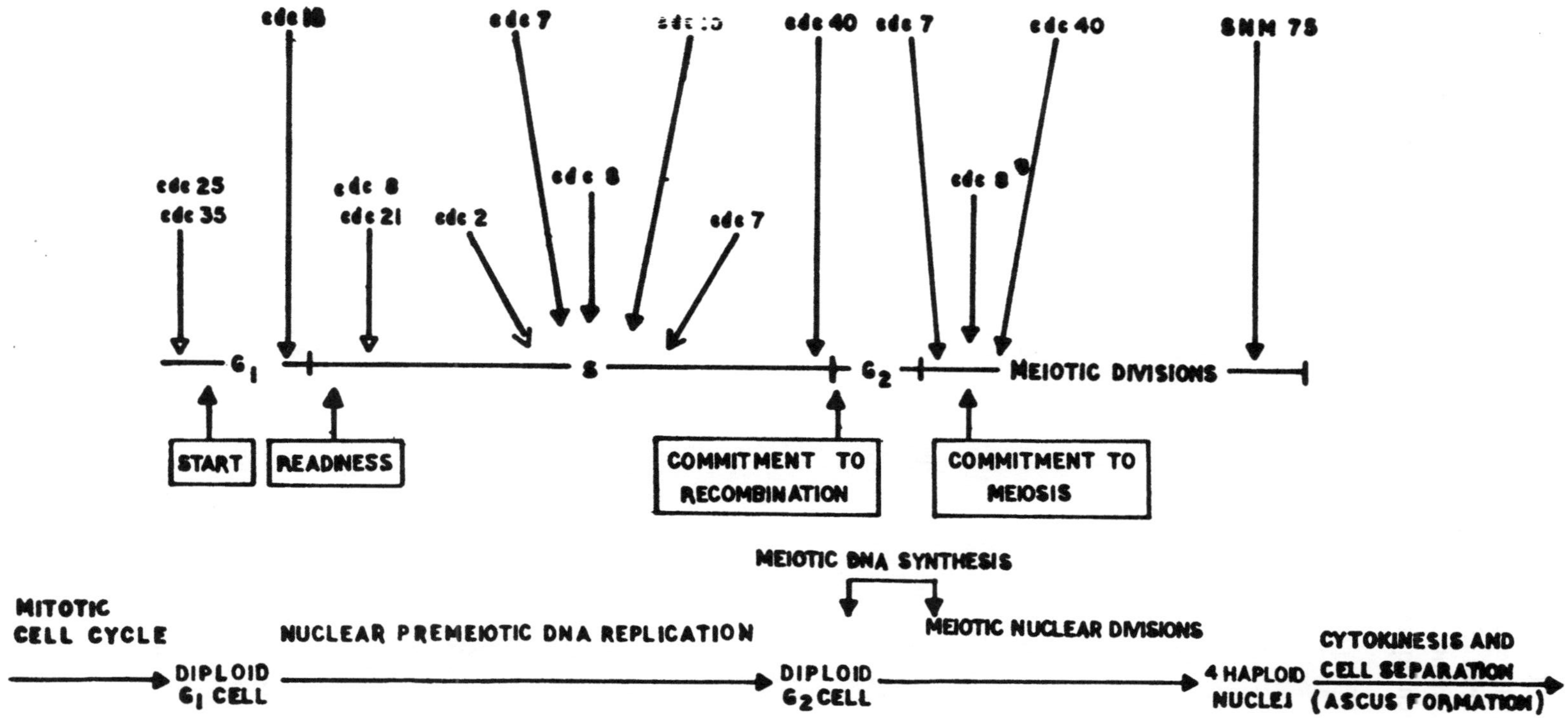

FIGURE 10: The state of knowledge on the genetic control of meiotic development of yeast cells is depicted. Liberty has been taken in using our observations (◄); observation of Shilo *et al.* 1978 (◄); and that of Schild & Byers, 1978 (←).

meiosis irrespective of SC formation. Hence, preservation of SC structure against 2 M NaCl treatment as already known (Solari, 1972) need not be valid for such elements. In our search for such NHC proteins, techniques of high resolution SDS polyacrylamide gel electrophoresis were adopted. The extraction of the chromatin proteins was carried out on the following rationality. It was earlier shown (Hadlaczky *et al.*, 1981) that extraction with HCl (0.25 N) and

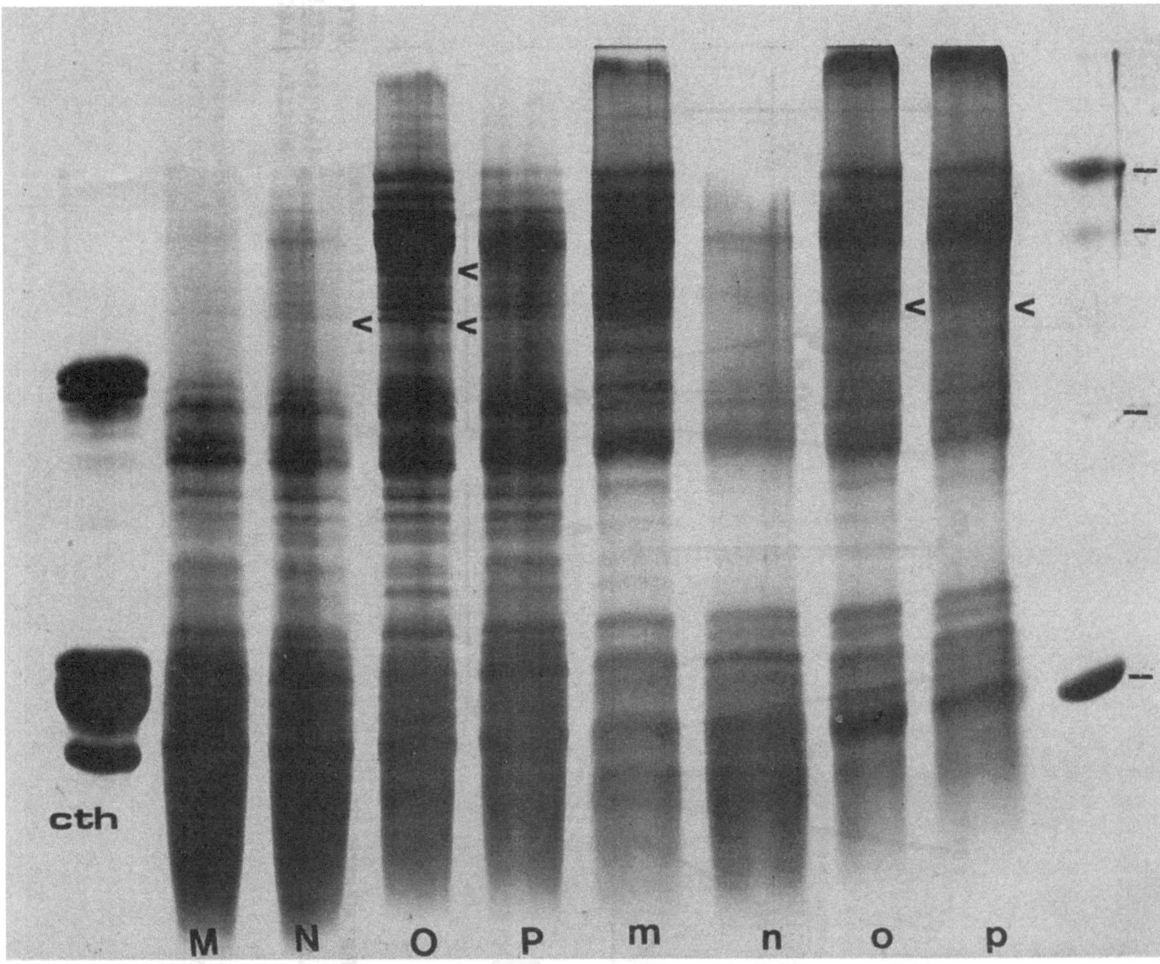

FIGURE 11: 0.25 N HCl and 0.6 M NaCl extracted chromatin protein patterns of NJS-2-1 cells grown for the following times in SPM.
M: HCl extraction of vegetative cell chromatin.
m: NaCl extraction of vegetative cell chromatin.
N: HCL extraction of 2, 3 h in SPM cell chromatin.
n: NaCl extraction of 2,3 h in SPM cell chromatin.
O: HCl extraction of 6,8 h in SPM cell chromatin.
o: NaCl extraction of 6,8 h in SPM cell chromatin.
P: HCl extraction of 8,9 h in SPM cell chromatin.
p: NaCl extraction of 8,9 h in SPM cell chromatin.
Molecular weight markers are indicated in Kilodaltons (68,45,25 and 12.5), < indicates the location of 40 kda and 32.5 kda proteins.

CTH: Calf thymus histones.

Coomassie Brilliant Blue R-250 staining.

NaCl (0.6 M) releases a significant amount of histones and that the extracted NHC proteins are very similar to those which are found in 2 M NaCl extraction of chromosomes. However the cross fibres of the central space of the SC as well as the lateral elements and central elements are preserved when treated with 2 M NaCl (Solari, 1972). Thus, treatment with 0.6 M NaCl for extraction of chromatin protein in the present case eliminates the possibility of inclusion of any SC proteins in our search.

Yeast chromatin at various phases of growth were monitored. One experimental line contained yeast chromatin at mitotic growth. The second line contained chromatin of cells which had been in sporulation regime for 6 to 8 h when the first batch of cells reached meiotic prophase. Thus the appearance of any specific kind of protein(s) in this line could be correlated with the attainment of meiotic prophase. In order to check that eventuality, two separate lines had been included to scan the chromatin at 2–3 h and 8–9 h growth during the sporulation regime of the same strain. Additionally chromatin of arrested growth stages of two cell division cycle mutants—cdc 15 and cdc 40—were also used. Cells of cdc 15 at restrictive temperatures cannot proceed beyond the first nuclear separation at mitotic growth, whereas cells of cdc 40 cease to grow beyond late meiotic prophase. Inclusion of these lines offered the opportunity of looking into the chromatin at specific stages of arrested growth.

Nuclei of all the cell lines were isolated and the chromatin prepared under identical conditions to maintain uniformity between the qualitative and quantitative virtues. Comparison of the NHC proteins of the above lines were carried out taking into consideration that factors like nuclear dense crescent and nuclear matrix proteins do not interfere in our evaluation. Proteins from the chromatin have been isolated by following the procedure noted below:

(A) The chromatin pellets were suspended in water and then kept on ice. To this an equal volume of ice-cold 0.5 N HCl was added to bring the final concentration of HCl to 0.25 N. The suspension was left on ice for 2 h with occasional stirring on a vortex mixer.

(B) 0.6 M NaCl extractions were made in 10 mM Tris HCl (pH 8.0), 1 mM $MgCl_2$ and 1 mM PMSF for 2 h on ice (Hadlaczky *et al.*, 1981).

(C) The particulate chromatin left after the above extractions were spun at 15,000 rpm at 4°C. These pellets were then dissociated in 2 per cent SDS at 25°C for about 2 h.

Protein extraction from the clean nuclei following treatment with HCl and NaCl, and also with SDS of the residues which could not be released with either HCl or NaCl, revealed that no evidence could be gathered for the presence of any meiosis specific proteins. However when a similar extraction procedure was followed with clean chromatin, the presence of two proteins could be identified in HCl extracted preparations, the appearance of which were specific to the 6–8 h stage of meiotic growth. They had MW of around 40 kd and 32.5 Kd. After marking the two proteins (40 kd and 32.5 kd) which appeared in 6/8 h growth in SPM only, we tried to follow their appearance and disappearance in the sporulation cycle. Figure 11 shows the results. In lanes N and O the presence of 32.5 Kd can be located, which however declines in P. In lane O we could also locate the 40 kd which did not appear in lane N and faded away in lane P. NaCl extracted proteins did show up these proteins as well. Since these proteins are extractable with 0.6 M NaCl, they are not believed to be part of the synaptonemal complex. Both the proteins could be extracted by 0.25 N HCl as well as 0.6 M NaCl (pH 8.0).

Furthermore it was realized that none of these proteins be enriched in the cdc 40 arrested cell lines. The experimental line of cdc 40 was arrested for 5 h during which the activity of proteinase could not be eliminated. Additionally it is always possible that cdc 40 gene activity had nothing to do with the relevant proteins.

Micrococcal nuclease is known to act on chromatin at the internucleosome DNA and when the digestion is prolonged it completely degrades the DNA in these regions. In the case of longer digestion the DNA of the nucleosome regions are attacked (Noll, 1974). On the other hand, the sensitivity of chromatin to DNase I depends on factors other than the sensitivity to micrococcal nuclease. DNase I nicks the two strands of both linker and core DNAs (Felsenfeld, 1978). It is known, however, that unlike chromatin from higher eukaryotes, transcribed yeast chromatin and total yeast chromatin are equally sensitive to DNase I digestion (Lohr and Hereford, 1979). We used these two nucleases in our study to digest the yeast chromatin at various orders of compaction in meiotic as well as mitotic cells. It was expected to reveal the nature of compaction of chromatin through digestion kinetics as well as the release of the associated NHC proteins of the superstructures of the chromatin. The NHC proteins associated with the superstructure have been characterized through SDS polyacrylamide gel electrophoresis. The materials left after the digestion were also characterized as they were expected (Goetsch and Byers, 1982) to contain the elements of axial core and the synaptonemal complex in an enriched way in the meiotic prophase chromatin.

Absorbance of fractions across the gradient of micrococcal nuclease digested chromatin was studied (Banerjee, 1986). It was seen that the chromatin from the vegetative cells and 6/8 h growth during meiosis had been digested releasing smaller fragments. On the other hand, chromatin of cdc 40 and cdc 15 after digestion released relatively larger fragments. In case of DNase I digestion, release of fragments was maximum in case of mitotic chromatin followed by 6/8 h growing meiotic chromatin. The contents of DNA (A_{260}) and protein (A_{280}) in the fractions across the sucrose gradient of micrococcal nuclease, as well as DNase I digested chromatin from mitotically growing cells, 6/8 h in SPM cells and cdc 40 arrested cells, were scanned. In all the cases digestion was carried out for a limited period of time in order to release only those proteins which were responsible for the maintenance of the superstructure of the chromatin.

Figure 12a shows the released proteins where the near absence of the core histones is notable. By DNase I digestion, the release of a number of unique proteins (lane n) in the chromatin of 6/8 h SPM could be documented. At least one amongst these proteins (84 kd) could also be released by micrococcal nuclease digestion. These were specific to 6/8 h growth conditions as they were absent in mitotically growing yeast cell chromatin. The 40 kd and 32.5 kd proteins earlier identified, however, could not be found in such preparations. Interestingly the 32.5 kd protein could be found in the chromatin residues after digestion with micrococcal nuclease and DNase I (Figure 12b). As the 32.5 kd protein can only be extracted by 0.6 M NaCl, its location could quite likely be on the axial elements. The same protein could also be found in the cdc 40 arrested cultures, although no enrichment could be noticed. It is apparent from the preparation that core histones are present, signifying that the residues do contain chromatin fibres (200–300 Å) that associate with the synaptonemal complex in lanes, N, O and n and o. DNase I digested residues showed up another protein of 25 kd in 6/8 h growth stages.

Figure 13 shows the chromatin protein extracted by HCl from the micrococcal nuclease and DNase I digestion residues. Micrococcal nuclease treated chromatin at 6/8 h of meiotic growth revealed (lane P) the presence of at least seven polypeptides not present in mitotic chromatin (lane O). The MW of the polypeptides was estimated to be 138, 119, 113, 96, 47.5, 39 and 37 kd. Except for the 37 kd protein all others

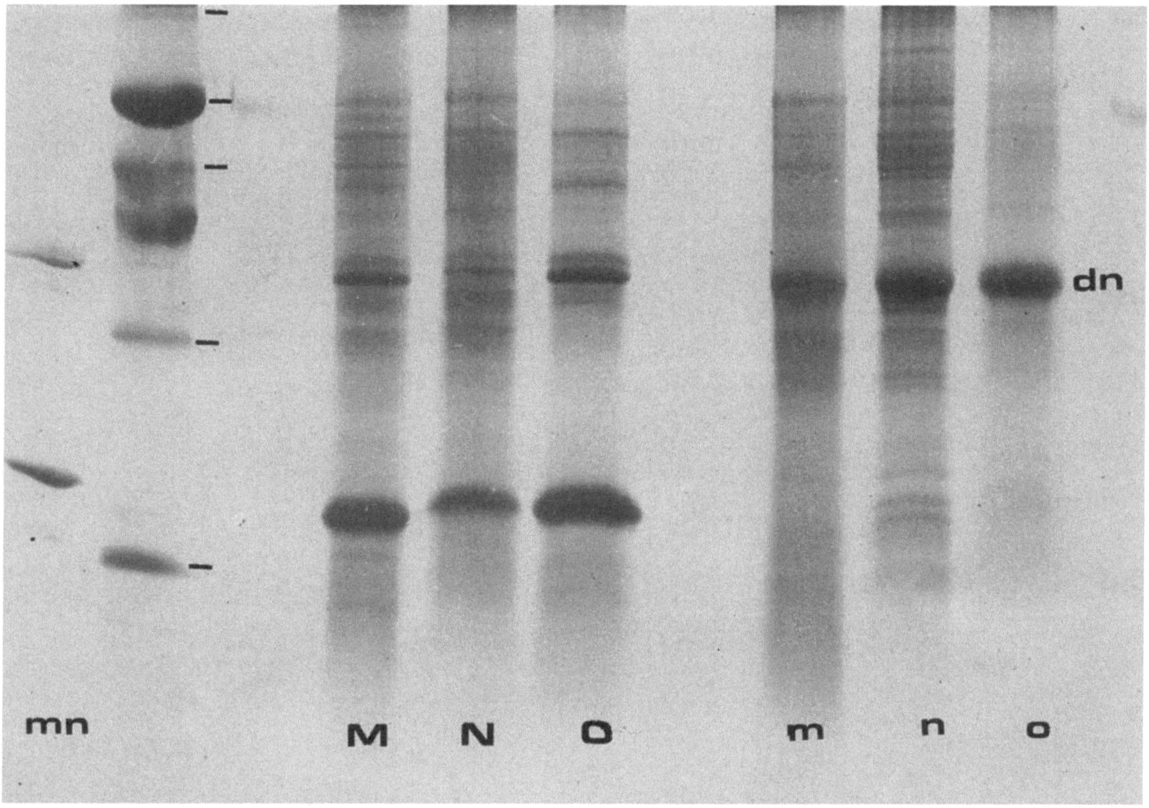

FIGURE 12a: Released proteins after micrococcal nuclease and DNase I digestions.
M: Micrococcal nuclease digestion of NJS-2-1 vegetative cell line chromatin.
N: Micrococcal nuclease digestion of 6,8 h (in SPM) of NJS-2-1 cell chromatin.
O: Micrococcal nuclease digestion of cdc 40 blocked cell line chromatin.
m: DNase I digestion of NJS-2-1 vegetative cell line chromatin
n: DNase I digestion of 6,8 h (in SPM) of NJS-2-1 cell chromatin.
o: DNase I digestion of cdc 40 blocked cell line chromatin molecular weight markers (2nd lane from left) are shown in Kilodaltons (158, 68, 45, 25, 12.5).
mn: Polypeptide band due to micrococcal nuclease, also present in lanes M, N, O.
dn: Polypeptide band due to DNase I in lanes m,n,o Coomassie Brilliant Blue R-250 staining.

could also be seen in case of cdc 40 chromatin (Lane Q). Three of the high molecular weight proteins could also be seen in cdc 40 chromatin (lane T) digested with DNase I residues. The seven polypeptides, as mentioned above, have the likelihood of being located on the axial elements of the synaptonemal complex. Their absence in lane S might have been due to the differential digestive properties of micrococcal nuclease and DNase I.

In the case of digestion of the chromatin samples with micrococcal nuclease, the added nuclease did not sediment with the chromatin. Micrococcal nuclease binds tightly to chromatin

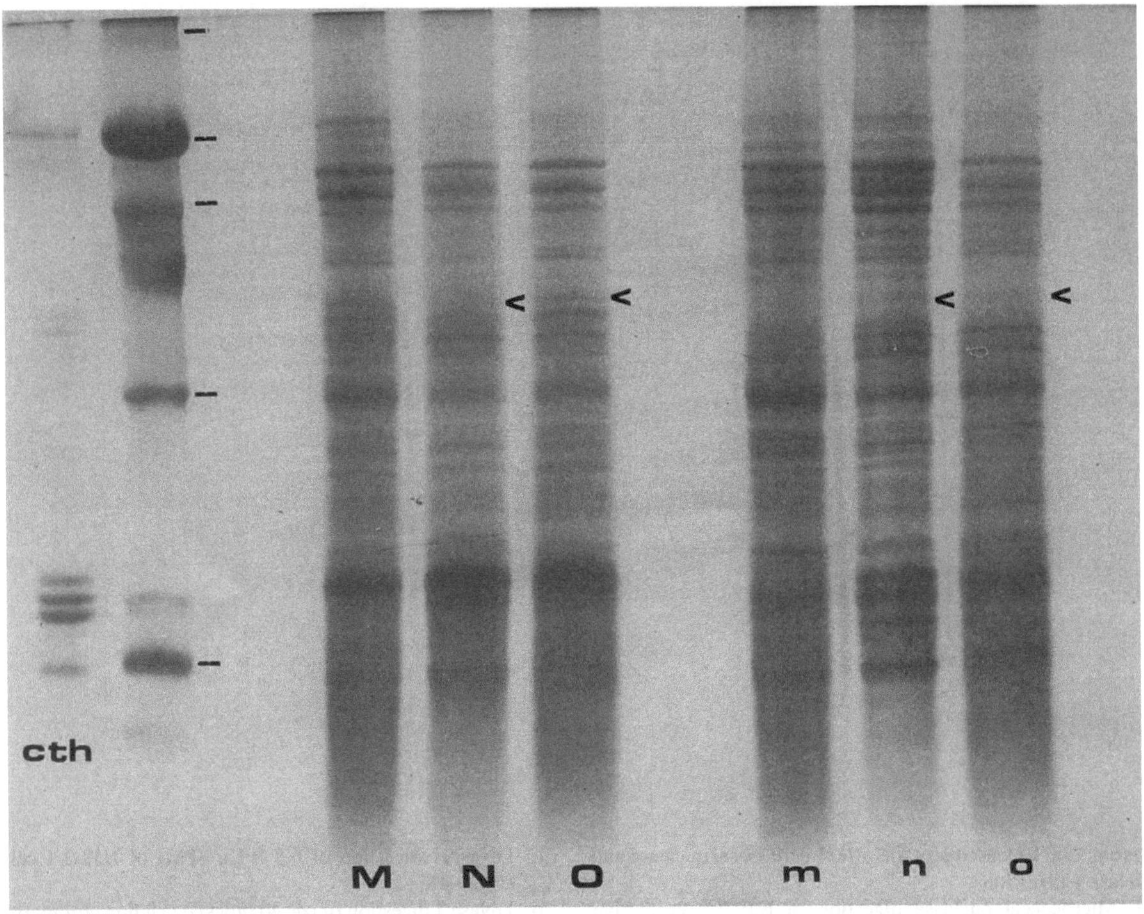

FIGURE 12b: Proteins extracted by HCl from the micrococcal nuclease and DNase I digestion residues of chromatins. Lanes have the same chromatin materials as for Figure 12a. Molecular weight markers are shown in Kilodaltons. < indicates the presence of 32.5 kDal protein Coomassie Brilliant Blue R-250 staining.

in the absence of Ca^{2+} at low ionic strength (Levinger and Carter, 1979). This is why our digestion buffer contained 0.5 mM $CaCl_2$ and 50 mM NaCl at a pH of 7.7. The protein bands due to free nuclease can be seen in Figure 12a (released proteins) and not in Figure 12b (residual structures). The concentration dependence on digestion rates represents a qualitative biochemical assay for states of higher order chromatin superstructures. This had come out very nicely in the digestion kinetics of mitotic cell chromatin, meiotic prophase (6/8 h in SPM) chromatin, cdc 40 arrested cell line chromatin and the most inaccessible cdc 15 blocked cell line. As the superstructures had been attacked differentially, we expected that only the NHC

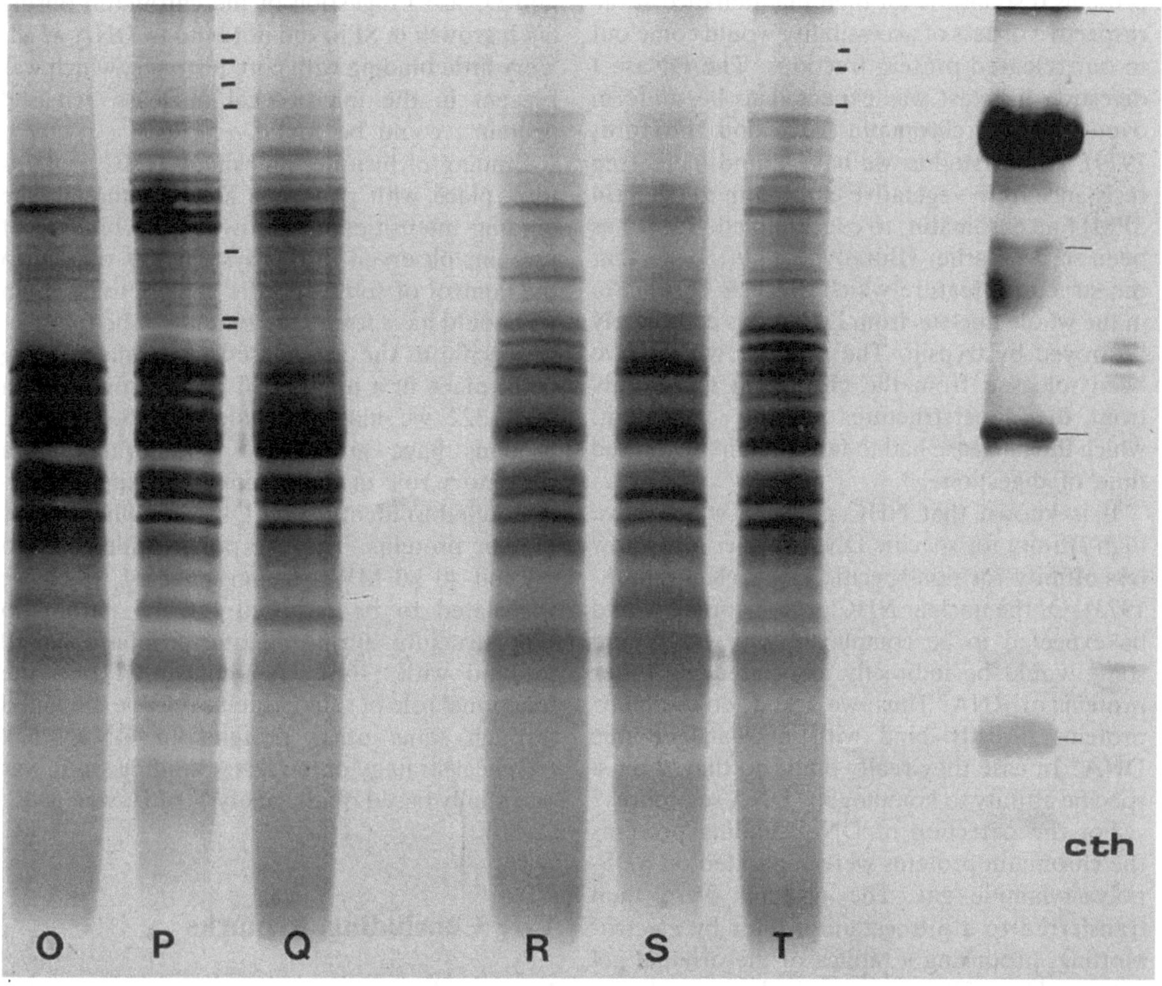

FIGURE 13: Proteins extracted by HCl from the micrococcal nuclease and DNase I digestion residues of chromatin, on silver staining.

Lanes have the following materials:

O: Micrococcal nuclease digestion of NJS-2-1 vegetative cell line chromatin.

P: Micrococcal nuclease digestion of 6,8 h (in SPM) of NJS-2-1 cell chromatin.

Q: Micrococcal nuclease digestion of cdc 40 blocked cell line chromatin.

R: DNase I digestion of NJS-2-1 vegetative cell line chromatin

S: DNase I digestion of 6,8 h (in SPM) of NJS-2-1 cell chromatin

T: DNase I digestion of cdc 40 blocked cell line chromatin.

Molecular weight markers (2nd lane from right) are shown in Kilodaltons

CTH: Calf thymus histones

proteins responsible for their maintenance at the respective orders of accessibility would come out in our released protein fractions. The DNase I digestion in yeast was expected to be uniform over the entire chromatin (Lohr and Hereford, 1979). In our studies we have found increasing resistance from vegetative chromatin to 6/8 h (in SPM) line chromatin, to cdc 40, to cdc 15. It has been shown earlier (Burgoyne *et al.*, 1976) that the structural feature which protects chromatin in the whole nucleus from DNase I is completely destroyed by trypsin. The proteins which have been released from the chromatin come only from the superstructures of the chromatin, which the nuclease had to face first in the limited time of digestion.

It is known that NHC proteins which have high affinity for specific DNA species also show less affinity for non-specific DNA (Klinesmith, 1973). Of the nuclear NHC proteins some would be expected to be complexed with DNA and some would be indirectly associated via other proteins or RNA. Thus, we asked, do the NHC proteins directly bind with nuclear genomic DNA? In case they really bind, do they show a specific affinity to homologous DNA sequences?

For the detection of DNA binding proteins the chromatin proteins were separated on SDS-polyacrylamide gel. The proteins were then transferred to a nitrocellulose filter by electroblotting, producing a replica of the original gel pattern. These filters were probed using radioactively labelled DNA produced by nick translation of pBR 322 carrying loads of yeast nuclear DNA ranging from 1 Kb to 5 Kb (Bowen *et al.*, 1980).

Unique sequences of yeast nuclear DNA ranging from 1.2 to 4.9 kb generated by Pst I digestion were cloned into pBR 322 following the standard methodology. Proteins of micrococcal nuclease and DNase I digested chromatin were separated in gel. The most striking feature to emerge from these experiments was that the released proteins from the micrococcal nuclease and DNase I digestion of the chromatin during 6/8 h growth in SPM did not bind to DNA at all. Very little binding with core histones, which was present in the micrococcal nuclease released proteins, could be found.

Binding of histones and many NHC proteins took place with pBR 322 DNA, although the binding intensities were low. If the binding of proteins observed in this system was related to the control of some specific gene activity, then we should have found the binding to be relatively specific to the DNA used. As some binding takes place in a number of NHC proteins with pBR 322 we may conclude that these NHC proteins have more of a structural than a regulatory role in the chromatin. Experimentation failed to identify any DNA binding meiosis specific proteins. The two polypeptides of 32.5 kd and 40 kd MW, the presence of which we speculated to be required for the chromatin superstructure during meiotic prophase, failed to bind with DNA. We concluded that the functional role of such proteins may be mediated through some other proteins or RNA. The technical strategy of the DNA binding study was essentially based on the analysis of Bowen *et al.*, (1980).

VII. Concluding Remarks

Functional regulation of meiotic processes has remained largely inaccessible to experimental investigation. The remark made by Taylor (1974) still remains valid. He opined that '... meiosis is still a potential battleground where dead hypotheses litter the field or rest uneasily in shallow graves, ready to emerge and haunt any conscientious scientist who tries to consolidate a victory for any particular thesis'. We are conscious of the fact that the frame of reference of this article is too broad and that none of our observations is likely to put to rest any of the old ghosts. Nevertheless, we intend to attract the attention of the reader to the fact that sufficient progress has been made in the cell biology of meiosis to point out the direction in which future studies must proceed in order to unravel the grand problems of the regulation of the meiotic process in causal terms. The present article has not been intended as a review but rather a collection of our thoughts on the subject, which we trust may prove useful.

To reach an understanding of the regulation of chromosome behaviour at meiosis, it is necessary to know how the internal organization of the chromosomes changes in meiosis through the intervention of gene products, the purely autonomous determination of chromosomal characters. The ultimate expression of any such character is therefore the product of both forms of regulation and their interaction. Our principal findings in this series of studies are indicated below. They may not be at all original in their nature of reference, nevertheless they appear to build a sense of rationality.

The more we are able to decipher intricate details of meiotic characteristics of development from diverse biological systems, the more are we getting convinced that the meiotic developmental course is controlled basically in the same way in all systems. This commonness is perhaps conserved in all eukaryotes in the course of evolution. Since the function of meiotic nuclear division remains the same in all biological systems, the common features are expected to carry out common functional roles. Our evidence for occurrence of meiotic DNA synthesis in a simple eukaryote like baker's yeast suggests that the lessons that we learnt from DNA metabolism in the lily may hold true in the case of yeast as well. Such a specialized control of DNA metabolism has been considered to be essential for reduction division (Sen and Kundu, 1977) as well as genetic recombination (Sen, 1969b). The role of gene products of cdc mutants indicate that the meiotic machinery is to a large extent dependent upon the differential control of already known genes (Simchen, 1974, 1978). Such control leads to differential results. Prolonged premeiotic DNA synthesis and completion of genomic DNA duplication at meiotic prophase may form examples of such a result. Prime factors that trigger off the course of things to happen in meiosis quite likely reside outside the target cells. Once the cells receive the signal and get committed, the developments follow, usually one after the other. Often we do find cases of disturbances. The developmental calendar does contribute towards such a rationality.

The high efficiency of the control mechanism of meiotic development is additionally dependent upon certain meiosis specific gene products. The present evidence that some NHC proteins do possibly function exclusively during meiosis indicates that there are requirements of meiosis specific functions to be carried out. Such specialized functions are mediated through elements that are developed *de novo* in the meiotic cells, obviously controlled by the same string of regulatory machinery of meiosis. Such meiosis specific functional components have been known to be present. The NHC proteins in the present case may play a role in the meiotic prophase specific organization (Hamkalo *et al.*, 1981) of the chromatin fibres. These proteins are contemplated to be not a part of the SC material though they may help in association of SC to the chromatid.

Thus the phenomenon of meiotic development prove to be an ideal case to understand how the factors, some of which otherwise carry out routine functions in mitosis, co-ordinate to bring about a remarkably different outcome in meiosis. Clearly, the final product is a result of switching on of such genes to function specially only once in the life cycle. Once we know the nature of these gene products during meiotic prophase development, the key to the understanding of one of the most remarkable and beautiful life processes will be within our reach.

Acknowledgements

The authors are thankful to Dr S.C. Hawthorne, Dr L.H. Hartwell, Dr G. Simchen, Dr P.B. Moens, Dr D.H. Williamson and Dr. D. von Wettstein for providing strains of yeast as and when required and also for helpful discussions. The association of Dr P.K. Srivastava and Dr A. Chatterjee at certain stages of this study is acknowledged. Technical assistance from Bishnupada Pradhan and Aloke Chatterjee during the work is thankfully recorded. The help rendered by Dr Sudhamoy Ghosh and Sumantra Sen during the preparation of the MS is acknowledged gratefully.

References

Baker BS, Carpenter ATC, Esposito MS, Esposito RE, Sandler L (1976). Genetic control of meiosis. Ann Rev Genetics 10; 53.

Banerjee A (1986). Characterisation of chromatin proteins in meiotic prophase. Ph. D. Thesis, University of Calcutta.

Banerjee A, Chatterjee A, Das S, Sen SK (1983). Role of histone and nonhistone chromatin in meiotic chromatin organisation. 15th Int Cong Genetics (Abstract) pp 350.

Behal A. Prakash K, Rao MRS (1987). Identification of a meiotic prophase specific nuclear matrix protein in the rat. J Biol Chem 262; 10898.

Bennett MD, Chapman VC, Riley R (1971). The duration of meiosis in pollen mother cells of wheat, rye and triticale. Proc Roy Soc Lond B, 178; 259.

Bouchard RA, Stern H (1980). DNA synthesised at pachytene in *Lilium*: A non divergent subclass of moderately repetitive sequences. Chromosoma 81; 349.

Bowen B, Steinberg J, Laemmli UK, Weintraub H (1980). The detection of DNA binding proteins by blotting. Nucl Acids Res 8; 1.

Burgoyne LA, Mobbs J, Marshall A (1976). Nucleic Acid Res 3; 3293.

Burgoyne LA, Marshall AJ, Mobbs J (1977). Proc Aust Biochem Soc 10; 65.

Callan HG (1967). The organisation of genetic units. J Cell Sci 2; 1.

Callan HG (1972). Replication of DNA in the chromosomes of eukaryotes. Proc Roy Soc Lond B 181; 19.

Chatterjee A, Ghosh S, Das S, Sen SK (1982). Isolation of yeast ascospores free of vegetative cell contamination. J Gen Microbiol 128; 2725.

Croes AF (1966). Duplication of DNA during meiosis in Baker's yeast. Exp Cell Res 41; 452.

Culotti J. Hartwell LH (1971). Genetic control of the cell division cycle in yeast III. Seven genes controlling nuclear division. Exptl Cell Res 67; 389.

Das S (1980). Cell biological aspects of sporulation in *Saccharomyces cerevisiae* : Calendar of meiotic developmental events relevant to genetic crossingover. Ph.D. Thesis, University of Calcutta.

Denhardt DT (1966). A membrane filter technique for the detection of complementary DNA. Biochem Biophys Res Comm 23; 641.

Dresser ME (1987). The synaptonemal complex and meiosis: An immunocytochemical approach. In meiosis (ed. P.B. Mones, Academic Press Inc., New York.

Esposito MS, Esposito RE, Arnaud M, Halvorson HO (1970). Conditional mutants of meiosis in yeast. J Bact 104; 202.

Esposito RE, Esposito MS (1974a). Genetic recombination and commitment to meiosis. Proc Natl Acad Sci USA 71; 3172.

Esposito MS, Esposito RE (1974b). Genes controlling meiosis and spore formation in yeast. Genetics 78; 215.

Felsenfeld G (1978). Chromatin. Nature 271; 115.

Fogel S, Mortimer RK (1971). Recombination in yeast. Ann Rev Genet 5; 219.

Gaddipati JP, Sen SK (1978). DNA replication studies in in genus *Vicia* through fibre autoradiography J Cell Sci 29; 85.

Goetsch L, Byers B (1982). Meiotic cytology of *Saccharomyces cerevisiae* in protoplast lysates. Mol Gen Genet 187; 54.

Grivell AR, Jackson JF (1968). Thymidine kinase: Evidence for its absence from *Neurospora crassa* and some other microorganisms and the relevance of this to the specific labelling of Deoxyribonucleic Acid. J Gen Microbiol 54; 307,

Hadlaczky G, Sumner AT, Ross A (1981). Protein depleted chromosomes I. Structure of isolated protein depleted chromosomes. Chromosoma 81; 537.

Hamkalo BA, Goldsmith MR, Rattner JB (1981). In Proceedings of the Second International Congress on Cell Biology, Berlin, (ed. HG Schweiger), Springer Verlag, pp 152.

Hartwell LH (1971). Genetic control of the cell division cycle in yeast II Genes controlling DNA replication and its initiation. J Mol Biol 59; 183.

Hartwell LH (1973). Three additional genes required for deoxyribonucleic acid synthesis in *Saccharomyces cerevisiae*. J Bact 115; 966.

Heyting Christa, Moens, Peter B, Raamsdonk, Willem Van, Dietrich Axel JJ, Vink, Agnes CG (1987). Identification of two major components of the lateral elements of synaptonemal complexes of the rat. Eur J Cell Biol 43; 148.

Hirota Y Ryter A, Jacob F (1968). Thermosensitive mutants of *E. coli* affected in the process of DNA synthesis and cellular division. Cold Spring Harb Symp Quant Biol 33; 677.

Hotta Y, Stern H (1971). Analysis of DNA synthesis during meiotic prophase in *Lilium*. J Mol Biol 55; 337.

Hotta Y, Stern H (1978), DNA unwinding protein from meiotic cells of *Lilium*. Biochemistry 17; 1872.

Hotta Y, Tabata S, Stern H (1984). Replication and nicking of zygotene DNA sequences: Control by a meiosis specific protein. Chromosoma 90; 243.

Hotta Y, Tabata S, Bouchard RA, Pinon R, Stern H (1985). General recombination mechanism in extracts of meiotic cells. Chromosoma 93; 140.

Howell SH (1974). An analysis of cell cycle controls in temperature sensitive mutants of *Chlamydomonas reinhardii.* In Cell Cycle Controls (eds. GM Paddila, LL Cameron, A Zimmermann), Academic Press, New York.

Ieradi LA, Moss SB, Bellve AR (1983). Synaptonemal complexes are integral components of the isolated mouse spermatocyte nuclear matrix. J Cell Biol 96; 1717.

Iyengar GAS, Deka PC, Kundu SC, Sen SK (1977). DNA synthesis in course of meiotic development in *Neurospora crassa*. Genet Res, Camb 29; 1.

Jasani N (1981). Genetic control of meiosis in yeast. Ph.D. Thesis, University of Calcutta.

Johnston LH, Williamson DH, Johnston AL, Fennell DJ (1982). On the mechanism of premeiotic DNA synthesis in the yeast *Saccharomyces cerevisiae.* Exp Cell Res 141; 53.

Kassir Y, Simchen G (1978). Meiotic recombination and DNA synthesis in a new cell cycle mutant of *Saccharomyces cerevisiae*. Genetics 90; 49.

King RC (1970). The meiotic behaviour of the *Drosophila oocyte*. Int Rev Cyt 28, 125.

Klinesmith LJ (1973). Specific binding of phosphorylated nonhistone chromatin proteins to deoxyribonucleic acid. J Biol Chem 248; 5648.

Kofman Alfaro S, Chandley AC (1970). Meiosis in the male mouse. An autoradiographic investigation. Chromosoma 31; 404.

Kumaroo KK, Jahnke G, Irvin JL (1975). Changes in basic chromosomal proteins during spermatogenesis in mature rat. Arch Biochem Biophysics 168; 413.

Levinger LF, Carter CW Jr (1979). Superstructural differences between chromatin in nuclei and in solution are revealed by kinetics of micrococcal nuclease digestion. J Biol Chem 254; 9477.

Lohr D, Hereford L (1979). Yeast chromatin is uniformly digested by DNase I. Proc Natl Acad Sci 76; 4285.

Lu BC, Jeng DY (1975). Meiosis in Coprinus VII. The prekaryogamy S Phase and postkaryogamy DNA replication in *C. lagopus*. J Cell Sci 17; 461.

Mather J, Hotta Y (1977). A phosphorylatable DNA binding protein associated with a lipoprotein fraction from rat spermatocyte nuclei. Exp Cell Res 109; 181.

McCusker JH, Haber JE (1977). Efficient sporulation of yeast in media buffered near pH 6. J Bact 132; 180.

Meistrich ML, Brock WA, Grines SR Jr, Platz RD, Hnilica LS (1978). Fed Proc 37; 2522.

Mills D (1972). Effect of pH on adenine and aminoacid uptake during sporulation in *Saccharomyces cerevisiae.* J Bact 112; 519.

Mitchison JM (1974). In Cell Cycle Controls (ed. JM Padilla, IL Cameron, A Zimmermann, Academic Press, New York, pp 125.

Moens PB, Rapport E (1971). Synaptic structure in the nucleus of sporulating yeast. J Cell Sci 9; 665.

Moens PB, Heyting C, Dietrich AJJ, Raamsdonk WV, Chen Q. Synaptonemal complex antigen location and conservation (in press).

Nelson WG, Pienta KJ, Barrack ER, Coffey DS (1986). The role of the nuclear matrix in the organisation and function of DNA. Ann Rev Biophys Biophys Chem 15; 457.

Noll M (1974). Subunit structure of chromatin. Nature 251; 249.

Pukkila PJ (1977). Biochemical analysis of genetic recombination in eukaryotes. Heredity 39; 193.

Rao BJ, Rao MRS (1987). DNase I site mapping and micrococcal nuclease digestion of pachytene chromatin reveal novel structural features. J Biochem 262; 4472.

Roscoe DH, Robinson H, Carvonell AW (1973). DNA synthesis and mitosis in a temperature sensitive Chinese Hamster cell line. J Cell Physiol 82; 333.

Schindler D, Davis J (1975). Inhibitors of macromolecular synthesis in yeast. Methods in Cell Biology. Vol 12 (ed. David M Prescott), Academic Press, New York, p 17.

Schild D, Byers B (1978). Meiotic effects of DNA defective cell division cycle mutations of *Saccharomyces cerevisiae*. Chromosoma 70; 109.

Scyedin SM, Kistler WS (1980). Isolation and characterization of rat testis. J Biol Chem 255; 5949.

Sen SK (1969a). Chromatin organisation during and after synapsis in culture microsporocytes of *Lilium* in presence of mitomycin C and cycloheximide. Exp Cell Res 55; 123.

Sen SK (1969b). Regulation of chiasma frequency in *Lilium* microsporocytes *in vitro*. Nature 224; 178.

Sen SK (1970a). Axial elements in plants meiotic chromosomes. Nature 228; 78.

Sen SK (1970b). Synaptonemal complexes in haploid *Petunia* and *Antirrhinum* sp. Naturwissenschaften 11; 550.

Sen SK (1971). Chromatin organisation and recombination in Eukaryotes. Z Pflanzenzuchtg 65; 15.

Sen SK, Kundu SC (1977). Formation of chiasmata in the synchronous meiotic cycle of *Lilium longiflorum*. Chromosoma 64; 167.

Sen SK, Kundu SC, Gaddipati JP (1977). High speed scintillation autoradiography of DNA fibres undergoing DNA synthesis at zygotene and pachytene in Lily. Exp Cell Res 108; 471.

Sen SK, Hazra SK, Iyengar GAS, Banerjee RS (1981). Comparison of intra and inter gene recombination in the meiotic mutant C(3)G of *Drosophila melanogaster.* Genetica 55; 47.

Shah M (1981). Isolation and characterisation of yeast mutants in relation to meiosis and/or recombination. Ph.D. Thesis, University of Calcutta.

Sherman F, Roman HL (1963). Evidence for two types of allelic recombination in yeast. Genetics 48; 255.

Shilo V, Simchen G, Shilo B (1978). Initiation of meiosis in cell cycle initiation mutants of *Saccharomyces cerevisiae.* Exp Cell Res 112; 241.

Shires A, Carpenter MP, Chalkely R (1976). A cysteine containing H2B like histone found in mature mammalian testis. J Biol Chem 251; 4155.

Simchen G (1974). Are mitotic functions required in meiosis? Genetics 76; 745.

Simchen G (1978). Cell cycle mutants. Ann Rev Genet 12; 161.

Simchen G, Hirschberg J (1977). Effects of the mitotic cell cycle mutation cdc 4 on yeast meiosis. Genetics 86; 57.

Simchen G, Pinon R, Salts Y (1972). Sporulation in *Saccharomyces cerevisiae:* Premeiotic DNA synthesis, readiness and commitment. Exp Cell Res 75; 207.

Slater ML (1973). Effect of reversible inhibition of deoxyribonucleic acid synthesis on the yeast cell cycle. J Bact 113; 263.

Sobel HM (1972). Molecular mechanism for genetic recombination. Proc Natl Acad Sci, USA 69; 2483.

Solari AJ (1972). Ultrastructure and composition of the synaptonemal complex in spread and negatively stained spermatocytes of the golden hamster and the albino rat. Chromosoma 39; 237.

Stern H, Hotta Y (1973). Biochemical controls of meiosis. Ann Rev Genet 7; 3766.

Stern H, Hotta Y (1978). Regulatory mechanism in meiotic crossing over. Ann Rev Plant Physiol 29; 415.

Stern H, Hotta Y (1980). The organisation of DNA metabolism during recombination phase of meiosis with special reference to humans. Mol Cell Biochem 29; 145.

Taylor JH (1974). Chairman's remarks. Symposium 4: Meiosis. Genetics 78(1); 187.

Trostle-Weige PK, Meistrich ML, Brock WA, Nishioka K, Bremer JW (1982). Isolation and characterization of TH2A a germ cell specific variant of histone 2A in rat testis. J Biol Chem 257; 5560.

von Wettstein D (1971). The SC and fourstrand crossing over. Proc Natl Acad Sci, USA 68; 851.

von Wettstein D, Rasmussen SW, Holm PB (1984). The synaptonemal complex in genetic segregation. Ann Rev Genet 18; 331.

Williamson DA, Johnston LH, Fennell DJ, Simchen G (1983). The timing of the S-phase and other nuclear events in yeast meiosis. Exp Cell Res 145; 209.

12 Aneuploidy in Man and Mechanisms of Nondisjunction Inferred from Studying Djungarian Hamster Oocytes

INGO HANSMANN, FRIEDRICH BEERMANN, EDITH HUMMLER AND FRANZ THEURING*

Institut für Humangenetik
Universität Göttingen
Gosslerstrasse 12d, D 3400 Göttingen, FRG

Introduction

Genome mutations detrimentally affect the developmental capacity of an individual as well as of individual cells. Constitutional aneuploidy arises from parental germ cells and causes pregnancy wastage, neonatal deaths, malformation and mental retardation. Some types of heteroploidy acquired during prenatal or postnatal development seem to cause or predispose to malignant transformation. Notwithstanding that constitutional and acquired aneuploidy causes severe health problems we are still ignorant about causes of aneuploidy and the respective disordered events of one of the most relevant biological processes of cell division.

In the first part of this chapter we summarize briefly some features of constitutional aneuploidy in man. We are aware that this brief review has to remain fragmentary and we apologize for not having considered so many relevant contributions from several colleagues.

*Present address: Institut für Biochemie, Schering AG; Berlin, FRG
Address correspondence to I. Hansmann

In the second part we refer to our own work with Djungarian hamster oocytes. This animal model has been used by us for the last several years for studying mechanisms of nondisjunction.

Considering the data obtained so far it appears to us that, although being certainly heterogeneous, a significant proportion of constitutional trisomy arises by a disordered molecular control of germ cell division. The controlling factors and the respective genes have yet to be identified. In future it may be important to study these events in the context of failures of cell proliferation causing acquired aneuploidy, e.g. in tumorigenesis.

I. Aneuploidy in Man

Incidence of Aneuploidy

The human species is, compared to other mammals, affected by an extremely high frequency of constitutional aneuploidy (for review see Bond and Chandley, 1983; Hansmann, 1983). The reasons for this surprising species specificity are still unknown.

A large set of data from cytogenetic analysis on newborns, spontaneous and induced abortions and even on human sperm metaphases has accumulated during the past years (Tables 1–3). Approximately 60,000 live born children were studied and about 0.5 per cent of them (Table 1) are aneuploid, either for an autosome (trisomy 13, 18 or 21) or for the gonosomes (47,XXX; 47,XXY; 47,XYY; 45,XO). In fact, these anomalies represent only the very tip of an iceberg:

TABLE 1

Heteroploidy in liveborns, perinatal deaths and in spontaneous and induced abortions (data from Boúe *et al.*, 1985*; Hook and Hamerton, 1977)**

	Spontaneous abortions*	Induced abortions*	Perinatal deaths*	Live-borns**
No. of cases	2085	1250	1176	56952
% heteroploid	50–60	6.5	5.5	0.55

TABLE 2

Frequency and type of heteroploidy in spontaneous and induced abortions (*: Boué *et al.*, 1985 by tissue culture; **: Eiben *et al.*, 1987 by direct preparation of chorionic villi)

	Spontaneous abortions I*	Spontaneous abortions II**	Induced abortions
No. of cases	2085	140	1250
% heteroploid			
total	50–60	48.6	6.5
polyploidy	22	21.2	17
monosomy X	20	7.4	11
trisomy	52	68.2	53
structural anomaly	4	4.6	3

TABLE 3

Frequency of aneuploid human sperm analysed after fertilization and culture of hamster oocytes

Authors	No. of sperm analysed	% hyper-haploid	% hypo-haploid	% with non-disjunction
Martin *et al.*, 1983	1000	2.4	2.7	4.8
Brandriff *et al.*, 1985	2468	0.7	0.9	1.5
Kamiguchi and Mikamo, 1986	1091	0.5	0.5	0.9
All	4559	1.0	1.2	2.1

aneuploidy is by an order of magnitude more frequent during pregnancy (see Hassold and Jacobs, 1984; Boué *et al.*, 1985). Aneuploid karyotypes were found among 6.5 per cent of abortions at the first trimester induced because of social reasons (Table 1). The most frequent type of anomaly was trisomy (53 %), followed by polyploidy (17 %) and monosomy X (11 %; Table 2). Aneuploidy adversely affects the outcome of pregnancy and causes perinatal deaths and pregnancy wastage (Table 1).

Altogether 50–60 per cent of early spontaneous abortions are due to chromosomal aneuploidy and polyploidy. Considering the incidence of spontaneous abortions as well as the frequency and type of chromosome anomaly one can estimate that at least 10 per cent of all human conceptions begin their development with an aneuploid karyotype (Hansmann, 1983). No conclusive data on the frequency and type of cytogenetic abnormalities in human zygotes/oocytes are available so far.

Several groups, however, analysed human sperm metaphases obtained by *in vitro* fertilization and culture of hamster oocytes (Table 3). According to the results from more than 4,500 such metaphases about 1 per cent of all human sperm carry one or several surplus chromosomes, yielding a conservative estimation of 2.1 per cent for the nondisjunction frequency during spermatogenesis. Malsegregated chromosomes from all groups (A–G) were detected (Martin *et al.*, 1983; Brandriff *et al.*, 1985; Kamiguchi and Mikamo, 1986). Thus, the preferential occurrence of specific chromosomes in trisomies at birth or in spontaneous abortions does not reflect the primary nondisjunctional events at least during paternal meiosis.

Parental origin of trisomy

The parental origin of additional chromosomes in trisomic offspring was first analysed by using chromosome specific markers like red-green colour blindness or the Xg-blood group (for review see Hassold and Jacobs, 1984). These markers are located rather distally from the centromere of the X-chromosome. Due to the high recombination frequency they are only useful for discriminating between maternal and paternal errors but not for allocating nondisjunction to meiosis I or II.

Sanger *et al.* (1977) determined the origin of the supernumerary gonosome in more than 500 individuals with the Klinefelter's syndrome. According to the results of the Xg-blood group analysis ⅓ of meiotic errors were attributed to spermatogenesis and ⅔ to oogenesis. No conclusive information was obtained in families with an XXX-daughter, although, the majority seems to have arisen by an error during oogenesis (Sanger *et al.*, 1977). Later on chromosomal heteromorphisms, visualized by Q- and C-banding, were used to trace nondisjunction in autosomal trisomies. These heteromorphic chromosomal regions are located pericentromerically and crossing-over between the phenotypic marker and the centromere is more unlikely to occur. Other pitfalls, like postzygotic nondisjunction and secondary nondisjunction in germ cells from mosaic parents, have to be considered and may bias the basic events of spontaneous meiotic nondisjunction.

The largest numbers of reports tracing the origin of nondisjunction are available from the analysis of families having a liveborn individual with trisomy 21 (Table 4). The majority of errors (80 %) were assigned to oogenesis and there preferentially (68 %) to meiosis I (Hassold and Jacobs, 1984). For spermatogenesis (20 %) meiosis I contributed nearly twice as often (13%) to trisomy 21 in liveborns than meiosis II (7%). Results from trisomy 21 in abortuses are not significantly different from that pattern: maternal errors contributed to 75 per cent of all trisomies 21 (meiosis I: 67 %; meiosis II: 8 %). While the remaining 25 per cent of paternal errors were all allocated to meiosis I (Table 4).

TABLE 4

Parental origin of additional chromosomes in trisomic abortions and liveborn individuals determined by chromosomal heteromorphisms (according to Hassold and Jacobs, 1984)

Population investigated	Trisomy for chromosome	No. of cases	Origin					
			Mat			Pat		
			I	II	I/II	I	II	I/II
liveborn	21	391	60.9	13.0	6.6	11.5	8.9	1.0
abortions	21	17	52.9	5.9	17.6	11.8	—	11.8
	13	14	57.1	14.3	14.3	7.1	7.1	—
	14,15	8	75.0	—	25.0	—	—	—
	16	34	73.5	5.9	11.8	5.9	2.9	—
	22	28	75.0	—	21.4	3.6	—	—

With respect to the other autosomal trisomies, e.g. for chromosome 3,4,9,13,14,15,16 and 22, the paternal contribution seems to be significantly lower and less than 12 per cent (Table 4). Evidently, the majority of those trisomic conceptions which develop up to the end of the first trimester or even up to birth is characterized by two maternally inherited chromosomes. This pattern could reflect an extreme parental difference in nondisjunction frequencies with oogenesis being more vulnerable.

The obviously less frequent involvement of spermatogenesis might on the other hand be derived from a less favourable developmental capacity of trisomic conceptions with two paternal chromosomes. It was shown recently that uniparental disomy for chromosomes or even only chromosome regions in otherwise chromosomally normal mice could result in prenatal lethality or in developmental alterations (for review see Monk, 1988). So far 6 different chromosomes/chromosome regions of the mouse karyotype have been identified for which one paternal and one maternal copy must be present for an undisturbed prenatal development (Mouse News Letter, 81, 1988). Mouse chromosomes showing a defective complementation are chromosome 2, 6, 7, 8, 11 and 17. If the so-called genomic imprinting is functioning also in man, then at least those human chromosomes/chromosome regions syntenic to the respective mouse chromosomes could show defective complementation as well. Methylation has been proposed as the molecular basis for imprinting and so far paternally inherited loci showed hypomethylation (for review see Monk, 1988). It is unresolved whether the relation of two paternally versus one maternally imprinted chromosome/s has a more detrimental effect on embryonic development or vice versa.

According to the data summarized by Hassold and Jacobs (1984) meiosis I errors outnumber those at meiosis II. Among abortuses trisomic for chromosome 13, 16, 21 and 22, errors at meiosis I occur more frequently than at meiosis II by a factor of 5.3, 10.1, 13.3, and 32.3, respectively. In liveborn individuals with trisomy 21, however, the factor is only 3.9 (compared to 13.3 in trisomic 21 abortions). The apparent difference in the allocation of meiotic errors between liveborns and abortions could reflect a negative influence of di-isotrisomy (a consequence of meiosis II errors, Engel, 1980) on embryonic survival.

With the advent of chromosome and centromere specific RFLPs (Restriction Fragment

TABLE 5

Parental origin in autosomal and gonosomal aneuploidy determined by RFLP-analysis

Karyotype	No. of cases	Origin of aneuploidy maternal	paternal	Authors
47, XXY	35	15	20	Jacobs *et al.*, 1988
47, XXX	9	8	1	May *et al.*, 1988
45, X0	35	28*	7**	Hassold *et al.*, 1988
Trisomy				
13	8	5	3	Hassold *et al.*, 1987
18	3	3	—	Kondoh *et al.*, 1988
21	53	37	16	***)

*) $45,X^{M}0$; **) $45,X^{P}0$; ***) Davies *et al.*, 1984; Hassold, 1985; Stewart *et al.*, 1985; Rudd *et al.*, 1988; Bricarelli *et al.*, 1988

Length Polymorphisms, Table 5) in combination with the simpler cytogenetic analysis of spontaneous abortions directly from chorionic villi (Eiben *et al.*, 1987) the question about the parental origin of trisomies can be approached now on a broader scale.

Maternal age

Maternal age is the best characterized etiological factor associated with trisomy in man. The association was shown in liveborn individuals, fetuses analysed by amniocentesis as well as in spontaneous and induced abortions (Penrose, 1933; Taylor, 1968; Polani, 1969; Carothers *et al.*, 1978; Hassold *et al.*, 1980; Hansmann, 1983; Hassold and Jacobs, 1984). Maternal age is, however, inversely related with the gonosomal aneuploidies of 45, X0 and 47,XYY (Fujita and Hirano, 1975; Kajii and Omaha, 1979; Warburton *et al.*, 1980).

The association of different types of heteroploidy with maternal age is exemplified in Tables 6 and 7, where data from two studies on spontaneous abortions are summarized. The data demonstrate furthermore that with the so-called direct preparation of chorionic villi, a method which can more easily be used for analysing spontaneous abortions (Eiben *et al.*, 1986), comparable results to the more laborious tissue culture technique are obtained. From Tables 6 and 7 it is evident that trisomies significantly increase with maternal age, polyploidy and 45,X0 in contrast decrease.

The association of trisomy with maternal age was presumed to be either due to a decreased ability for aborting trisomic conceptuses (Golbus, 1981; relaxed-selection: Aymé and Lippman-Hand, 1982) or due to an increasing frequency of chromosomal malsegregation during oogenesis in older women. So far, no evidence for the hypothesis of a 'relaxed-selection' was found (Hassold and Jacobs, 1984).

Recently, the age-related frequencies of individual autosomal trisomies in spontaneous abortions were analysed by two groups (Hassold and Chiu, 1985; Risch *et al.*, 1985). Hassold and Chiu (1985) calculated the relation of 641 autosomal trisomies with maternal age and concluded that trisomies for different chromosome groups behave very heterogeneously. Trisomies for the A and B group did not change in frequency with age and the frequency for trisomy 4 even decresased at higher maternal ages (inverse relation). Trisomy 16 differed from

TABLE 6

Frequency and type of chromosome anomalies in spontaneous abortions ($n = 2364$) from different maternal age groups analysed by tissue culture (according to Hassold and Chiu, 1985)

Maternal age (years)	No. of abortions	% abnormal abortions	% abnormal abortions with		
			Polyploidy	45,X0	Trisomy
15–23	602	38.0	9.1	12.1	16.8
24–29	765	48.6	12.8	11.0	24.8
30–35	637	50.2	9.7	6.1	34.4
36–39	185	55.7	3.2	8.7	43.8
⩾40	75	72.0	4.0	1.3	66.7

TABLE 7

Frequency and type of chromosome anomalies in spontaneous abortions ($n = 130$) from different maternal age groups analysed by direct preparation of chorionic villi (according to Eiben *et al.*, 1987)

Maternal age (years)	No. of abortions	% abnormal abortions	% abnormal abortions with		
			Polyploidy	45,X0	Trisomy
16–24	18	50.0	22.2	11.1	16.7
25–29	41	36.6	4.9	2.4	26.8
30–34	34	47.0	5.9	2.9	26.5
35–39	29	55.2	6.9	3.5	37.9
⩾40	8	87.5	0.0	0.0	87.5

the other trisomies by demonstrating a more steady and linear increase up to high maternal ages. The other trisomies like that for chromosome D, 17–20 and 21, 22 show a rather steep increase above 35 years of age.

Risch *et al.* (1986) subsequently published data on 515 trisomic abortions and they also concluded that autosomal trisomies seemed to be of heterogeneous origin with a maternal age-related factor affecting the smaller chromosomes preferentially. In principle, their data do not differ from that obtained by Hassold and Chiu (1985). Taken together, each chromosome is characterized by individual maternal age-specific risks. The possible implication of these findings for the mechanisms of maternal age-related nondisjunction have been discussed recently (Hummler *et al.*, 1987).

Morton *et al.* (1988) analysed data on trisomies in live births, amniocentesis and in spontaneous abortuses and concluded that trisomies seem to increase monotonically with maternal age. They also differentiated between age-dependent (trisomy 16) and age-independent trisomies (A, B and C trisomies). Accordingly, trisomies for the smaller chromosome are considered to be in part age-dependent as well as independent. Their data on the frequency of

TABLE 8

Frequency of trisomies in spontaneous abotions ($n=2292$) in relation to maternal age (according to Morton *et al.*, 1988)

Maternal age (years)	No. of abortions	% trisomy	% double trisomy
14–24	663	21.7	0.0
25–29	616	33.6	1.1
30–34	625	40.5	1.3
35–39	300	50.0	2.3
≥40	88	64.5	9.1

TABLE 9

Frequency (%) of individual trisomies in spontaneous abortions (see also Table 8, from Morton *et al.*, 1988)

Maternal age (years)	Trisomy for chromosome(s)*							
	1–3,5	4	C	D	16	17–20	21	22
14–24	3.0	5.6	3.1	3.5	24.5	1.6	11.2	8.4
25–29	3.2	5.4	2.1	3.8	35.3	0.9	10.0	8.1
30–34	1.4	3.4	1.7	3.7	36.4	2.0	8.1	15.7
35–39	1.7	2.5	1.5	6.4	26.8	1.8	7.0	20.4
≥40	1.5	0.0	2.3	7.2	16.9	4.2	10.8	10.8

* Calculated for individual chromosomes from the respective group

trisomies, double trisomies and on trisomies for individual chromosomes are summarized in Tables 8 and 9. Double trisomies seem to increase more steeply with maternal age than individual trisomies (Table 6) which is in accordance to our model for nondisjunction at meiosis I in mammalian oocytes in which the phase of inhibiting spindle function is determining the pattern and frequency of nondisjunction (Hummler *et al.*, 1987).

In fact, we were able to show recently that the inhibition of the early meiosis I spindle causes nondisjunction of several bivalents, whereas late inhibition results in preferential malsegregation of single so-called late segregators (Hummler and Hansmann, 1988). Chromosome 16 is most often involved in trisomies at all maternal age groups followed by trisomy 21 and 22 (Table 7). The acrocentric chromosomes with their multiple copies of r-DNA genes and the chromosome 16, characterized by a relatively large amount of constitutive heterochromatin are represented in 55 to 73 per cent of all trisomies. This high contribution is (on the basis of the relatively small number of chromosomes (6) in comparison to the remaining 17 chromosomes) an overproportional contribution to trisomy. According to our model (Hansmann *et al.*, 1985; Hummler *et al.*, 1987) these chromosomes are suspected to be late segregators.

Brook *et al.* (1984) have shown by comparing aneuploidy in early embryos from unilaterally ovariectomized, sham operated and aging control mice that the biological and not the chronological age of a female might be relevant.

Seasonal variation

Studies on the seasonal variation in the frequency of Klinefelter's syndrome gave controversial results; some authors reported such a variation, whilst others found none (see Nielsen *et al.*, 1973). Jacobs (1979), cited by Videbeck and Nielsen, 1984) analysed data from the literature and found an excess of births for children with sex chromosome anomalies in April, May and June, without any characteristic seasonality over a period of several years. This corresponds with data from the Danish Cytogenetic Central Register (Videbck and Nielsen, 1984) which showed accordingly, no seasonal variation in the birth of children with any type of sex chromosome abnormality in the period after 1945.

With respect to Down syndrome, several investigations on seasonality have been performed. Some groups found such a correlation, others did not. Janerich and Jacobsen (1977) averred that the seasonal variation in estradiol receptors in mammary tumours (Hughes *et al.*, 1976) shows the same variation as that for Down syndrome reported by Harlap (1974). It was proposed that endocrine factors which are responsible for changes in estradiol-receptor concentration also affect chromosome segregation during the late phases of meiosis.

Jongbloet (1971), amongst others, described a significant seasonality and proposed the so-called seasonal preovulatory over-ripeness oovopathy (SPOO-) hypothesis. This hypothesis is based on the presence of an evolutionary vestige of 'ovulatory' alternating with 'anovulatory' seasons, a phenomenon which is known to exist in non-human primates and other mammals. Such a mono-dioestric system is considered to be influenced by exogenous factors, whereas the phylogenetically younger polyoestric pattern appears to be an autonomous endogenous system. Jongbloet *et al.* (1982) predicted an association of ovulation behaviour and the incidence of Down syndrome. Accordingly, ovulation should be delayed during the transitions from anovulatory to ovulatory seasons and vice versa. Down syndrome conceptions indeed appear to occur more frequently during the seasonal transitions of increasing and decreasing ovulation rate, and disproportionally less during the season in which the ovulation is stabilized.

Later, Jongbloet *et al.* (1982) analysed births of 287 patients with Down syndrome whose extra chromosome 21 has been attributed to an error of either one of the 4 meiotic divisions (Mat I or II, Pat I or II). Interestingly, he found a significant clustering of those conceptions with a Mat I- error around May–July. The authors also suggested that the circannual periodicity of prolactin serum concentrations is related to this seasonality. Higher levels of prolactin interfere with the preovulatory development of granulosa cells and cause a protracted preovulatory phase (McNeilly, 1980). Jongbloet *et al.* (1982) conclude that hyperprolactinemia may play a central role in the etiology of delayed ovulation and chromosomal nondisjunction. This is a new aspect for the understanding of nondisjunction, although there are no data about the involvement of prolactin in nondisjunctional events. Hansmann (1983) argued that the possibility has to be envisaged that the seasonal variation might reflect an altered selection during pregnancy.

Sexual behaviour

For a long while it has been suggested that the occurrence of Down syndrome is correlated with sexual behaviour and may play a role in association with parental age (German, 1968). This role in age effects is questionable, because only 10–20 per cent of nondisjunctional events seem to occur in maternal meiosis II, which might be the only stage susceptible for alterations induced by sexual behaviour. In attempting to test this hypothesis, Milstein-Moscati and Becak (1978, 1981) found, that mothers of Down syndrome children report significantly longer intercoital intervals at the time of conception compared to control mothers. This relationship was independent of maternal age. Different results were reported by others (Sever *et al.*, 1970) and there was some discussion about the relationship between unusually low coital rates and the risk of having a child with Down syndrome (James, 1978, 1983; Milstein-Moscati and Becak, 1983). It cannot be ignored that there is some systematic malfunction (e.g. thyroid dysfunction or premature aging) which affects both coital rates and Down syndrome.

The higher incidence of Down syndrome in illegitimacy (Shokeir, 1968) or amongst Catholic women in Western Australia (Mulcahy, 1978) and the Netherlands (Jongbloet *et al.*, 1978) seems to be connected with the interrelation of ovulation and fertilization. Contraceptive failures and premarital conceptions may have also favoured delayed fertilization (Juberg, 1983).

Preconceptional aging of sperm could also be the result of sexual behaviour. The effect of aged sperm was investigated mainly in rabbit blastocysts and an increased frequency of chromosomally abnormal conceptuses was observed (for review Shaver and Martin-DeLeon, 1975). Martin-DeLeon and Boice (1982) reported increased frequencies of chromosome abnormalities in mouse zygotes after fertilization with aged sperm.

Preovulatory overripeness

Studies on spontaneous and artificially induced preovulatory overripeness in *Xenopus laevis* (Mikamo, 1968) and in the rat (Butcher and Fugo, 1976; Fugo and Butcher, 1971; Mikamo and Hamaguchi, 1975; Kamiguchi *et al.*, 1979) demonstrated its potential role in producing chromosomal and developmental anomalies.

Eggs from *Xenopus* are prevented from spawning for over one year, if the stimulus to induce ovulation is not sufficient. In such overripe oocytes, Mikamo (1968) observed unusual behaviour of meiotic chromosomes and degeneration of the spindle. Fugo and Butcher (1976) investigated the occurrence of intrafollicular overripeness of the oocytes in rats by delaying ovulation with sodium pentobarbital treatment to suppress the mid-cycle surge of LH. A 48-hour delay in rats with 4-day estrus cycles resulted in the resumption of meiosis postponed by 2 days (Freeman *et al.*, 1970).

Butcher and Fugo (1976) demonstrated a detrimental effect of pentobarbital induced delayed ovulation on embryonic development in rats due to, e.g., an increased incidence of chromosome anomalies on day 11 of gestation (1.3 per cent). Mikamo and Hamaguchi (1975) observed in rat blastocysts an aneuploidy rate about twice as high as that of the study by Butcher and Fugo (1976). Kamiguchi *et al.* (1979) cytogenetically studied the effect of preovulatory overripeness of primary oocytes after pentobarbital treatment in 2-cell rat embryos and found a significantly increased aneuploidy rate in the experimental group (5.2 versus 1.8 per cent).

Fugo and Butcher (1971) showed that follicular overripeness due to spontaneously occurring 6-days cycles in aged female rats produced similiar anomalies, i.e. degeneration of fertilized ova, a reduced implantation rate and an increased embryonic death rate compared to the pentobarbital treated group.

It cannot be excluded, however, that the observed anomalies are exclusively caused by pentobarbital. Uilenbroek *et al.* (1984) recently found a rapid decrease in follicular oestradiol synthesis in rats after pentobarbitone treatment at proestrus. This effect was due to a decrease in the activity of the enzyme converting progesterone to androstenedione. Laing *et al.* (1984) inhibited, by a single injection of LH-RH antiserum in 4- and 5-day cyclic mice, the ovulatory surge of gonadotropins. Cytogenetical analysis of oocytes which were delayed in ovulation for 2 days revealed no increased incidence of nondisjunction in comparison to oocytes ovulated normally.

In man, intrafollicular ageing of the oocyte might occur as a result of a prolonged menstrual cycle. Such irregular and often extended cycles are known at puberty and during the climacteric period associated with changes in the secretion of gonadotropic hormones (Butcher, 1975). Such hormonal alteration might influence the environment of the maturing oocyte. However, no conclusive data are available on the association of prolonged cycles and the frequency of chromosome anomalies.

Postovulatory overripeness

Previous studies with amphibia and mammals also suggest that overripeness during the time of the second meiotic division could lead to numerical chromosomal anomalies (Szollosi, 1975). The postovulatory capacity of mammalian oocytes to fertilize is estimated to last approximately 6 to 36 h (Juetten and Bavister, 1983). Before this capacity is lost, cytological changes take place. Rotation and migration of the spindle to the central portion of the oocyte (Szollosi, 1971) and disturbance of the spindle followed by dispersal of chromosomes (Austin, 1976) occur within a few hours after ovulation. Rodman (1971) suggested that precocious disjunction of chromatids in mouse oocytes could be responsible for errors of chromosome segregation during meiosis II after delayed fertilization. He described 'ageing' oocytes as characterized by a dispersed or sometimes absent cumulus cell mass. One has to take into account, however, that all these observations of precocious separation have only been made after treatment of oocytes with hyaluronidase (Hansmann, 1983). It is known that hyaluronidase is a potent inducer of *in vitro* parthenogenetic development in mice (Kaufman, 1973). Thus, such oocytes with less or even without cumulus cells are more extensively exposed to the enzyme and may have responded with parthenogenetically activated 2nd meiotic division.

Polyploidy is known to be a main consequence of fertilizing postovulatory overripe oocytes as observed by Shaver and Carr (1969) in rabbits, and by Yamamoto and Ingalls (1972) and Adachi and Ingalls (1976) in hamsters. Moreover, these authors described also monosomies, which may result from anaphase lagging, but no cases of trisomy. German (1968) suggested that a decreased frequency of coitus might lead to aged oocytes and thus might contribute to the frequency of trisomy 21. However, direct evidence for the occurrence of delayed fertilization is difficult to obtain in man.

α-1-Antitrypsin

α-1-antitrypsin is an inhibitor of proteolytic enzymes in human serum. It is produced by liver and inhibits cellular proteases like collagenase, elastase, trypsin and thrombin. More than 30 variants of α-1 AT have been identified and they are all believed to be codominant and determined by alleles at a single locus, referred to as Pi (protease inhibitor). Compared to the common Pi M type, some of the non-M variants, e.g., S and Z, show lower serum levels. There are some reports suggesting a role of antitrypsin variants in chromosome behaviour during cell division. Aarskog and Fägerhol (1970) and Kueppers *et al.* (1975) found more cases of sex chromosome mosaicism in probands heterozygous for Pi variants. Fineman *et al.* (1976) studied the Pi types of patients with sex chromosome mosaicism and Down syndrome. They confirmed the previous results and reported also on an increased frequency of α-1 AT heterozygotes among Down syndromes born to older mothers. However, such an association of non-M variants with Down syndrome was not found by others (Arnaud *et al.*, 1976; Guanti and diLoreto, 1980; McPhee *et al.*, 1980) and no increased incidence of Pi variants was observed in patients born to older or younger mothers (Breg *et al.*, 1981).

Jongbloet *et al.* (1981), however, found a significantly increased frequency of the deficiency variants MS and MZ in mothers of trisomic 21 children where nondisjunction was allocated to the first meiotic division in oocytes. Aarskog *et al.* (1978) argued that the presence of certain alleles at the Pi locus could decrease crossover frequency which might lead to chromosome malsegregation by an increase of univalent formation.

Other hypotheses are based on the decreased inhibitory activity of some α-1-antitrypsin variants, specially M and Z either by affecting the stability of the spindle or by accelerating the breakdown of the nucleoprotein connections between the NOR bearing chromosomes. Proteases, however, are also involved in the processes of ovulation and implantation, and any alteration of proteolytic activities could have secondary effects, e.g., on the selection of chromosomally abnormal oocytes and embryos.

Virus infection

Epidemiological investigations on the incidence of Down syndrome in the State of Victoria, Australia, led Collmann and Stoller (1962) to suggest that the frequency of this congenital anomaly is influenced by a virus infection preferentially affecting oocytes in ageing females. Three years later, the same authors (Stoller and Collmann, 1965) reported that between 1953 and 1954 the pattern of Down syndrome in this region of Australia followed with a 9-month lag that for infectious hepatitis. There are numerous discussions about this correlation (see ref. in Cohen *et al.*, 1977) but many were unable to confirm these reports in other countries, in, e.g., Britain (Leck, 1966), Italy (Ceccarelli and Torbidoni, 1967) and in the United States (Stark and Fraumeni, 1966; Kogon *et al.*, 1967).

For long it has been known that virus infection could increase the frequency of structural aberrations of chromosomes. Such aberrations, predominantly breaks, have been observed in peripheral leucocytes of patients with measles (Nichols *et al.*, 1962; Koskull and Aula, 1977), rubella (Kuroki *et al.*, 1966), infectious hepatitis (Aya *et al.*, 1966; Mella, 1968) and in the bone-marrow cells of patients with infectious hepatitis (Matsaniotis *et al.*, 1966). Mitotic suppression and increased aneuploidy and stickiness of the chromosomes have also been described in infectious hepatitis (Mella, 1968). These changes were demonstrable up to nine months after the appearance of jaundice. It was suggested that altered mitotic divisions are the effect of virus-induced spindle-abormalities (Nichols, 1966).

Evans (1967) argued that any increase in the incidence of Down syndrome that may be associated with viral infection has little or nothing to do with the induction of chromosome breakage. He favoured the idea of a virus effect on nucleoli of human oocytes, which should result in a reduced capacity for nucleolus dissolution during meiosis. Accordingly, such a persistence of nucleoli would lead to failures in the segregation of those bivalents sharing nucleoli.

There are several investigations on mammalian cells infected by intranuclear replicating DNA viruses and they have all shown that the nucleolus plays an important role in the development of the virus. This often results in specific virus-induced changes of the nucleolus, e.g. fragmentation and progressive destruction of the nucleolus after herpes simplex infection, enlargement of nucleoli due to increased DNA synthesis (SV 40 virus) and increased storage of rRNA (adenovirus) (Evans, 1967; Bouteille *et al.*, 1982). Since nucleolar persistence and satellite association seem to be proportional to the size of the nucleolus (Miller, 1981) these findings support the suggestion of Evans (1967). No information is available, however, on the behaviour of nucleoli in germ cells after virus infections.

Ross (1981) reported the birth of 5 trisomic children whose mothers had been all exposed to a communicable disease (rubella, infectious hepatitis, measles, mumps or varicella) during the five months preceding pregnancy. No trisomic child has been observed from mothers who were affected earlier or later than that preconceptional period.

Thyroid disease and antithyroid antibodies

Altered thyroid function has long been suspected as a risk factor for nondisjunction. As early as 1921, Dollinger reported an increased thyroid activity in mothers of Down syndrome cases, and in 1938 (Myers) and 1949 (Benda) similar reports followed. Ek (1959) investigated 41 mothers of Down syndrome children and found that their serum protein bound iodine was significantly increased. This was not confirmed in other studies (Coppen and Cowie, 1960; McDonald, 1972). Van Wijck *et al.* (1961) reported on a trisomy 18 child of a 41-year-old hyperthyroid mother and Migeon (quoted in Merz *et al.*, 1966) described a mongoloid born to a 26-year-old mother who had thyroid enlargement for at least 1 year prior to conception. Other investigations supported these findings (Fialkow *et al.*, 1971; McDonald, 1972) and showed that mothers of mongoloid children had a higher incidence of thyroid disease.

Interestingly, an increased frequency of nondisjunction as a consequence of altered thyroid levels was reported for male pituitary deficient Snell dwarf mice (Wauben-Penris and van Buul-Offers, 1982). The untreated dwarf mice (hypothyroid), as well as dwarfs artificially made hyperthyroid by treatment with thyroxine, showed higher nondisjunction levels than the controls.

In 1964, Fialkow published a hypothesis on the relation between thyroid autoantibodies and aneuploidy. He suggested that a parental immunological abnormality should predispose to nondisjunction in the germ cells. Although, such an association was not confirmed by Wren *et al.* (1967) and McDonald (1972), many studies in favour of that hypothesis appeared (Fialkow *et al.*, 1965; Burgio *et al.*, 1966; Vanhaelst *et al.*, 1970; Schuler *et al.*, 1970; Fialkow *et al.*, 1971; see ref. in Cohen *et al.*, 1977).

A different aspect of this problem has been studied by other workers. The supposed significance of satellite association in the pathogenesis of Down syndrome raised the question as to whether a correlation exists between satellite association and thyroid conditions. Merz *et al.* (1966) studied the effect of thyroxine on the number of nucleoli per cultured testicular tumour cell. They found a lower number of nucleoli and concluded that thyroxine might have caused coalescence of these organelles followed by an increased frequency of satellite associations. However, in discordance with these results another *in vitro* study showed an increased frequency of nucleoli after thyroxine treatment of human kidney epithelial T-1 cells (Siegel and Tobias, 1966).

Nankin *et al.* (1968) compared the satellite association tendency in patients with thyrotoxicosis and in non-affected control probands and found no significant difference between these groups. Schuler *et al.* (1970) examined individuals with thyroxine autoantibodies and showed that these antibodies did not have increased satellite associations either in chromosomally normal or in patients with trisomy 21. There are contradicting results (Hanson, 1970; Nilsson *et al.*, 1975; Zankl *et al.*, 1980) and there is no definitive evidence that satellite associations or thyroxine levels have any influence on the frequency of meiotic nondisjunction in man.

Double nucleolar organizer region (dNOR variants)

The acrocentric D- and G-group chromosomes are characterized by specific rDNA containing regions (NOR) in the very short arm, which can be visualized by $AgNO_3$-staining or by *in-situ* hybridization with labelled rRNA- or rDNA-probes. Some individuals have been reported with NOR-variants on one or several chromosomes. These variants, termed dNOR, seem to consist of a duplicated area of rDNA and segregate as Mendelian traits (for review see Jackson-Cook *et al.*, 1988). These authors observed, by analysing the parental origin of trisomy, an extraordinary high frequency of dNOR variants in 15 of the 50 families studied. In 13 families they were able to allocate the meiotic error and in 12 of the informative families the additional chromosome was indeed derived from the dNOR-carrier parent. No preferential association of the dNOR with a specific acrocentric chromosome was detected. None of the 50 controls and only one of the 41 spouses not having contributed to the trisomy were shown to have such a dNOR variant.

Considering the significantly different frequencies of dNOR variants in controls versus Down syndrome families, these authors estimated that families with dNOR variants would contribute to approximately 96 per cent of all non-translocation/non-cryptic trisomy 21 mosaic families with several cases of Down Syndrome. Indeed, Jackson-Cook *et al.* (1988) detected in 9 of 10 such families a dNOR variant in one or even both parents. In 10 trisomic offspring from these 9 families, nondisjunction could be allocated to one of the parental meiotic divisions by comparing the heteromorphisms on chromosome 21. The additional chromosome was derived in all cases not only from the parent with the dNOR but also from an error at meiosis I.

These data, if corroborated, would suggest that dNOR variants represent, in addition to maternal age, a considerable risk factor for the occurrence of trisomy 21. Variants of NORs seem to play an important role in the etiology of Down syndrome; the respective mechanisms have yet to be determined.

Influence of chiasmata

Four major events take place during the mammalian meiotic prophase: 1. pairing of the homologous chromosomes, 2. recombination by crossing-over followed by chiasma formation, 3. co-orientation, and 4. the segregation of homologous half bivalents at meiosis I (John, 1976). An interrelationship between recombination/chiasma frequency and nondisjunction has been assumed for many years. Darlington (1929) was the first to suggest that homologous chromosomes must be tied together by their chiasmata at diakinesis for a regular segregation at the first meiotic division. This was confirmed by Dobzhansky (1933) and Mather (1938) who envisaged that half-bivalents require at least one common chiasma for a normal segregation.

Henderson and Edwards (1968) put forward the 'production line' hypothesis which linked sequential oocyte maturation with diffrences in the frequency of chiasmata. The hypothesis was based on observations of a declining chiasma frequency/increasing univalency in oocytes from aging mice. The authors postulated a gradient in the fetal ovary at the time of chiasma formation. Accordingly, they suggested that oocytes are ovulated along this production line, i.e. the early fetal oocytes with a higher chiasma frequency are ovulated early during the reproductive life span, whereas those formed later and having less chiasmata, are ovulated only at higher maternal ages.

In man, several groups have studied chiasma frequencies in males (McDermott, 1973; Hulten, 1974; Lange *et al.*, 1975). The data summarized by Polani (1981) suggest a mean chiasma number per cell at diakinesis of about 50, with a range of 44 to 56. Hulten (1974) showed a decline in chiasma frequency with decreasing chromosome size. Aging seems not to affect the mean chiasma frequency in human spermatocytes (Edwards, 1970; Lange *et al.*, 1975). Male mice showed a slight but insignificant increase in chiasma frequency with age (Henderson and Edwards, 1968; Jagiello and Fang, 1979), whereas an increase in the frequency of bivalents with two chiasmata at the expense of bivalents with a single bivalent was observed by Speed (1977).

Female mice showed in contrast a decrease in chiasma frequency at advanced ages. The effect was reported to be strain specific (Henderson and Edwards, 1968; Luthardt *et al.*, 1973; Polani and Jagiello, 1976; Speed, 1977; Jagiello and Fang, 1979). These results are in accordance with findings in the Chinese hamster (Sugawara and Mikamo, 1983) in which the average number of chiasmata per oocyte was also significantly lower at higher maternal ages.

De Boer and van der Hoeven (1980), using a mouse strain homozygous for a reciprocal translocation, showed in aged females an increase of true univalents for their short marker chromosome accompanied by an increase in nondisjunction of the same chromosome. Furthermore, the presegregation, i.e. the separation of the homologous chromatids, at meiosis I occurred more frequently in oocytes from aged females.

In constrast to findings of Henderson and Edwards (1968), Luthardt *et al.* (1973) and Polani and Jagiello (1976), Speed (1977) was not able to confirm an increased frequency of univalents in aged female mice. He suggested that the increased number of univalents seen at metaphase I in aged oocytes of certain mouse strains may be attributable to a more contracted state of chromosomes. Polani and Jagiello (1976) found no relation between the frequency of univalents at metaphase I and the percentage of aneuploid metaphase II oocytes in female mice. They argued that the observed univalents in older females are not true univalents derived from asynaptic or desynaptic bivalents, but that they mostly arise as artefacts. In Chinese hamsters, Sugawara and Mikamo (1983) showed that the incidence of both univalents at metaphase I and hyperhaploidy at metaphase II oocytes were significantly increased in the older females. However, they discovered no correlation between univalency and nondisjunction of the respective chromosomes. Almost all univalents were found in the smallest metacentric subgroup, whereas nondisjunction affected more or less all bivalents. Luthardt *et al.* (1973) and Speed (1977) also confirmed such a non-random univalent formation of both true and ambiguous univalents involving especially the smaller chromosomes.

One aspect of the production line hypothesis proposed by Henderson and Edwards (1968) concerns the timing of meiotic prophase in the fetal mouse ovary. They concluded from the observed variation of 4–5 days with which mouse oocytes enter dictyotene that there is sufficient time for a gradient to develop. Jagiello and Fang (1979) supported this production line by observing that chiasma frequency from early diplotene oocytes is significantly greater than from later (diplotene) oocytes. However, others have questioned the existence of such a production line (Crone *et al.*, 1965; Baker, 1972; Speed, 1982). Speed and Chandley (1983), who analysed in detail the developmental sequence during meiotic prophase, failed to corroborate the difference in chiasma frequency between day 16 and 18 of gestation in the mouse. Dietrich and Mulder (1983) analysed sequentially these stages of oogenesis using different preparation techniques. They found a considerable variation between different litters in the population of oocytes at various meiotic stages on the same day of gestation (e.g. oogonia were found until day 17). Fetuses from the same litter, however, showed no significant variation.

Darlington (1937) developed the concept of chiasma terminalization to account for the apparent shift in chiasma distribution between diplotene and metaphase I in certain species, i.e. chiasmata are changed in position by moving towards the ends of bivalents. Movements of chiasmata might also take place in mammalian oocytes during the prolonged germinal vesicle stage with progressive loss of chiasmata. Speed (1977) observed in aged mice, as indicated by Henderson and Edwards (1968), a reduction in the number of bivalents with two chiasmata which was accompanied by a corresponding increase of single terminal chiasma. Terminal chiasmata, involving especially the smaller chromosomes, were also found more frequently in aged Chinese hamsters (Sugawara and Mikamo, 1983).

Precocious desynapsis of bivalents that have recombined normally could also result from age dependent environmental effects. Maguire (1974), for example, suggested that aging promotes the decay of a chiasma binder. The possibility cannot be excluded, however, that some observed age effects on the number of chiasmata result from artifacts due to a higher sensitivity to experimental manipulations in oocytes from older females. Our own investigations in aged Djungarian hamster females with a significantly higher frequency of nondisjunction revealed no increased incidence of univalents (Hummler *et al.* 1987).

Sugawara and Mikamo (1986) analysed the frequency of univalents in oocytes from aged Chinese hamsters prepared by the air-drying technique (Tarkowski, 1966) and by their own gradual fixation method (Kamiguchi and Mikamo, 1986). An increased frequency of univalents was found only in air-dried preparations, and these authors concluded that the higher frequency of univalents reported for oocytes from aged females might be derived by preparational artifacts.

Beermann *et al.* (1987) investigated cross-over frequencies within a specific pairing region of the T(2;4)1Gö translocation in oocytes from young and old heterozygous translocation mice. Cross-over frequencies were calculated from chromosome configurations at metaphase II. For both marker chromosomes, i.e. the small 2^4 and the large 4^2, the frequency was higher in old than in young T(2; 4)1Gö/+females. This observation does not support the production line hypothesis (Henderson and Edwards, 1968) which predicts a decreased crossover frequency in aged oocytes.

Summarizing the experiments with oocytes from aged rodents we conclude:

1. The higher number of univalents and the decreased frequency of chiasmata at diakinesis/metaphase I are most probably due to preparational artifacts. These arise, e.g., from an altered chromosome configuration/rate of compaction in preovulatory oocytes resuming meiosis in follicles from old females.

2. Univalency is not a predominant cause for the specific types of maternal age-dependent nondisjunction. True univalency might, however, predispose to presegregation.

II. Aneuploidy in Djungarian Hamster Oocytes

The Djungarian hamster (*Phodopus s. sungorus*, Rodentia, Cricetinae) is an Asiatic hamster inhabitating the steppes between the Ural Mountains, Mongolia, Djungaria and Altai. This species has mainly been used for research in carcinogenesis, mutagenesis (Pogosianz *et al.*, 1970; Pogosianz, 1975) and for studying the influence of photoperiodicity on reproduction in male hamsters (Hoffmann, 1973). The karyotype was analysed by conventional chromosome staining and by different banding techniques (Schmid *et al.*, 1986). The diploid chromosome number of *Phodopus s. sungorus* is

2*n*-28 with an extreme heterogeneous chromosome morphology, varying from large metacentric to small acrocentric chromosomes. All chromosomes, with the exception of 12, 13, and the Y which are acrocentric, are meta- or submetacentric. By silver staining NORs were demonstrated at the telomeres of the short arms of chromosome 5, 7, 12 and 13. Interestingly, five chromosomes (i.e. chromosome 1,2,4,6 and 7) do contain interstitial blocks of constitutive heterochromatin.

Gonadotropin induced nondisjunction in primary oocytes

In the course of studying the spontaneous frequency of nondisjunction during the first meiotic division in mammalian oocytes (Hansmann and Probeck, 1979; Hansmann and El-Nahass, 1979) we observed that the injection of gonadotropins for stimulated ovulation induced a significant incidence of nondisjunction in oocytes from these hamsters (Hansmann *et al.*, 1980). In those experiments females received intraperitoneally pregnant mares serum (PMS) and 48 h later human chorionic gonadotropin (HCG). Usually 19 h after HCG application ovulated oocytes were isolated from the oviducts, and were prepared cytogenetically. The frequency of chromosomal malsegregation at the first meiotic division was established by counting chromosomes from individually fixed oocytes at metaphase II.

Gonadotropin (GTH) stimulation induced in a dose-dependent manner the ovulation of diploid (2n = 28) and hyperhaploid ($\geqslant$14) oocytes. At high doses (e.g. 10 IU) up to 10 per cent heteroploid oocytes were ovulated. We do observe this phenomenon of GTH induced heteroploidy in our breeding colony now for approximately 10 years.

Influence of maternal age on nondisjunction

The influence of maternal age on chromosome segregation during the first meiotic division was analysed in ovulated occytes from three different groups of females (Hummler *et al.*, 1987). From Table 10 it can be seen that the frequency of hyperhaploidy is significantly increased in oocytes from middle-aged (mean: 36.7 weeks) and old females (mean: 62.4 weeks). In contrast, the frequency of diploid oocytes (most of them being arrested at metaphase I) decreased with maternal age.

Calculating the nondisjunction frequency per bivalent, it became evident that the most pronounced increase occurred in oocytes of middle-aged females. This effect was due to nondisjunction of several bivalents in 7 of 20 hyperhaploid oocytes from that group. In one oocyte even 6 bivalents were malsegregated.

To analyse a possible association of univalents and nondisjunction, metaphase I oocytes were prepared from two age groups, i.e. from young (8–14 weeks) and old (46 ± 2.7 weeks) females. For each group of chromosome (A–D, E, F–G) the frequency of chiasmata (C1), terminal chiasmata (C2 and A) and univalents were determined (Table 11).

The number of chiasmata was not significantly different between oocytes from young (13.2 ± 3.2) and old (12.2 ± 36) females. Only a few univalents of the E, F–G group chromosomes could be detected and there was no significant age effect. The only age effect observed was that on the number of terminal (C2) or terminalized (A) chiasmata. The high frequency of C2 and A pairing configuration in oocytes from old (56, 0%) compared to young females (48.8 %) was due to an increase of C2 and A bivalent configurations of the middle-sized E chromosomes. These chromosomes, however, were only rarely involved in nondisjunction. Thus, the results of our study can be used as a further argument against the validity of the production-line hypothesis (Henderson and Edwards, 1968).

TABLE 10

Influence of maternal age on chromosome segregation during the first meiotic division in oocytes from the Djungarian hamster (according to Hummler *et al.*, 1987)

Age of females	No. of ovulated oocytes analysed	% diploid oocytes	% hyperhaploid oocytes	% of oocytes with nondisjunction*	Nondisjunction* per bivalent in 100 oocytes
young	482	1.0	0.6	1.2	1.1
middle age	1054	0.3	2.8	5.6	10.8
old	499	0.0	3.6	7.2	7.3

* calculated from hyperhaploidy x 2

TABLE 11

Chiasmata and univalents in metaphase I oocytes from young and old Djungarian hamsters

Age of females	Chromosome group									
	A–D				E			F–G		
	No. and type of pairing configuration									
	Total	C1	C2,A	U	C1	C2,A	U	C1	C2,A	U
young	346	109	78	0	40	28	1	27	62	1
old	352	99	92	0	28	42	1	26	61	3

Preferential nondisjunction

Due to the favourable morphology of Djungarian hamster chromosomes we were able to identify the specific malsegregated bivalent in hyperhaploid oocytes, even without the application of banding techniques. In the case of gonadotropin induced nondisjunction we observed a preferential involvement of the large A–D chromosomes (Hansmann *et al.*, 1980). A similar pattern of nondisjunction was observed later on by injecting colchicine or MBC (methylbenzimidazole-carbamate) during the late preovulatory period and in hyperhaploid oocytes from middle aged and old females (Table 12). Altogether 65 nondisjunctional events were allocated and most of them (73.9 per cent) occurred in the group of the large A–D chromosomes. The E and F–G group chromosomes contributed less often to hyperhaploidy (13.9 and 12.3 per cent respectively).

To evaluate the possible causes of preferetial nondisjunction we compared the observed and theoretically expected frequencies of nondisjunction according to two theoretically expected rates (Hummler and Hansmann, 1985):

First expectation: Each bivalent has the same chance to be involved (Chance, Table 13)

TABLE 12

Preferential involvement of bivalents in nondisjunction causing hyperhaploidy in oocytes from old females (Hummler *et al.*, 1987) and after colchicine or MBC treatment (Hummler and Hansmann, 1985; 1988)

Chromosome group	Maternal age		Colchicine	MBC	Total
	middle	old	at 5.5 h post HCG	at 6 h post HCG	
A–D	17	4	12	15	48
E	3	1	1	4	9
F–G	2	2	1	3	8
All	22	7	14	22	65

TABLE 13

Factors determining susceptibility for nondisjuction during meiosis I in oocytes of Djungarian hamsters. Calculation for 65 malsegregated bivalents.

Chromosome group	No. of chromosomes	Relative length (%) in mitosis (Hummler and Hansmann, 1985)	No. of malsegregated bivalents		
			observed	expected from length	expected from chance
A–D	6	68.8	48	44.7	27.9
E	3	19.0	9	12.4	13.9
F–G	5	12.4	8	8.1	23.2
Total	14	100.2	65	65.2	65

Second expectation: Chromosomal properties such as length provide bivalent specific risks (Length, Table 13).

Evidently, the observed pattern differs significantly from that which can be expected from nondisjunction by chance. There is no significant difference from the second expectation, i.e. chromosome length as a determinator of bivalent specific risks.

The molecular basis for the specific involvement of the large chromosomes is still unknown. Furthermore, we do not know whether other factors like the interstitial constitutive heterochromatin of these chromosomes are involved.

Spindle phase

In our model on nondisjunction (Hummler *et al.*, 1987) we suggest that bivalents segregate asynchronously during meiosis I and that those bivalents which were preferentially affected by malsegregation were those segregating last. Furthermore, bivalents should function in this model more or less passively as individual targets for structural and functional alterations of the meiotic spindle. Accordingly, we suggested that preferential nondisjunction results from a time dependent susceptibility of polekinetochore microtubili which should differentiate between asynchronously segregating bivalents.

The spindle poison Carbendazim (MBC) which is known to bind reversibly to mammalian tubulin (Havercroft *et al.*, 1981) was used to investigate whether indeed the time of the inhibition spindle function determines the frequency and pattern of malsegregated bivalents. MBC was applied at an early (4.5 h post HCG) or a later stage (6 h post HCG) of preovulatory oocyte maturation and the effect on meiosis I was subsequently analysed at metaphase II (Tables 14 and 15). From Table 15 it is evident that the time at which spindle phase is inhibited is decisive for both the frequency and pattern of nondisjunction. An early inhibition results in malsegregation of several bivalents (Table 14) and all bivalents seem to have the same risk of being involved (Table 15). The pattern is compatible with an inhibitory activity throughout the spindle phase and characteristic for a chemical. Late spindle phase inhibition affects less oocytes, less bivalents per oocyte and specifically those chromosomes considered to be late segregators.

TABLE 14

Frequency and pattern of nondisjunction in Djungarian hamster oocytes after inhibiting the early or late meiosis I spindle by Carbendazim (according to Hummler and Hansmann, 1988)

Time of MBC application	No. of oocytes analysed	% diploid oocytes	% hyper-haploid oocytes	malsegregated bivalents per oocyte
early	120	3.3	40.6	0.89
late	200	1.5	22.5	0.28

TABLE 15

Influence of spindle phase inhibition on the pattern of malsegregation at meiosis I in Djungarian hamster oocytes (Hummler and Hansmann, 1988)

Chromosome group	% malsegregated bivalents	
	early phase $n=55$	late phase $n=22$
A–D	54.6	68.2
E	16.4	18.2
F–G	29.1	13.6
Total	100.1	100

Rate of meiosis

It has been proposed earlier that an altered rate of meiosis in preovulatory oocytes could be related to nondisjunction (Crowley *et al.*, 1979; Hansmann *et al.*, 1983). This presumes that the rate at which oocytes resume meiosis from dictyotene to metaphase II would not follow a timely fixed programme, but may be variable and characteristically altered in oocytes becoming aneuploid later on. Evidence for an altered rate of meiosis in oogenesis and also spermatogenesis was discussed recently (Hansmann *et al.*, 1988).

We analysed the meiotic rate in preovulatory oocytes from young and old Djungarian hamsters and found a typically altered meiotic rate in oocytes from old females. As discussed earlier, oocytes from these females do have a significantly higher frequency of nondisjunction. Maturation is characterized by a premature germinal vesicle breakdown, a long lasting metaphase I stage and an accelerated segregation phase (Hummler *et al.*, 1987; unpublished results).

In conclusion, we suggest that failures in the endocrine control of follicular maturation are primarily responsible for chromosome malsegregation at meiosis. We presume that these failures alter the signalling which is required for inducing the maturation of the preovulatory oocyte. The quantitatively/qualitatively altered signal(s) seem to cause an abnormal programme for the resumption of meiosis within the oocyte which, amongst other possible effects on the cytoplasm, is followed by an altered meiotic rate. The meiotic spindle serves as a mediator between an altered programme of cytoplasmic maturation and the nucleus. The spindle responds to these unfavourable intracellular conditions with structural and/or functional alterations which do affect asynchronously segregating bivalents. The frequency and pattern of malsegregated half-bivalents is assumed to depend on the stage of meiosis at which pole-kinetochore microtubuli are inhibited. The factors controlling oocyte division and spindle function have to be elucidated.

Acknowledgements

The expert assistance of Marita Heinze-Sprotte in preparing the manuscript is gratefully acknowledged. The research on the Djungarian hamster is financed by grants from the Deutsche Forschungsgemeinschaft.

References

Aarskog D, Fägerhol MK (1970). Protease inhibitor (Pi) phenotypes in chromosome aberrations. J Med Genet 7; 367–70.

Aarskog D, Aarseth P, Fägerhol MK (1978). Alpha-1-antitrypsin (Pi)-types in recurrent miscarriages. Clin Genet 13; 81–4.

Adachi K, Ingalls TH (1976). Ovum ageing and pH imbalance as a cause of chromosomal anomalies in the hamster. Science 194; 946–8.

Arnaud F, Burdash NM, Winson GB, Fudenberg HH (1976). Alpha-1-antitrypsin (Pi)-types in Down's syndrome. Clin Genet 10; 238–43.

Austin CR (1976). Chromosome detoriation in ageing eggs of the rabbit. Nature 213: 1018–9.

Aya T, Makino S, Hirayama A (1966). Some chromosomal studies on patients with infectious hepatitis. Proc Jap Acad 42; 1088–93.

Aymé S, Lippman-Hand A (1982). Maternal-age effect in aneuploidy: does altered embryonic selection play a role? Am J Hum Genet 34; 558–65.

Baker TG (1972). Oogenesis and ovulation. In Reproduction in mammals. I. Germ cells and fertilization. (eds Austin CR, Short RV), Cambridge University Press, London; 14–45.

Beermann F, Bartels I, Franke U, Hansmann I (1987). Chromosome segregation at meiosis I in female T(2;4)1Gö/+ mice: No evidence for a decreased crossover frequency with maternal age. Chromosoma 95; 1–7.

Benda CE (1949). Prenatal factors in mongolism. J Am Med Assoc 139; 979.

Bond DJ, Chandley AC (1983). Aneuploidy. Oxford Monographs on Medical Genetics, No. 11, Oxford University Press, Oxford.

Boué A, Gropp A, Boué J (1985). Cytogenetics of pregnancy wastage. Adv Hum Genet 44; 1–57.

Bouteille M, Hernandez-Verdun D, Dupuy-Con AM, Bourgeors CA (1982). Nucleoli and nucleolar-related structures in normal infected and drug-treated cells. In The nucleolus (eds Jordan EG, Culls CA), Cambridge University Press: 179–211.
Brandriff B, Gordon C, Ashworth L, Watchmaker G, Moore D, Wyrobek AJ, Carrano AV (1985). Chromosomes of human sperm: Variability among normal individuals. Hum Genet 70; 18–24.
Breg WR, Fineman RM, Johnson AM, Kidd KK (1981). The current status of alpha-1-antitrypsin and other factors in Down syndrome. In Trisomy 21 (Down Syndrome) (eds de la Cruz FF, Gerald PS), University Press, Baltimore: 205–14.
Bricarelli FD, Pierluigi M, Perroni L, Grasso M, Arslanian A, Sacchi N (1988). High efficiency in the attribution of parental origin of nondisjunction in trisomy 21 by both cytogentic and molecular polymorphosims. Hum Genet 79; 124–7.
Brook JD, Gosden RG, Chandley AC (1984). Maternal ageing and aneuploid embryos. Evidence from the mouse that biological and not chronological age is the important influence. Hum Genet 66; 41–5.
Burgio GR, Severi F, Rossori R, Vaccaro R (1966). Autoantibodies in Down's Syndrome. Lancet I; 497–8.
Butcher RL (1975). The role of intrauterine environment and intrafollicular ageing of the oocyte on implantation rates and development. In Ageing gametes, their biology and pathology (ed. Balandau RJ), S Karger, Basel: 201–18.
Butcher RL, Fugo NW (1976). Overripeness and mammalian ova. Delayed ovulation and chromosome anomalies. Fertil Steril 18; 297–302.
Carothers AD, Collyer S, de Mey R, Frackiewicz A (1978). Parental age and birth order in the aetiology of some sex chromosome aneuploidies. Ann Hum Genet 41; 277–87.
Ceccarelli G, Torbidoni L (1967). Viral hepatitis and Down's syndrome. Lancet I; 438.
Cohen BH, Lilienfeld AM, Kramer S, Hyman LC (1977). Parental factors in Down's Syndrome—Results of the second Baltimore case-control study. In Population cytogenetics: Studies in humans (eds. Hook SB and Porter IH), Academic Press, New York: 301–52.
Collmann RD, Stoller A (1962). A survey of mongoloid births in Victoria, Australia, 1942–1957. Am J Public Health 52; 813–29.
Coppen A, Cowie V (1960). Maternal health and mongolism. Brit Med J 1; 1843–7.
Crone M, Levy E, Peters H (1965). The duration of the premeiotic DNA synthesis in mouse oocytes. Exp Cell Res 39; 678–88.
Crowley TH, Gulati DK, Hayden TC, Lopez P, Dyer R (1979). A chiasma-hormonal hypothesis relating Downs syndrome and maternal age. Nature 280; 417–8.
Darlington CD (1929). Chromosome behaviour and structural hybridity in the Tradescantiae. J Genet 21; 207–86.
Darlington CD (1937). Recent advances in cytology. 2nd ed, London, Churchill.
Davies KE, Harper K, Bonthron D, Krumlauf R, Polkey A, Pembrey ME, Williamson R (1984). Use of a chromosome 21 cloned DNA probe for the analysis of nondisjunction in Down syndrome. Hum Genet 66; 54–6.
De Boer P, van der Hoeven A (1980). The use of translocation-derived 'marker-bivalents' for studying the origin of meiotic instability in female mice. Cytogenet Cell Genet 26; 49–58.
Dietrich AJJ, Mulder RJP (1983). A light and electron microscopic analysis of meiotic prophase in female mice. Chromosoma 58; 377–85.
Dobzhansky T (1933). Studies on chromosome conjugation II. The relation between crossing-over and disjuction of chromosomes. Z Indukt Abstamm Vererb 64; 269–309.
Dollinger A (1921). Zur Aetiologie des Mongolismus. Ztschr Kinder, 27; 332.
Edwards JH (1970). The operation of selection. In: Jacobs PA, Price WH, Can P (eds). Human Population Cytogenetics. Pfizer Medical Monographs. Vol 5. Edinburgh University Press, Edinburgh; 241–62.
Eiben B, Borgmann S, Schübbe I, Hansmann I (1987). A cytogenetic study directly from chorionic villi of 140 spontaneous abortions. Hum Genet 77; 137–41.
Eiben B, Schübbe I, Borgmann S, Hansmann I (1986). Rapid cytogenetic diagnosis of early spontaneous abortions. Lancet I; 1273–4.
Ek JI (1959). Thyroid function in mothers of mongoloid infants. Acta Faediat (Uppsala) 48; 33–41.
Engel E (1980). A new genetic concept: uniparental disomy and its potential effect, isodisomy. Am J Med Genet 6; 137–43.
Evans HW (1967). The nucleolus, virus infection, and trisomy in man. Nature 214; 361–3.
Fialkow PJ (1964). Autoimmunity: A predisposing factor for chromosomal aberrations? Lancet I; 474–5.
Fialkow PJ, Thuline HC, Hecht F, Bryant J (1971). Familial predisposition to thyroid disease in Down's syndrome. Controlled immunoclinical studies. Am J Hum Genet 23; 67–86.
Fialkow PJ, Uchida IA, Hecht F, Motulsky AG (1965). Increased frequency of thyroid autoantibodies in mothers of patients with Down's Syndrome. Lancet II; 868–70.
Fineman RM, Kidd KK, Johnson AM, Breg WR (1976). Increased frequency of heterozygotes for alpha-1-antitrypsin variants in individuals with either sex chromosome mosaicism or trisomy 21. Nature 260; 320–1.

Freeman ME, Butcher RL, Fugo NW (1970). Alteration of oocytes and follicles by delayed ovulation. Biol Reprod 2; 209–15.

Fugo NW, Butcher RL (1971). Effects of prolonged oestrous cycles on reproduction in aged rats. Fertil Steril 22; 98–101.

Fugo NW, Butcher RL (1976). Overripeness in mammalian ova I. Overripeness and early embryonic development. Fertil Steril 17; 804–14.

Fujita H, Hirano Y (1975). Maternal age of patients with Down's syndrome and Turner's syndrome according to birth date and their karyotypes. Jpn J Hum Genet 20; 269.

German J (1968). Mongolism, delayed fertilization and human sexual behaviour. Nature 217; 516–18.

Golbus MS (1981). The influence of strain, maternal age, and method of maturation on mouse oocyte aneuploidy. Cytogenet Cell Genet 31; 84–90.

Guanti G, di Loreto M (1980). Alpha-1-antitrypsin quantitative and qualitative Pi phenotyping characterization in Down's syndrome subjects and in their parents. Am J Hum Genet 32; 174–8.

Hansmann I (1983). Factors and mechanisms involved in nondisjunction and X-chromosome loss. In: Cytogenetics of the mammalian X chromosome. Part A. Progress and Topics in Cytogenetics, Vol. 3A (ed. AA Sandberg), Alan R Liss, New York; 131–70.

Hansmann I, El-Nahass E (1979). Incidence of nondisjuction in mouse oocytes. Cytogenet Cell Genet 24; 115–21.

Hansmann I, Probeck HD (1979). Chromosomal imbalance in ovulated oocytes from Syrian hamsters (*Mesocricetus* auratus) and Chinese hamsters (*Cricetulus griseus*). Cytogenet Cell Genet 23; 70–6.

Hansmann I, de Boer P, Speed RM (1988). Aneuploidy-related delay of meiotic development in the mouse and the Djungarian hamster. In The cytogenetics of mammalian autosomal rearrangements (ed A Daniel), Progress and Topics in Cytogenetics, Vol. 8, Alan R Liss, New York; 295–314.

Hansmann I, Jenderny J, Probeck HD (1980). Gonadotropin-induced aneuploidy in oocytes from *Phodopus sungorus*. Eur J Cell Biol 22; 24.

Hansmann I, Jenderny J, Probeck HD (1983). Low doses of X-rays decrease the risk of diploidy in mouse oocytes. Muta Res 109; 99–110.

Hansmann I, Bartels I, Beermann F, Caspari D, Franke U, Hummler E, Theuring F (1985). Mechanism of nondisjunction: Facts and perspectives. In Aneuploidy (eds Dellarco FL, Voytek PE, Hollaender A), Plenum Press, New York; 417–32.

Hansson A (1970). Differences in the satellite association pattern on the human population. Hereditas 66; 21–30.

Harlap S (1974). A time-series analysis of the incidence of Down's Syndrome in West Jerusalem. Am J Epidemiol 99; 210–217.

Hassold T (1985). The origin of aneuploidy in humans. In Aneuploidy: etiology and mechanisms (eds Dellarco VL, Voytek PE, Hollaender A), Plenum Press, New York; 103–15.

Hassold T, Chiu D (1985). Maternal age-specific rates of numerical chromosome abnormalities with special reference to trisomy. Hum Genet 70; 11–17.

Hassold TJ, Jacobs PA (1984). Trisomy in Man. Ann Rev Genet 18; 69–97.

Hassold T, Benham F, Leppert M (1988). Cytogenetic and molecular analysis of sex-chromosome monosomy. Am J Hum Genet 42; 534–41.

Hassold T, Jacobs PA, Leppert M, Sheldon M (1987). Cytogenetic and molecuar studies of trisomy 13. J Med Genet 24; 725–32.

Hassold T, Jacobs PA, Kline J, Stein Z, Warburton D (1980). Effect of maternal age on autosomal trisomies. Ann Hum Genet 44; 29–36.

Havercroft JC, Quinlin RA, Gull K (1981). Binding of carbendazole to tubulin and its influence by immunofluorescence microscopy. J. Cell Sci 49; 195–204.

Henderson SA, Edwards RG (1968). Chiasma frequency and maternal age in mammals. Nature 218; 22–8.

Hoffman K (1973). The influence of photoperiod and melatonin on testis size, body weight and pelage colour in the Djungarian hamster (*Phodopus sungorus*). J Comp Physiol 85; 267–82.

Hook EB, Hamerton JF (1977). The nonrandomness of human chromosome abnormalities. In Population cytogenetics: studies in humans (eds Hook EB and Porter H), Academic Press, New York; 237–50.

Hughes A, Jacobsen HI, Wagner RK, Jungblut PW (1976). Ovarian-independence fluctuations of estradiol receptor levels in mammalian tissues. Molec Cell Endocr 5; 379–88.

Hulten M (1974). Chiasma distribution of diakinesis in the normal human male. Hereditas 76; 55–78.

Hummler E, Hansmann I (1985). Preferential nondisjunction of specific bivalents in oocytes from Djungarian hamsters (*Phodopus sungorus*) following colchicine treatment. Cytogenet Cell Genet 39; 163–7.

Hummler E, Hansmann I (1988). Pattern and frequency of nondisjunction in oocytes from the Djungarian hamster are determined by the stage of first meiotic spindle inhibition. Chromosoma 97; 224–30.

Hummler E, Theuring F, Hansmann I (1987). Meiotic nondisjunction in oocytes from aged Djungarian hamsters correlates with an alteration in meiosis rate but not in univalent formation. Hum Genet 76; 357–64.

Jackson-Cook CK, Nance WE, Brown JA (1988). Role of

the double nucleolar organizer region (dNOR) variant in the etiology of trisomy 21. In: The cytogenetics of mammalian autosomal rearrangements (ed. A. Daniel), Progress and Topics in Cytogenetics, Vol 8, A Liss, New York; 697–721.

Jacobs PA, Hassold TJ, Whittington E, Butler G, Collyer S, Keston M, Lee M (1988). Klinefelter's syndrome: an analysis of the origin of the additional sex chromosomes using molecular probes. Ann Hum Genet 52; 93–109.

Jagiello G, Fang JS (1979). Analysis of diplotene chiasma frequencies in mouse oocytes and spermatocytes in relation to ageing and sexual dimorphism. Cytogenet Cell Genet 23; 53–60.

James WH (1978). Down syndrome and parental coital rate. Lancet II; 895.

James WH (1983). Parental coital rates and Down syndrome. Am J Med Genet 14; 783–4.

Janerich DT, Jacobsen HI (1977). Seasonality in Down Syndrome. An endocrinological explanation. Lancet I; 515–6.

John B (1976). Myths and mechanisms of meiosis. Chromosoma 54; 295–325.

Jongbloet PH (1971). Month of birth and gametopathy. An investigation into patients with Down's, Klinefelter's and Turner's syndromes. Clin Genet 2; 315–30.

Jongbloet PH, Frants RR, Hamers AJ (1981). Parental alpha-1-antitrypsin (Pi) types and meiotic nondisjunction in the aetiology of Down's Syndrome. Clin Genet 20; 304–9.

Jongbloet PH, Mulder AM, Hamers AV (1982). Seasonality of pre-ovulatory non-disjunction and the aetiology of Down's Syndrome. A European collaborative study. Hum Genet 62; 134–8.

Jongbloet PH, Pastkoke AJM, Hamers AJH, van Erkelens-Zwets JHJ (1978). Down Syndrome and religious groups. Lancet II; 1310.

Juberg RC (1983). Origin of chromosomal abnormalities: Evidence for delayed fertilization in meiotic nondisjunction. Hum Genet 64; 122–7.

Juetten J, Bavister BD (1983). Effects of egg ageing on in vitro fertilization and first cleavage division in the hamster. Gamete Research 8; 219–30.

Kajii T, Omaha K (1979). Inverse maternal age effect in monosomy X. Hum Genet 51; 147–51.

Kamiguchi Y, Mikamo K (1986). An improved, efficient method for analyzing human sperm chromosomes using zona-free hamster ova. Am J Hum Genet 38; 724–40.

Kamiguchi Y, Funaki K, Mikamo K (1979). Chromosomal anomalies caused by preovulatory overripeness of the primary oocyte. Proc Japan Acad 55; 398–402.

Kaufman MH (1973). Parthenogenesis in the mouse. Nature 242; 475–6.

Kogon A, Krommal R, Peterson DR (1967). Viral hepatitis and Down's Syndrome. Lancet I; 615.

Kondoh T, Tonoki H, Matsumoto T, Tsukahara M, Niikawa N (1988). Origin of the extra chromosome in trisomy 18. Hum Genet 79; 377–8.

Koskull HV, Aula P (1977). Distribution of chromosome breaks in measles, fanconi's anemia and controls. Hereditas 87; 1–10.

Kueppers F, O'Brien F, Passarge E, Rüdiger HW (1975). Alpha-1-antitrypsin phenotypes in sex chromosome mosaicism. J Med Genet 12; 263–4.

Kuroki Y, Makino S, Aya T, Nagayama T (1966). Chromosome abnormalities in cultured leucocytes from rubella patients. Jap J Hum Genet 11; 17–23.

Laing SC, Gosden RG, Fraser HM (1984). Cytogenetic analysis of mouse oocytes after experimental induction of follicular overripening. J Reprod Fert 70; 387–93.

Lange K, Page BM, Elston RC (1975). Age trends in human chiasma frequencies and recombination fractions. I, Chiasma frequencies. Am J Hum Genet 27; 410–8.

Leck I (1966). Incidence and epidemicity of Down's Syndrome. Lancet II; 457–60.

Luthardt FW, Palmer CG, Yu PL (1973). Chiasma and univalent frequencies in aging female mice. Cytogenet Cell Genet 12; 68–79.

Maguire MP (1974). The need for a chiasma binder. J Theor Biol 48; 485–7.

Martin RH, Balkan W, Burns K, Rademaker AW, Lin CC, Rudd NC (1983). The chromosome constitution of 1000 human spermatozoa. Hum Genet 63; 305–9.

Martin-DeLeon PA, Boice ML (1982). Sperm aging in the male and cytogenetic anomalies. An animal model. Hum Genet 62; 70–7.

Mather K (1938). Crossing-over. Biol Rev 13; 252–92.

Matsaniotis N, Kiossoglou K, Maounis F, Anagnostatus DE (1966). Chromosomes in infections hepatitis. Lancet II; 1421.

May KM, Hassold T, Jacobs PA, Lee M, Robinson A (1988). The triple X syndrome: Parental origin of nondisjunction. Am J Hum Genet 43; 114A.

McDermott A (1973). The frequency and distribution of chiasmata in man. Ann Hum Genet 37; 13–20.

McDonald AJ (1972). Thyroid disease and other maternal factors in mongolism. Can Med Assoc J 106; 1085–9.

McNeilly AS (1980). Paradoxical prolactemia. Nature 20; 212.

McPhee H, Ananthakrishnan R, Raft Li (1980). Antiproteases and Down's Syndrome in an Australian population. J Med Genet 17; 170–3.

Mella B (1968). Chromosomal damage associated with infectious hepatitis and its possible teratogenic significance. Neurology 18; 741–4.

Merz T, Prempree T, Migeon BR (1966). Thyrosin, satellite association and trisomy. Nature 209; 1198–1200.

Mikamo K (1968). Mechanism of nondisjunction of meiotic chromosomes and of degeneration of maturation spindles in eggs affected by intrafollicular overripeness. Experientia 24; 75–8.

Mikamo K, Hamaguchi H (1975). Chromosomal disorder caused by preovulatory overripeness of oocytes. In Ageing gametes, their biology and pathology (eds. Blandau RJ), S Karger, Basel, München, Paris, London, New York, Sydney; 72–97.

Miller OJ (1981). The nucleolus. In Trisomy 21 (Down Syndrome). (eds de la Cruz FF, Gerald PS), University Park Press, Baltimore; 205–14.

Milstein–Moscati I, Becak W (1978). Down syndrome and frequency of intercourse. Lancet II; 629–30.

Milstein-Moscati I, Becak W (1981). Occurrence of Down syndrome and human sexual behaviour. Am J Med Genet 9; 211–7.

Milstein-Moscati I, Becak W (1983). Occurrence of Down syndrome and human sexual behaviour. Am J Med Genet 14; 785–6.

Monk M (1988). Genomic imprinting. Genes Develop 2; 921–5.

Morton NE, Jacobs PA, Hassold T, Wu D (1988). Maternal age in trisomy. Ann Hum Genet 52; 227–35.

Mouse News Letter 1988, 81.

Mulcahy MT (1978). Down syndrome and parental coital rate. Lancet II; 895.

Myers CR (1938). An application of the control group method to the problem of the etiology of mongolism. Proc Am Assoc Ment Deficiency 62; 142.

Nankin HR, Sode J, Ball MP (1968). Satellite association and trisomy in thyrotoxicosis. J Med Genet 5; 45–7.

Nichols WW (1966). The role of viruses in the etiology of chromosome abnormalities. Am J Hum Genet 18; 81–92.

Nichols WW, Levan A, Hall B, Ostergren G (1962). Measles-associated chromosome breakage. Hereditas 48; 367–70.

Nielsen J, Peterson GB, Therkelsen AJ (1973). Seasonal variation in the birth of children with aneuploid chromosome abnormalities. Hum Genet 19; 67–74.

Nilsson C, Hansson A, Wilsson G (1975). Influence of thyroid hormones on satellite association in man and the origin of chromosome abnormalities. Hereditas 80; 157–66.

Penrose LS (1933). The relative effects of paternal and maternal age in mongolism. J Genet 27; 219–24.

Pogosianz HE (1975). Djungarian hamster—a suitable tool for cancer research and cytogenetic studies. J Natn Cancer Inst 54; 659–64.

Pogosianz HE, Sokova OI, Prigogina EL (1970). A new animal for experimental cancer research—Djungarian hamster. Vopr Onkol 16; 90–8.

Polani PE (1969). Autosomal imbalance and its syndromes, excluding Down's. Br Med Bull 25; 81–92.

Polani PE (1981). Chiasmata, Down's syndrome, and nondisjunction. An overview. In Trisomy 21 (Down's syndrome)—Research perspectives (eds de la Cruz FF, Gerald PS), University Park Press, Baltimore; 111–30.

Polani PE, Jagiello GM (1976). Chiasmata, meiotic univalents, and age in relation to aneuploid imbalance in mice. Cytogenet Cell Genet 16; 505–29.

Risch N, Stein Z, Kline J, Warburton D (1986). The relationship between maternal age and chromosome size in autosomal trisomy. Am J Hum Genet 39; 68–78.

Rodman TC (1971). Chromatid disjunction in unfertilized ageing oocytes. Nature 233; 191–3.

Ross LC (1981). Trisomy after pregestational contact with communicable disease. Am J Epidem 113; 529–32.

Rudd NL, Dimnik LS, Greentree C, Mendes-Crabb K, Hoar DI (1988). The use of DNA probes to establish parental origin in Down syndrome. Hum Genet 78; 175–8.

Sanger R, Tipett P, Gavin J, Tusdale P, Daniels G (1977). Xg groups and sex chromosome abnormalities in people of northern European ancestry: An addendum. J Med Genet 14; 210–3.

Schmid M, Haaf T, Weis H, Schempp W (1986). Chromosomal homoeologies in hamster species of the genus *Phodopus* (Rodentia, Cricetinae). Cytogenet Cell Genet 43; 168–73.

Schuler D, Koos R, Siegler J, Horvarth L (1970). Thyroid auto-antibodies and satellite associations in Down's syndrome. Human Heredity 20; 13–8.

Sever JL, Gilkeson MR, Chen TC, Ley AC, Edmonds D (1970). Epidemiology of mongolism in the Collaborative Project. Ann NY Acad Sci 170; 328–40.

Shaver EL, Carr DH (1969). The chromosome complement of rabbit blastocysts in relation to the time of fertilization and ovulation. Can J Genet Cytol 11; 287–93.

Shaver EL, Martin-DeLeon PA (1975). Effects of aging of sperm in the female and male reproductive tracts before fertilization on the chromosome complement of the blastocysts. In Aging gametes (ed. Blandau RJ), S. Karger, Basel; 151–65.

Shokeir M (1968). Coital frequency and Down's Syndrome. Lancet II; 1081.

Siegel E, Tobias C (1966). Actions of thyroid hormones on cultured human cells. Nature 212; 1318–21.

Speed RM (1977). The effects of ageing on the meiotic chromosomes of male and female mice. Chromosoma 64; 241–54.

Speed RM (1982). Meiosis in the foetal mouse ovary. I. An

analysis at the light microscope level using surface spreading. Chromosoma 85: 427–37.
Speed RM, Chandley AC (1983). Meiosis in the foetal mouse ovary. II. Oocyte development and age-related aneuploidy. Does a production line exist. Chromosoma 88, 184–9.
Stark CR, Fraumeni JFJ (1966). Viral hepatitis and Down's syndrome. Lancet I; 1036.
Stewart G, Harris P, Galt J, Ferguson-Smith MA (1985). Cloned DNA probes regionally mapped to human chromosome 21 and their use in determining the origin of nondisjunction. Nucleic Acids Res 13; 4125–32.
Stoller A, Collmann RD (1965). Virus aetiology for Down's syndrome. Nature 208; 903–4.
Sugawara S, Mikamo K (1983). Absence of correlation between univalent formation and meiotic nondisjunction in aged female chinese hamsters. Cytogenet Cell Genet 35; 34–40.
Sugawara S, Mikamo K (1986). Maternal ageing and nondisjunction: A comparative study of two chromosomal techniques on the formation of univalents in first meiotic metaphase oocytes of the mouse. Chromosoma 93; 321–5.
Szollosi D (1971). Morphological changes in mouse eggs due to ageing in the fallopian tube. Am J Anat 130; 209–26.
Szollosi D (1975). Mammalian eggs ageing in the fallopian tubes. In Ageing gametes, their biology and pathology (ed. Blandau RJ), S Karger, Basel; 98–121.
Tarkowski AK (1966). An air-drying method for chromosome preparations from mouse eggs. Cytogenet 5; 394–400.
Taylor AI (1968). Autosomal trisomy syndromes: A detailed study of 27 cases of Edward's syndrome and 27 cases of Patau's syndrome. J Med Genet 5; 227–52.
Uilenbroek JTJ, van der Linden R, Woutersen PJA (1984). Changes in oestrogen biosynthesis in preovulatory rat follicles after blockage of ovulation with pentobarbitone. J Reprod Fertil 70; 549–55.
Vanhaelst L, Hayez F, Bonnyns M, Bastenie PA (1970). Thyroid autoimmune disease and thyroid function in families of subjects with Down syndrome. J Clin Endocr Metab 30; 792–7.
Van Wijck JAM, Stolte LAM, Van Kessel HIAM, Tijdink GAJ (1961). A trisomic child of a hyperthyroid mother. Lancet I; 887–8.
Videbeck P, Nielsen J (1984). Chromosome abnormalities and season of birth. Hum Genet 65; 221–31.
Warburton D, Kline J, Stein Z, Susser M (1980). Monosomy X: A chromosomal anomaly associated with young maternal age. Lancet I; 167–9.
Wauben-Penris PJJ, van Buul-Offers SC (1982). Meiotic nondisjunction in male Snell dwarf mice. J Hered 73; 365–9.
Wren PJ, Evans DAP, Velters JM, Chew A (1967). Autoimmune antibodies in mongol families. Lancet II; 186–8.
Yamamoto M, Ingalls TH (1972). Delayed fertilization and chromosome anomalies in the hamster embryo. Science 176; 518;–21.
Zankl H, Mayer C, Zang KD (1980). Association frequency and silver staining of nucleolus organizing regions in hyperthyroid parents of patients with trisomy 21. Hum Genet 54; 111–14.

13 Chromosome Alterations and Oncogenes in Human Neoplasia

AVERY A. SANDBERG[1] AND SURABHI KAKATI[2]

[1]The Cancer Center of Southwest Biomedical Research Institute
6401 East Thomas Road, Scottsdale, AZ 85251, USA*
[2]Roswell Park Memorial Institute
666 Elm Street, Buffalo, New York, 14263, USA

Introduction

Cytogenetic analyses have revealed a large number of chromosome rearrangements characterizing various human neoplastic diseases (Rowley, 1984; Sandberg, 1986a, b; Dal Cin and Sandberg, 1988). These chromosome abnormalities are usually represented by translocations, deletions, insertions, inversions, gains or losses of chromosome(s). In Tables 1–5 are listed those rearrangements (translocations and other morphologic chromosome changes) and numerical changes thought to be specific (primary) in nature and to characterize the neoplastic diseases listed.

When a leukaemia or tumour is associated with a sole karyotypic change, it can be assumed that this change represents the primary (or specific) chromosome change characterizing the particular leukaemia or tumour. The fact that one of these primary karyotypic changes is consistently shared by a neoplastic entity from unrelated patients indicates that the primary chromosomal event plays a key role in the malignant transformation of the cells. The presence of the primary chromosome change as the sole abnormality in the neoplastic cells is much more commonly encountered at diagnosis in leukaemia than in solid tumours. This is possibly due to the fact that the clinical (and hence biological) manifestations of leukaemia generally become apparent at a much earlier stage of the disease than do those of solid tumours and, therefore, the neoplastic (leukaemic) cells are often examined at a time when they contain only the primary karyotypic change. It is also possible that karyotypic elevation may occur more readily in solid tumours than in in leukaemias.

The identification of specific cytogenetic events in solid tumours has lagged behind that of the leukaemias for several reasons, including among others the frequent encounter of multiple and complex chromosomal changes. These complex karyotypic changes may confuse the picture as they render difficult the identification of the primary chromosome changes associated with specific tumour types (Figure 1). Except in some leukaemias (i(1 q) in CML, +8 in AL, −Y or

* Address correspondence to Dr. Sandberg.

TABLE 1

Common chromosome changes in ANLL (Classified by the FAB System)

M1—Acute Myeloblastic Leukaemia without Differentiation	M2—Acute Myeloblastic Leukaemia with Differentiation	M3—Acute Promyelocytic Leukaemia
−7	t(8;21)	t(15;17)
+8	−7	i(17q)
5q−	+8	t(1;17)
−5	5q−	
Ph+	7q−	
7q−; del or t with 3	−5	
	del or inv (3)	
	t(6;9)	
M4—Acute Myelomonocytic Leukaemia	**M5—Acute Monocytic Leukaemia**	**M6—Acute Erythroleukaemia**
del or t or inv(16)	del or t(11q)	−7
+8	+8	+8
−7	t(9;11)	5q−
del or t(11q); t(6;9);		−5
5q−/−5; 7q−; +4; Ph+		7q−
t(5;7) (q11.2; p11.2) (secondary)		−3 or t with 3
		dup (1)
M7—Acute Megakaryocytic Leukaemia		
inv or del (3)		
ring		
+8		
+21		

TABLE 2

Consistent and primary structural chromosome abnormalities in ALL

Rearrangement	Typical morpho-logy	Typical immunophenotype
t(1;11)(p32: q23)	L1	Pre-B ALL
t(1;19)(q23; p13)	L1	Pre-B ALL
t(2;8)(p12; q24)	L3	B ALL
t(4;11)(q21; q23)	L1, L2	Early B-precursor ALL, mixed phenotype
del (6q)	L1, L2	Common ALL
t(8;14) (q24; q11)	L1, L2	T ALL
t(8;14) (q24; q32)	L3	B ALL
t(8; 22)(q24; q11)	L3	B ALL
del/t(9p)	L1, L2	T ALL or early T-precursor ALL
t(9;22)(q34;q11)	L1, L2	Early B-precursor, common or pre-B ALL
t(10;14)(q24;q11)	L1, L2	T ALL
t(11;14)(p13;q11)	L1, L2	T ALL
t(11;14)(q23;q32)	L1, L2	non-B, non-T ALL
t(11;19)(q23;p13)	L1, L2	B ALL, mixed phenotype
del/t(12p)	L1, L2	Common ALL
Near-haploid	L1, L2	B ALL

−X in AML with t(8;21)), the additional (secondary) chromosome changes have not been found to follow a non-random pattern and invariably appear to be associated with the biological progression of the disease, such as increases in proliferative activity, drug resistance, tumour invasiveness, metastatic spread and aggressiveness. In other words, a leukaemia or tumour is at its lowest level of aggressiveness when it is associated with the primary karyotypic change only, the leukaemia or tumour showing more aggressive behaviour with the acquisition of additional chromosome changes (Nowell, 1986).

The findings of recurrent chromosome changes in human neoplasias, i.e., leukaemia and lymphoma, and recently in some solid

TABLE 3

Specific chromosome changes in lymphoma

t(2;8)(p12;q24)	Burkitt lymphoma
t(2;5)(p23;q35)	Non-Hodgkin lymphoma
+3	Non-Hodgkin lymphoma
+5	Non-Hodgkin lymphoma
6q−	CLL and non-Hodgkin lymphoma
6p−	Non-Hodgkin lymphoma
+7	Non-Hodgkin lymphoma
t(8;14)(q24;q32)	Burkitt lymphoma
t(8;22)(q24;q11)	Burkitt lymphoma
t(11;14)(q13;q32)	CLL and diffuse small cell lymphoma
11q−	CLL and non-Hodgkin lymphoma
+12	CLL and diffuse small cell lymphoma
14q11	T-Cell
t(14;18)(q32;q21)	Follicular lymphoma
14q+	Non-Hodgkin lymphoma
+18	Non-Hodgkin lymphoma
t(2;5) (p 23;q35)	Malignant histiocytosis

TABLE 4

Single chromosome changes in patients with primary myelodysplastic syndromes

−7
+8
del (5) (q12 − q34) and translocations involving 5q
del (7q)
del (11q)[a]
del (12) (p11p13)
del (13q)[b]
del (20) (q11q13)

t(1;3)(p36;q21)
t(2;11)(p21;q23)
t(6;9)(p23;q34)
t(11;21)(q24;q11.2)
i(17q)

[a] 11q 14 is always involved in either interstitial or terminal deletions.
[b] 13q 14 is always involved in interstitial deletions of variable size.

TABLE 5

Specific (primary) chromosome changes in Tumours

Tumours	Chromosome changes
Benign	
Meningioma and acoustic neuroma	– 22 or 22q –
Mixed tumours of salivary glands	t(3; 8) (p21; q12)
	t(9; 12) (p; 13–22; q13–15)
Lipoma	t with 12q14
Colonic adenoma	12q – and/or +7
	12q – and/or +8
Leiomyoma	t(12; 14) (q14–15; q22–24)
	7q –
Cortical adenoma of kidney	+7, +7, +17, – Y
Carcinomas	
Bladder	i(5p)
	+7
	– 9/9q –
	11p –
Prostate	del (10) (q24)
Lung (SCLC)	del (3) (p14p23)
Colon	12q –*
	+7*
	+8*
	+12*
	17 (q11)*
	17p –*
	– 18*
Kidney	del (3) (p11→p21)
Uterus	1q –*
Ovary	6q –*
	t(6; 14) (q21; q24)
Endometrium	trisomy 1q
	+10
Sarcomas	
Liposarcoma (myxoid)	t(12; 16) (q13; p11)
Synovial sarcoma	t(X; 18) (p11.2; q11.2)
Rhabdomyosarcoma (alveolar)	t(2; 13) (q37; q14)
Extraskeletal myxoid chondrosarcoma	t(9; 22) (q31; q12.2)
Leiomyosarcoma	– 22*
Embryonal and other	
Testicular (germ cell tumors)	i(12p)
Retinoblastoma	del(13) (q14)**
	i(6p)
Wilms tumour	del(11) (p13)**
Neuroblastoma	del(1) (p32p36)
Malignant melanoma	del(6) (q11q27)*
	i(6p)*
	del(1) (p11p22)*
	t(1; 19) (q12; q13)
Mesothelioma	del(3) (p13→p23)
Ewing sarcoma and peripheral neuroepithelioma	t(11; 22) (q24; q12)
Glioma	– 22*

* Not yet proved to be primary. ** Associated with a constitutional chromosome change.

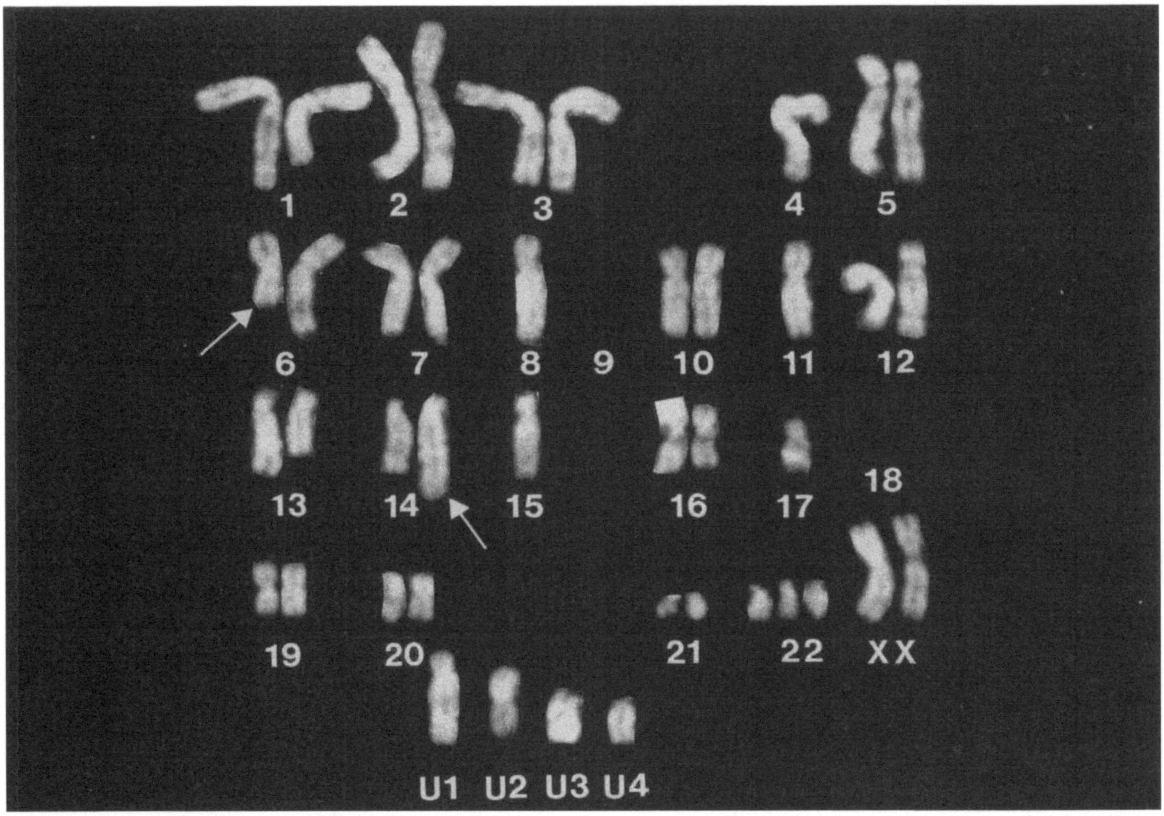

FIGURE 1: Karyotype (Q-banded) of an ovarian cancer cell showing a t(6;14), which was present in all of the cells of the tumour, and a number of additional (secondary) changes (−4,−8,−9,−9,−11,−13q,−15,−17,−18,−18,+22). Secondary changes are common in cancers, particularly those of epithelial origin. U1−U4 represent abnormal chromosomes whose origin could not be ascertained.

tumour support the idea that chromosome rearrangements play an important role in tumour development. In the last few years some of these non-random karyotypic changes have been molecularly investigated and it has been demonstrated that: (1) structural chromosome changes can alter specific genes that are located at the breakpoints or they can affect gene dosage, and (2) numerical changes result in either gene amplification or loss of heterozygosity for complete chromosomal sets of genes (Dal Cin and Sandberg, 1988).

Leukaemias and Lymphomas

A major contribution of cytogenetics to the understanding of the acute leukaemias has been the demonstration that recurrent karyotypic anomalies characterize certain of these leukaemias (Tables 1−4) and serve as independent prognostic factors.

When a leukaemia is associated with a sole karyotypic change (translocation, loss or gain of a chromosome, insertion, deletion), it can be assumed that this represents the primary or specific chromosome change characterizing the particular leukaemia or tumour. When a group of leukaemias of similar origin and pathology contain the same primary karyotypic change, it can be safely assumed that they constitute a subset or subtype within the leukaemia or tumour type.

Though the primary chromosome change may play a key role in the process of leukomogenesis or carcinogenesis, and in some cases the evidence is more than suggestive (e.g., the abnormal *PHL/ABL* gene in CML) there is the possibility that the primary karyotypic change reflects events at the molecular level which may have preceded the chromosome change and, in fact, be the cause of it. Thus, however essential the karyotypic change is in the development of the neoplastic state, there may exist an underlying molecular event necessary for the genesis of the specific chromosome change.

Though emphasis naturally has been placed on the specific chromosome changes and their meaning, it must be kept in mind that significant percentages (and in some conditions, the majority) of cases have not had a primary karyotypic change established. Furthermore, the wide array of additional (secondary) cytogenetic changes which have so far defied classification or organization must be kept in mind when evaluating the role of the chromosome changes in various conditions.

The characterization of the various leukaemias has been for a number of years based on the so-called FAB systems (Bennett *et al.*, 1981; Atkin and Baker, 1982; Bennett *et al.*, 1985) which have relied heavily on the characteristics of cellular morphology and immunology. The cytogenetic demonstration that FAB categories thought to be relatively homogeneous contained subsets characterized by specific chromosome changes ultimately led to MIC Workshops (First, Second, Third in 1986, 1988 and 1988 respectively, MIC Cooperative Study Group) in which these changes were related to the various clinical and cytological aspects of each leukaemic subset. In these MIC workshops members of the FAB group, immunologists, histochemists, clinicians and cytogeneticists critically evaluated the information available in each area in an attempt to to define as accurately and reliably as possible the nature of the conditions associated with specific chromosome changes. It is hoped that these results will find use in the diagnostic and clinical aspects of various hematological neoplastic disorders.

Solid Tumours

Although progress in solid tumour cytogenetics has not been as rapid or clinically meaningful as that in the leukaemias, advances in methodology have made possible the detailed examination of the karyotypes of a number of tumour entities (e.g., sarcomas, testicular and kidney cancers, neuroblastoma), with most of the carcinomas remaining to be elucidated cytogenetically (Table 5). These advances have led to the description of a number of specific chromosome changes in solid tumours, as well as to two solid tumour workshops (First and Second International Workshops on Chromosomes in Solid Tumours, 1986, 1987) dealing with the karyotypic aspects of such tumours. Future solid tumour workshops will undoubtedly combine progress at the cytogenetic as well as at the molecular and other levels. As in the case of the leukaemias, the combination of cytogenetic and molecular results is very likely to lead to the description of subtypes within existing tumour entities and the utilization of the findings in the diagnosis, classification and development of therapeutic approaches and prognostic aspects of these tumours.

The number of solid tumours characterized by a consistent specific chromosome marker is still relatively small (Table 5), as is the number of cases of chromosomally investigated tumours of each subtype. However, several conclusions can be inferred:

1. Consistent chromosome changes in tissue proliferations are no longer a criterion of malignancy, since such changes have been described in tumours considered benign on histological and behavioural bases.
2. Lipoma and myxoid liposarcoma (LPS), both tumours of adipose tissue but one benign and the other malignant, share a consistent chromosomal translocation involving very close, if not an identical, region on chromosome 12, i.e., band 12q13 in myxoid LPS and region 12q13 or q14 in typical lipomas. This observation, though it may speak in favour of a possible common basis, in at least one of the steps responsible for the neoplastic process in adipose tissue, will have to be confirmed or refuted by appropriate molecular studies.
3. An identical chromosomal change may be observed in a tumour derived from cells of the same lineage but at different levels of differentiation, e.g., in germ cell tumours of the testis, an identical i(12p) can be found in histologically different tumours.
4. A number of different primary cytogenetic changes may be found within a previously defined tumour entity, each possibly characterizing a subtype within the entity, e.g., as in bladder cancer in which subtypes may be characterized by one of the following changes: +7, −9, 5q− or i(5p) and 11p−.
5. Marker chromosomes have the highest degree of specificity since they are associated with a precise tumour subtype and support the histological definition. For example, within the liposarcoma type, the myxoid subtype is the only one to exhibit t(12;16) (q13;p11), whereas other types of chromosome changes were observed in well-differentiated and pleomorphic subtypes.
6. Cytogenetic findings can be used by the pathologist as guides for the correct diagnosis of tumours lacking specific cellular differentiation or characteristic histology.

Molecular Correlations

The molecular applications of cytogenetic findings in leukaemia and solid tumours have been more informative in the former than the latter. Thus, it has been shown that the translocation (9;22) (q34;q11) leading to the genesis of the Philadelphia (Ph) chromosome in chronic myelocytic leukaemia (Figure 2) is associated with translocation of the oncogene *ABL* to an abbreviated chromosome 22 in which the break occurs within a 5.8 kb segment of the *PHL* (or *BCR*) known as bcr. The formation of the near chimeric gene, *ABL/PHL* leads to the production of an abnormal mRNA and thus to an abnormal protein product. The latter has been implicated as the possible cause of the leukaemia. Chromosomal changes leading to rearrangements of the oncogene *MYC* have also been described in Burkitt lymphoma and related conditions with t(8;14) (q24;p32). Though similar results have been observed in a few leukaemias with specific chromosome changes, the list is not very large and much remains to be done molecularly in the leukaemias and lymphomas and even more so in solid tumours.

Chromosome Changes in Benign Tumours

The chromosomal studies in human malignant neoplasms have shown that most, if not all, malignant cells exhibit chromosome abnormalities, often occurring in a non-random fashion (Sandberg, 1985). In contrast, the chromosomal status of so-called benign neoplasms has been far less defined, due essentially to the paucity of the studies reported to-date. Though it had been generally accepted that benign tumours have a

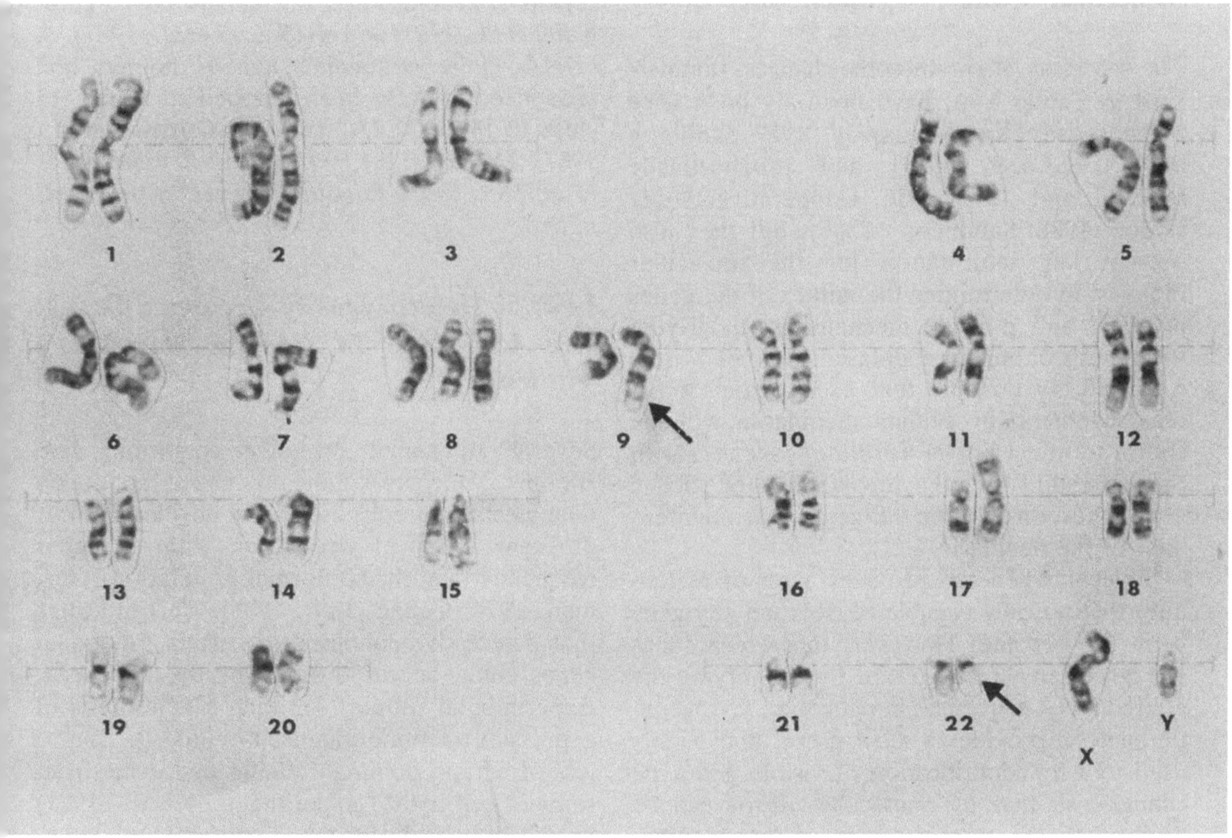

FIGURE 2: G-banded karyotype of a leukaemia cell from a patient with CML in blastic phase showing in addition to the Ph translocation, a +8 and an i(17q). The latter two changes are common in the blastic phase and, in fact, their appearance may precede or herald the onset of that phase.

normal chromosome constitution (Sandberg, 1980), non-random chromosome abnormalities have been described in a number of benign tumours, i.e., meningiomas, mixed tumours of the salivary glands, lipomas, leiomyomas, adenomas of the colon and cortical adenomas of the kidney. Such data highlight the fact that the concept of an abnormal chromosome pattern being related to a malignant or premalignant state versus a normal diploid karyotype related to a benign state may have to be qualified. On the other hand, cytogenetic data on benign tumours may provide clues to an understanding of the biological basis for the distinctive behaviour of benign versus malignant neoplasms.

Molecular Cancer Cytogenetics

The specificity of chromosome changes, summarized in Tables 1–5, have not only been used diagnostically (Karakousis *et al.*, 1987; Sandberg and Turc-Carel, 1987) and prognostically (Second and Third MIC Cooperative Study Group, 1988; Sandberg, 1986b), but they also present key information for the molecular biologist in determining the nature of the genes involved and possible mechanisms underlying the genesis of human neoplasia. In other words, it is entirely possible that each chromosomal rearrangement in human neoplasia will be shown to affect a gene (proto-oncogene?) with the possibility that the involvement of such a gene is responsible for the genesis or maintenance of the neoplasia.

Molecular genetics provides a set of techniques theoretically capable of isolating any gene from the genome. However, these techniques are not often directly useful for identifying the altered DNA sequences in neoplasia. Cytogenetic analysis provides a most useful and widely used tool for identification of visible genotype changes, so that the molecular efforts can be directed to the correct region of the genome.

An exciting challenge for molecular genetics has been recently presented by non-random chromosome changes in tumours considered benign on histologic and behavioural bases (Sandberg, 1988). The fact that benign as well as malignant proliferations, e.g., lipoma and myxoid liposarcoma, respectively, both adipose tissue tumours, share the same chromosome region (12q13–q14) in consistent rearrangements (Truc-Carel *et al.*, 1986a and b) may serve as an excellent model for the study of molecular events underlying benign versus malignant proliferation. One of the first steps is to investigate the proximity of the breakpoints in these two tumours at the molecular level and to ascertain whether these breakpoints fall within the same gene. Two genes have been localized to this region of chromosome 12, namely int-1 (Turc-Carel *et al.*, 1987) and gli (Kinzer *et al.*, 1987). A cDNA clone containing human homeo box sequences has also been mapped to 12q12–13 close to the COL2A1 gene (Cannizzaro *et al.*, 1987). However, the involvement of these genes in adipose tissue tumours has yet to be determined.

Loss of Heterozygosity Creates a Recessive Condition in Certain Hereditary Tumours

Several autosomal recessive conditions like Bloom syndrome, Fanconi anaemia, ataxia telangiectasia predispose to the development of different types of neoplasms with a higher frequency than in the normal population (German, 1974; Cohen *et al.*, 1975). This indicates that a recessive condition of certain functional genes could be responsible for the neoplasia. Experimental support for such a hypothesis of expression of tumourigenecity could be due to loss of certain dominant alleles first came from somatic cell hybrid experiments. Harris (1971) produced hybrid cells by: (1) normal and malignant cell fusion and, (2) malignant and malignant cell fusion. The hybrid cells derived from normal and malignant cell fusion initially seemed to lose their malignant potential, but later regained their malignant potency after loss of some chromosomes of the normal non-malignant parent. This indicated that malignancy can also be brought about by recessive rather than dominant mutation or by loss of wild type alleles.

Loss of a wild type locus may be achieved either through non-disjunction, non-disjunction and subsequent reduplication, mitotic recombination, deletion, gene inactivation and a second mutation inactivating the functional gene.

Knudson (1971) first hypothesized: (1) that

sporadic and inherited forms of a particular cancer could both result from mutations in the same gene, (2) that these mutations act recessively at the cellular level, and (3) that both copies of the gene must be lost for cancer to develop.

This hypothesis proved to be true in few inherited childhood tumours, specifically in retinoblastoma, and Wilms' tumour and in dominantly inherited conditions such as familial adenomatous polyposis.

Retinoblastoma

Retinoblastoma (RB) is a malignant tumour arising in the eyes of children. Its incidence is approximately one in every 20,000 live births (Devesa, 1975). Most children with retinoblastoma are first identified by parents, who generally consult an ophthalmologist for one or more of several presenting complaints. In order of frequency, these are leukokoria (56%), strabisums (20%), painful eye with glaucoma (7%) and poor vision (5%). Less common presenting findings are orbital cellulitis, unilaterally dilated pupil, heterochromia, hyphema, nystagmus and even failure to thrive (Grabowski and Abramson, 1987).

Three forms of retinoblastoma have been noted: hereditary, non-hereditary, and chromosomal deletion form:

1. The hereditary form of RB refers to clinical situations in which there is a positive family history of the tumour or in cases of sporadic disease where the individual is bilaterally affected. No chromosomal defect is found in the peripheral lymphocytes in this hereditary form of RB,
2. The non-hereditary form of RB is applied to individuals with unilateral disease with no family history of the tumour and no chromosomal abnormality in peripheral lymphocytes. It is, however, estimated that approximately 15 per cent of patients with sporadic unilateral disease actually have the hereditary form of RB.
3. In the chromosomal deletion form of RB, the lymphocytes have a deletion at 13q14, to which band the locus for RB has been assigned (Sparkes *et al.*, 1983).

The possible genotypes at the RB locus are as follows: If only one eye is involved and there is no family history of RB, the tumour is usually non-hereditary and the constitutional genotype is presumed to be Rb+/Rb+. When retinoblastoma develops bilaterally or when there is a positive family history of RB, a mutant or inactive allele of germinal origin is assumed with a genotype of Rb+/Rb− and dominant inheritance is the rule. Approximately 2 to 3 per cent of patients with retinoblastoma have the chromosome 13 deletion form of the tumour and a genotype Rb+/−. In addition, a significant percentage of patients with RB may have a chromosome deletion mosaicism (Motegi, 1982).

In the chromosome deletion form of RB, activity of the enzyme esterase D was studied in great detail (Sparkes *et al.*, 1980). The locus for esterase D is closely linked to the Rb locus. This enzyme shows an electrophoretic polymorphism with two common alleles being types 1 and 2. Only 50 per cent esterase D activity was found in various tissues in all the patients with chromosome deletion form of RB. This indicates that the normal and non-deleted 13 chromosomes were lost from the tumour. This observation was further supported by studies using several gene loci with restriction fragment length polymorphism. Loss of heterozygosity was also indicated in these tumours. This loss of heterozygosity creates a recessive condition for RB.

Various studies indicated that there is a close relation between hereditary RB and osteosarcoma. Children who are treated successfully for hereditary RB have a 10–15 per cent risk of developing osteosarcoma as a secondary tumour. Hansen *et al.* (1985) indicated that osteosarcoma and RB share a common pathogenic mechanism. Using recombinant DNA probes which are homologous to loci on chromosome 13, they showed loss of constitutional heterozygosity in the majority of secondary osteosarcomas and several sporadic osteosarcomas.

Wilms' Tumour

Wilms' tumour (WT) is the second most common abdominal tumour of childhood. The tumour arises in the kidney and usually originates from three tissue elements: blastoma of the embryonic kidney, epithelium and stroma. Incidence of WT is about 7.6 per 1,000,000 live births.

WT may occur in an autosomal dominant form. A fraction of WT patients have a constitutional deletion of 11p13 (Riccardi *et al.*, 1978; Franke *et al.*, 1979). These patients have mental retardation, retinal abnormalities and abnormal genitalia, very commonly associated with aniridia.

As in RB, in WT also a germinal or somatic mutation may occur, this time on one chromosome 11 and predispose to WT, with a subsequent somatic genetic event on the normal homolog at 11p13 permitting tumour development (Knudson, 1976; Riccardi *et al.*, 1978). That the tumour-specific somatic genetic event may involve 11p has been supported by the demonstration of loss of tumour-specific heterozygosity for 11p loci even in the presence of an apparently normal chromosome 11 (Fearon *et al.*, 1984; Koufos *et al.*, 1984; Orkin *et al.*, 1984; Reeve *et al.*, 1984). To study the frequency and mechanism of tumour-specific genetic events, Dao *et al.* (1987) have examined the karyotype and chromosome 11 genotype of normal and tumour tissues from 13 childhood renal tumour patients with different histologic tumour types and associated clinical conditions. Tumours of 8 of the 12 WT patients showed molecular evidence of loss of 11p DNA sequences by somatic recombination (4 cases), chromosome loss (2 cases), and recombination or chromosome loss and duplication (2 cases). One malignant rhabdoid tumour in a patient heterozygous for multiple 11p markers did not show any tumour-specific 11p alterations. These findings confirm the critical role of 11p sequences in WT development and reveal that mitotic recombination may be the most frequent mechanism by which such a tumour develops.

Dominantly Inherited Conditions

Beckwith-Wiedemann syndrome (BWS) is autosomal dominant in inheritance and shows multiple congenital malformations coupled with a concurrent high risk for the development of specific rare tumours. More than 10 per cent of all individuals with BWS develop WT, hepatoblastoma, rhabdomyosarcoma and adrenal carcinoma. All these three childhood tumours share a common genetic event which is loss of heterozygosity of an allele at the 11p as in WT. Genetic events associated with these three childhood tumours is similar to WT (Koufos *et al.*, 1985).

Familial adenomatous polyposis (FAP) which was previously known as familial polyposis coli is a genetic disorder transmitted as an autosomal dominant trait, and characterized by the development of numerous adenomatous polyps in the colon and rectum within the first decades of life. The patients are at an extremely high risk of colorectal carcinoma if left untreated.

In a recent case report, we showed a constitutional interstitial deletion in the long arm of chromosome 5 at band between 5q15 – q22 in a mentally retarded individual with multiple developmental abnormalities and FAP (Herrera *et al.*, 1986). Following this report Bodmer *et al.*

TABLE 6

Chromosome oncogene locations in the human genome

Chromosome	Location	Oncogene
1	1p36.1 – p36.2	FGR
1	1p35 – p32	LCK
1	1p32	MYCL
1	1p32	BLYM
1	1p31 – 32	JUN
1	1p22 and/or lp12-cen	NRAS
1	1q22 – q24	SKI
2	2p24	MYCN
2	2p13–cen	REL
3	3p21.33 – p22 or chr17	ERBA2
3	3p25 – p24	RAFI
4	4q11 – q22	KIT
5	5p14 – p13	MLVI2
5	5q33 – q34	CSF1R (FMS)
6	6p23 – q12	YES2
6	6p21	PIM
6	6p12 – p11	KRAS1P
6	6q21	SYR
6	6q16 – 22	ROS1
6	6q22 – q23	MYB
6	6q24 – 27	MAS1
7	7p22 – p15	RAL
7	7p14 – q21	ARAF2
7	7p13 – p12	EGFR
7	7q31	MET
8	8q11 or q22	MOS
8	8q13–qter	LYN
8	8q24	MYC
8	8q24	PVT1
9	9q34	ABL
11	11p15.5–pter	HRAS1
11	11q13	INT2
11	11q13	SEA
11	11q13.3	BCL1
11	11q23 – q24	ETS1
12	12p12.1	KRAS2
12	12q12–q14	INT1
12	12q13–q14.3	GLI
13	13q12	FLT
14	14q24.3 – q31	FOS
14	14q32	AKT1
15	15q25 – q26	FES
17	17q11 – q21	ERBA1
17	17q11 – q12	NGL
17	17q21 or chr3	ERBA2
18	18q21.3	YES1
18	18q21.3	BCL2
18	18q21	BCL3
18	18q22–qter	ERV1
19	19p13.2–q13.2	MEL
20	20q12–q13	HCK
20	20q11–q13	SRC
21	21q22.3	ETS2
22	22q12.3–q13.1	PDGFB (SIS)
X	Xp21–q11	ARAF1
X	Xq27	MCF2

(1987) localized the gene for FAP at band 5q21–q22. A polymorphic DNA probe which is localized at this region of chromosome 5q produces two alleles of 3.9 and 4.4 kilobases. The affected individuals show heterozygosity with this polymorphic probe. This finding was later confirmed by linkage analysis (Leppert *et al.*, 1987). Solomon *et al.* (1987) using a highly polymorphic mini-satellite probe which maps to chromosome 5q showed that at least 20 per cent of highly heterogeneous sporadic tumours of the colon lose one of the alleles present in matched normal tissues. Thus, recessiveness for a particular gene at 5q15-q22 may be critical in the progression of a relatively high proportion of colorectal cancers. However, in a recent report frequent allele loss of chromosome 22, with some additional losses on chromosomes 5, 6, 12q and 15 were noticed in colorectal tumours from FAP patients and some sporadic colorectal tumours (Okamoto *et al.*, 1988). The discrepancy of the findings indicates that FAP may be heterogeneous in the population.

Cancer and Oncogenes

As their name indicates, *oncogenes* are thought in essence to be capable of inducing and maintaining malignant transformation and proliferation. To-date, only a few of the known oncogenes (Table 6) have been shown to be directly involved in and affected by the karyotypic events in human leukaemia and cancer, e.g., the c-*abl* in CML, c-*myc* and *bcl*-1 in lymphoma. Though, as shown in Table 6, many oncogenes have been located to bands involved in various karyotypic changes, their direct involvement by the chromosomal rearrangements has not been demonstrated, possibly indicating that the chromosomal breaks affect areas of the bands removed from the oncogene loci. There is a possibility of indirect involvement of oncogenes in the initiation of the neoplastic process and certainly in the subsequent biology of the neoplasia, and activation expression or overexpression and/or amplification of a number of oncogenes in various leukaemias and cancers have been amply demonstrated. It is possible, as further primary karyotypic changes are established (particularly in solid tumours), that direct involvement of some oncogenes will be demonstrated or that these, as have the previously established cytogenetic changes, will point to the presence of oncogenes hitherto undercovered. It appears certain that the actual number of oncogenes, if the term is taken at its face value, greatly exceeds that known at present, considering the large array of specific chromosome changes already known in human cancer and leukaemia, and the multitude of types of human leukaemias and cancers.

References

Atkin NB, Baker MC (1982). Specific chromosome change, i(12p), in testicular tumours? Lancet ii; 1349.

Bennett JM, Catovsky D, Daniel MT, Flandrin G, Galton DAG, Gralnick HR, Sultan C (1981). Morphological classification of acute lymphoblastic leukemia: Concordance among observers and clinical correlations. Br J Haemat 47; 553–61.

Bennett JM, Catovsky D, Daniel MT, Flandrin G, Galton DAG, Gralnick HR, Sultan C (1985). Proposed revised criteria for the classification of acute myeloid leukemia: A report of the French-American-British Cooperative Group. Ann Intern Med 103; 620–9.

Bodmer WF, Bailey CJ, Bodmer J, Bussey HJR, Ellis A, Gorman P, Luceibello SH, Scambler P, Sheer D, Solomon E, Spurr NF (1987). Localization of the gene for familial adenomatous polyposis on chromosome 5. Nature 328; 614–6.

Cannizzaro LA, Croce CM, Griffin CA, Simeone A, Boncinelli E, Huebner K (1987). Human homeo box-containing genes located at chromosome regions 2q31–2q37 and 12q12–12q13. Am J Hum Genet 41; 1–15.

Cohen MM, Shahlam M, Dagan J, Shmueli E, Kohn G (1975). Cytogenetic investigations in families with ataxia telangiectasia. Cytogenet Cell Genet 15; 338–56.

Dal Cin P, Sandberg AA (1988a). Karyotypic analysis of solid tumours. In Pathology of the Nucleus (Ed. Underwood JCE). Springer-Verlag, Berlin (in press).

Dal Cin P, Sandberg AA (1988b). Chromosomal Aspects of Human Oncogenesis. CRC Crit Rev (in press).

Dao DD, Schroeder WT, Chao LY, Kikuchi H, Strong LC, Riccardi VM, Pathak S, Nichols WW, Lewis WH, Saunders GF (1987). Genetic mechanisms or tumor-specific loss of 11p DNA sequences in Wilms's tumour. Am J Hum Genet 40; 202–17.

Devesa SS (1975). The incidence of retinoblastoma. Am J Ophthalmol 80; 263–5.

Fearon ER, Vogelstein B, Feinberg AP (1984). Somatic deletion and duplication of genes on chromosome 11 in Wilms' tumour. Nature 309; 176–8.

First International Workshop on Chromosomes in Solid Tumours (1986). Cancer Genet Cytogenet 19; 1–197.

First MIC Cooperative Study Group (1986). Morphologic, immunologic and cytogenetic (MIC) working classification of acute lymphoblastic leukemias. Cancer Genet Cytogenet 23; 189–97 (1987).

Franke U, Holmes LB, Atkins L, Riccardi VM (1979). Aniridia-Wilms' tumour association: evidence for specific deletion of 11p13. Cytogenet Cell 24; 185–92.

German J (1974). Bloom's syndrome II. The prototype of human genetic disorders predisposing to chromosome instability and cancer. In Chromosomes and Cancer (Ed German J), John Wiley, New York; 601.

Grabowski EF, Abramson DH (1987). Intraocular and extraocular retinoblastoma. Hematology/Oncology Clinics of North America, I; 721–35.

Hansen MF, Koufos A, Gallie BL, Phillips RA, Fodstad O, Brogger A, Gedde-Dahl T, Cavenee WK (1985).

Osteosarcoma and retinoblastoma: A shared chromosomal mechanism revealing recessive predisposition. Proc Natl Acad Sci USA 82; 6216–20.

Harris H (1971). Cell fusion and analysis of malignancy. Proc Roy Soc Lond B 179, 1054; 1–20.

Herrera L, Kakati S, Gibas L, Pietrzak E, Sandberg AA (1986). Gardner syndrome in a man with an interstitial deletino of 5q. Am J Med Genet 25; 473–6.

Karakousis CP, Dal Cin P, Turc-Carel C, Limon J, Sandberg AA (1987). Chromosome changes in soft tissue sarcomas: a new diagnostic parameter. Arch Surg 122; 1257–60.

Kinzer KW, Bigner SH, Bigner DD, Trent JM, Law ML, O'Brien SJ, Wong AJ, Vogelstein B (1987). Identification of an amplified highly expressed gene in a human glioma. Science 236; 70–3.

Knudson AG (1971). Mutation and cancer: Statistical study of retinoblastoma. Proc Natl. Acad Sci USA 68; 820–3.

Knudson AG (1976). Genetics and etiology of childhood cancer. Ped Res 10; 513–7.

Koufos AM, Hansen F, Lampkin BC, Workman ML, Copeland NG, Jenkins NA, Cavenee WC (1984). Loss of alleles at loci on human chromosome 11 during genesis of Wilms' tumour. Nature 309; 170–2.

Koufos AM, Hansen MF, Copeland NG, Jenkins NA, Lampkin BC, Cavenee WK (1985). Loss of heterozygosity in three embryonal tumours suggests a common pathogenetic mechanism. Nature 316; 330–4.

Leppert M, Dobbs M, Scambler P, O'Connell P, Nakamura Y, Stauffer D, Woodward S, Burt R, Hughes J, Gardner E, Lathop M, Wasmuth J, Lalouel JM, White R (1987). Gene for familial polyposis coli maps to the long arm of chromosome 5. Science 238; 1411–4.

Motegi T (1982). High rate of detection of 13q14 deletion mosaicism among retinoblastoma patients using more extensive methods. Hum Genet 61; 95–7.

Nowell PC (1986). Mechanisms of Tumour Progression. Cancer Res 46; 2203–7.

Okamoto M, Sasaki M, Sugio K, Sato C, Iwama T, Ikenchi T, Tonomura A, Sasazuki T, Miyaki M (1988). Loss of constitutional heterozygosity in colon carcinoma from patients with familial polyposis coli. Nature 331; 273–7.

Orkin SH, Goldman DS, Sallen SE (1984). Development of hemizygosity for chromosome 11p markers in Wilms' tumor. Nature 309; 172–4.

Reeve AE, Housieux PJ, Gardner RJM, Chewing WE, Grindley RM, Millow LJ (1984). Loss of a Harvey ras allele in sporadic Wilms' tumor. Nature 309; 174–6.

Riccardi VM, Sujansky E, Smith AC, Francke U (1978). Chromosomal imbalance in the aniridia-Wilms' tumor association: 11p interstitial deletion. Pediatrics 61; 604–10.

Rowley JD (1984). Biological implications of consistent chromosome rearrangements in leukemia and lymphoma. Cancer Res 44; 3159–68.

Sandberg AA (1980). The chromosomes in human cancer and leukemia. Elsevier North-Holland. Inc., New York.

Sandberg AA (1985). Application of cytogenetics in neoplastic diseases. In CRC Critical Reviews in Clinical Laboratory Sciences, vol. 22, Ohio. CRC Press, Inc.

Sandberg AA (1986a). Chromosome changes in bladder cancer; clinical and other correlations. Cancer Genet Cytogenet 19; 163–75.

Sandberg AA (1986b). The chromosomes in human leukemia. Sem Hematol 23; 201–17.

Sandberg AA (1988). Editorial: The cytogenetic route of benign tumors. Cancer Genet Cytogenet 32; 11–2.

Sandberg AA; Turc-Carel C (1987). The cytogenetics of solid tumours. Relation to diagnosis, classification and pathology. Cancer 59; 387–95.

Second International Workshop on Chromosomes in Solid Tumors (1987). Cancer Genet Cytogenet 28; 1–54.

Second MIC Cooperative Study Group (1988). Morphologic, immunologic and cytogenetic (MIC) working classification of acute myeloid leukemias. Cancer Genet Cytogenet 30; 1–15.

Solomon E, Voss R, Hall V, Bodmer WF, Jass JR, Jeffrey AJ, Lucibello FC, Patel I, Rider SH (1987). Chromosome 5 allele loss in human colorectal carcinomas. Nature 328; 616–9.

Sparkes RS, Murphree AL, Lingua RW, Sparkes MC, Field LL, Funderburk SJ, Benedict WF (1983). Gene for hereditary retinoblastoma assigned to human chromosome 13 by linkage to esterase D. Science 219; 971–3.

Sparkes RS, Sparkes MC, Wilson MG, Towner JW, Benedict W, Murphree AL, Yunis JJ (1980). Regional assignment of genes for human esterse D and retinoblastomat to chromosome band 13q14. Science 208; 1042–4.

Third MIC Cooperative Study Group (1988). Morphologic, immunologic and cytogenetic (MIC) working classification of the primary myelodysplastic syndromes and therapy-related myelodysplasias and leukemias. Cancer Genet Cytogenet 32; 1–10.

Turc-Carel C, Dal Cin P, Rao U, Karakousis C, Sandberg AA (1986a). Cytogenetic studies of adipose tissue tumors. I. A benign lipoma with reciprocal translocation t(3;12) (q28;q14). Cancer Genet Cytogenet 23; 283–9.

Turc-Carel C, Limon J, Dal Cin P, Rao U, Karakousis C, Sandberg AA (1986b). Cytogenetic studies of adipose tissue tumors. II. Recurrent reciprocal translocation t(12;16) (q13;p11) in myxoid liposarcomas. Cancer Genet Cytogenet 23; 291–9.

Turc-Carel C, Pietrzak E, Kakati S, Kinniburgh AJ, Sandberg AA (1987). The human int-1 gene is located at chromosome region 12q12–12q13 and is not rearranged in myxoid liposarcoma with t(12;10) (q13;p11). Oncogene Res 1; 394–405.

14 Chromosome Alterations in Speciation and Neoplastic Transformation: A Parallelism

S. PATHAK

Department of Cell Biology
The University of Texas M.D. Anderson Hospital and
Tumor Institute at Houston
Houston, Texas 77030, USA

I. Introduction

One of the dogmas of classic cytogenetics is that the karyotype, that consists of the chromosome number and morphology, is a characteristic feature of a species. Any change that can be selected in the genetic makeup of a species might bring about speciation and thereby the origin of a new species. Evolutionary cytotaxonomists have already generated a wealth of information on the numerical and structural changes in the chromosome complements of diverse species of organisms. Different cytogenetic pathways, such as centric fusion, tandem translocation, inversion, addition of constitutive heterochromatin, and to a lesser extent, centric fission, have operated alone or in combination during the evolutionary history of vertebrate species. Thanks to the discovery of various chromosome banding techniques, we have a powerful tool to identify not only the entire chromosome but even a segment of a chromosome (Hsu, 1974, Sumner, 1982).

Ohno (1967) hypothesized that the mammalian X chromosome is conserved during the evolution of different species. His observation was based on findings that (a) the measurement and weight of cutouts of the X chromosomes indicated that the functional mammalian X chromosome, irrespective of its size and morphology, is 5 per cent of the female haploid genome, and (b) similar genes are mapped on the X chromosomes of different species. By G-banding analysis it was demonstrated that the X chromosome, irrespective of its size and morphology, showed homology based on two G-bands (A & B) in approximately 80 mammalian species (Pathak and Stock, 1974). The conservation of the X chromosome is therefore well established. Recent observations have shown that this conservation can be further extended to certain pairs of autosomes as well.

I would like to describe the circumstances under which the conservation of autosomes was discovered in this laboratory. One day in late 1973, while preparing a manuscript on the X chromosome homology, I entered the photographic darkroom of our laboratory and there I saw several G-banded metaphase photographs that were being dried by a graduate student. The

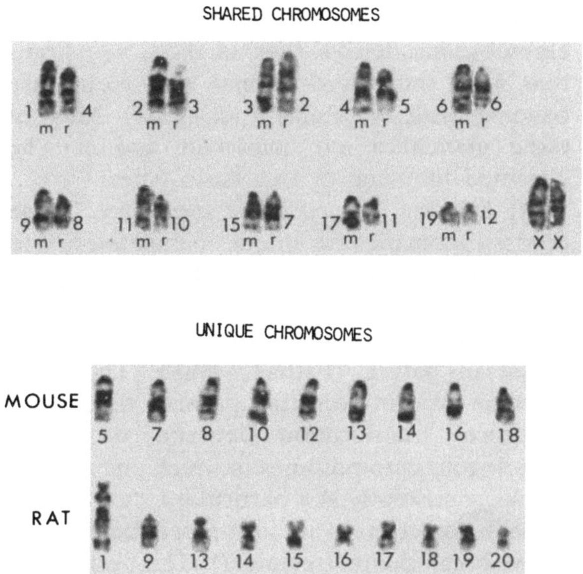

FIGURE 1: Giemsa-banded mouse, *Mus musculus*, and rat, *Rattus norvegicus*, karyotypes compared for their *shared* and *unique* chromosome complements. Autosomes are numbered following the standard nomenclature for mouse and rat karyotypes. A number of mouse (m) chromosomes show banding homology with rat (r) chromosomes (upper two rows). Unique chromosomes of these species are arranged in lower two rows.

G-banding patterns of some of those chromosomes were so familiar that I thought this graduate student had mistakenly made G-banded prints from *Peromyscus* negatives with which I was very familiar. 'Are these prints of *Peromyscus* chromosomes?' I asked this student. 'No, sir. These are G-banded plates of *Neotoma*' (another rodent genus on which the student was working for his Ph D). I could not believe my eyes and so took some of his G-banded pictures and the student to my office, where we compared the G-banded chromosome photographs of *Peromyscus* with those of *Neotoma*. To my amazement we found that there were a number of autosomes between these two genera with identical G-banding patterns. Later, a comparison of the G-banding patterns of several other species revealed that in even distantly related species some autosomes have identical G-banding patterns ('shared' chromosomes) and others have specific structural features in a particular species ('unique' chromosomes). This initial observation has been published in the *Journal of Mammalogy* (Mascarello *et al.*, 1974). Since then other authors have reported similar observations (Nash and O'Brien, 1982). The shared and unique chromosomes of the laboratory mouse, *Mus musculus*, and of the Norway rat, *Rattus norvegicus* are shown in Figure 1. There are 11 pairs of chromosomes shared by both species. However, their unique chromosomes vary in morphology as well as banding patterns.

A similar situation exists when chromosomes of human and murine tumours are analyzed carefully. There are two basic types of chromosome anomalies: (1) The *unique type* is causal, primary, and responsible for neoplastic transformation. An interstitial deletion of the 13q14 region in retinoblastoma, an interstitial deletion in the 11p13 region in Wilm's tumour, and involvement of the 3p14 region in renal cell carcinoma are examples of unique anomalies in these human neoplasms (Pathak, 1983). (2) The *shared type* of anomalies are acquired, secondary in nature, and responsible for maintenance, growth, heterogeneity, and invasiveness of the tumours. Such chromosomal anomalies, both structural and numerical, are present in a wide variety of pathologically distinct human neoplasms. Some of the shared anomalies are trisomy of chromosome 8, iso (17q), absence of the late-replicating X chromosome in the tumour of female origin, and absence of the Y chromosome in the tumour of male origin. Unique alterations in a particular tumour type must be present in the initial stage of the neoplastic transformation. Tumours of similar histopathologic types, when examined cytogenetically at different stages of progression, may not contain this unique marker chromosome at a later stage of tumour development. It is likely that during the progression of malignancy, the secondary types of genetic anomalies (shared chromosomes) are present and the primary types (unique) may be masked or totally missing. Once a cell has become neoplastic and has progressed to the next stage of tumour development the unique genetic alteration is no longer needed, because it has probably finished its job of neoplastic transformation. The function of a primary anomaly may be compared to the live battery of a car. A powerful battery is needed to start a car, but once the engine is started one does not need the battery power any more until one shuts the engine off and then tries to start it again.

The purpose of this brief review is to compare the similarity and uniqueness of chromosomal alterations, C-band distribution, and possible functions of the supernumerary chromosomes of different organisms, especially mammalian species with the chromosomal alterations, gene amplification, fragile sites, and other cytogenetic anomalies observed recently in human and other mammalian tumours.

II. Primary and Secondary Chromosome Anomalies

With the help of various chromosome banding techniques and particularly with high-resolution banding it has become possible to identify chromosomal defects even in those neoplasms that were considered normal with metaphase banding analysis (Yunis *et al.*, 1981). Most of these anomalies are consistent and can be classified into one of two basic types:

(a) *Primary chromosomal anomalies*. These primary anomalies are unique and are the first to be observed in any tumour cell. Such chromosomal defects have been thought to be responsible for neoplastic transformation and to be causal in nature (Pathak, 1983). The most common type of recurring primary defect is a reciprocal translocation between two nonhomologous chromosomes in which one partner breaks consistently at a particular locus (donor) while the other may vary in number and position of the break points (recipient). The best example to illustrate this point is the presence of a balanced translocation between chromosomes 8 and 14 in Burkitt's and non-Burkitt's lymphoma, representing a B-cell malignancy (Zech *et al.*, 1976). Variant translocations involving t(8;2) and t(8;22) have also been reported in Burkitt's lymphoma, indicating that chromosome 8 is a donor and chromosomes 2, 14 and 22 the recipients in Burkitt's lymphoma (Berger *et al.*, 1979a; Miyoshi *et al.*, 1979). Other examples of translocations in various myeloid and lymphoid series of leukemias and lymphomas are listed in Table 1.

Progress in the cytogenetic study of solid tumours has been scant, in spite of the fact that a significant number of leukemias and lymphomas have been studied cytogenetically and important observations have been made on primary and secondary anomalies. Recent introduction of tissue dissociating technique and the use of modified growth factors for short-term tumour cell cultures has provided evidence of recurrent anomalies in different sarcomas and carcinomas (First International Workshop on Chromosomes in Solid Tumours, 1986). A close examination of Table 1 indicates that in a limited number of

TABLE 1

Human neoplasms with a known unique chromosome alteration

Neoplasm	Chromosomal alteration	Reference
LEUKEMIA		
Chronic myelogenous leukemia	t(9;22)	Rowley, 1973a
Acute granulocytic leukemia	t(8;21)	Trujillo *et al.*, 1978
Chronic lymphocytic leukemia	+12	Autio *et al.*, 1979
	t(11;14)	Sandberg, 1981
Acute nonlymphocytic leukemia		
M1	t(9;22)	Bloomfield *et al.*, 1977
M2	t(8;21)	Rowley, 1973b
	+4	Mecucci *et al.*, 1986
M3	t(15;17)	Rowley *et al.*, 1977
M4	inv(16)	Yunis *et al.*, 1982
M5	t(9;11)	Hagemeizer *et al.*, 1982
Acute lymphocytic leukemia		
L1	t(9;22)	Bloomfield *et al.*, 1978
L2	t(4;11)	Van den Berghe *et al.*, 1979
L3	t(8;14)	Berger *et al.*, 1979a
LYMPHOMA		
Burkitt's and non-Burkitt's lymphoma	t(8;14)	Zech *et al.*, 1976
	t(2;8)	Miyoshi *et al.*, 1979
	t(8;22)	Berger *et al.*, 1979b
Follicular lymphoma	t(14;18)	Yunis *et al.*, 1982
Small cell lymphocytic lymphoma	t(11;14)	Yunis *et al.*, 1982
SARCOMA		
Liposarcoma	t(12;16)	Limon *et al.*, 1986
Rhabdomyosarcoma	t(2;13)	Truc-Carel *et al.*, 1986a
Ewing's sarcoma	t(11;22)	Aurias *et al.*, 1984
Synovial sarcoma	t(X;18)	Truc-Carel *et al.*, 1986b
CARCINOMA		
Retinoblastoma	del (13q)	Wilson *et al.*, 1973
Aniridia-Wilms' tumour	del (11p)	Riccardi *et al.*, 1978
Renal cell carcinoma	t(3;8)	Cohen *et al.*, 1979
	t(3;11)	Pathak *et al.*, 1982
Neuroblastoma	del (1p)	Brodeur *et al.*, 1981
Small cell lung carcinoma	del (3p)	Whang-Peng *et al.*, 1982
Breast tumour	t(1;3)*	Pathak *et al.*, 1979
Meningioma	del (22q)	Zankl and Zang, 1972
	t(14;22)	Bolger *et al.*, 1985
Colon tumour	del (12p)	Dhaliwal *et al.*, 1986
Malignant melanoma	del (6q)	Pathak *et al.*, 1983; Trent *et al.* 1983; Becher *et al.*, 1983
Ovarian tumour	t(6;14)	Wake *et al.*, 1980

* 1q translocation to chromosomes 5, 7, 6, 9, 11 and 12 were also present.

sarcomas a similar mechanism of translocation might be operating. In other words, all sarcomas thus far analysed chromosomally have shown specific translocations involving two non-homologous chromosomes. Whether these sarcomas would also show typical donor and recipient type chromosomes, as shown in leukemias and lymphomas, will depend on more cases studied in the future. A limited number of carcinomas studied have also revealed specific chromosomal defects (Table 1). Such anomaly is primarily of the deletion type. In several cases of meningiomas, Zankl and Zang (1972) reported a complete or partial monosomy of chromosome 22. Recently, a large family pedigree has been described by Bolger *et al.* (1985) with the incidence of meningioma and a constitutional translocation involving chromosomes 14 and 22. This particular translocation was not a reciprocal one and resulted in partial deletion of the long arm (q) of chromosome 22 and of the short arm (p) of chromosome 14. Some of the other examples are deletion 13q in retinoblastoma (Wilson *et al.*, 1973), deletion 11p in Wilms' tumour (Riccardi *et al.*, 1978), deletion 1p in neuroblastoma (Brodeur *et al.*, 1981), deletion 3p in small cell lung carcinomas (Whang-Peng *et al.*, 1982), and probably an interstitial deletion in 12p in patients with colon tumours (Dhaliwal *et al.*, 1986). Some of these anomalies are present in the normal lymphocyte or fibroblast cultures of these patients and therefore are constitutional in nature. In other cases, similar anomalies have been acquired in the target organ where the tumour has developed. It is apparent that primary (unique) anomalies present in leukemias and lymphomas are mostly acquired. Only recently has it been noticed that some of these anomalies could also be inherited (Pathak, 1986). On the other hand, specific anomalies reported in retinoblastomas, Wilm's tumours, renal cell carcinomas, meningiomas, colon carcinomas and breast tumours may be present as constitutional anomalies or as tissue-specific acquired anomalies.

Nevertheless, in many sarcomas, carcinomas, and other types of solid neoplasms, primary chromosomal defects are not yet known. With the knowledge of a primary (unique) chromosome anomaly, it would be possible to identify a person in a high-risk group and therefore maintain surveillance for any potential tumour occurrence. This situation would be analogous to the unique chromosomes of a given species by which that particular species can be easily identified.

(b) *Secondary chromosomal anomalies*. Secondary anomalies arise at a later stage of tumour development and are responsible for tumour progression, heterogeneity, and metastases. Also, in many cases these anomalies are shared by pathologically different neoplasms. For example, in chronic myelogenous leukemia (CML) the primary genetic anomaly is the translocation between chromosomes t(9;22), giving rise to the Ph chromosome. During the blastic phase, additional genetic anomalies in the form of trisomy 8 and isochromosome for the 17q have been reported. Isochromosome for the 17q has been reported in breast tumours, colon tumours (Pathak and Goodacre, 1986) and in many cases of lymphomas as well. The presence of multiple copies of chromosome 7 has been reported in a number of both hematologic and solid tumour neoplasms. Trisomy of chromosome 7 has been observed in many other solid tumours including malignant melanoma, brain tumours, breast tumours, colon cancer, and adenomas of the colon (Mitelman, 1985). In at least four head and neck tumour cell lines, multiple copies of chromosome 7 have been observed (our unpublished data). Absence of the late-replicating X chromosome from a tumour of female origin and absence of the Y chromosome from a tumour sample of male origin are additional examples of shared chromosome anomalies (Sandberg, 1980). In a kind of analogy, the Y chromosome tends to disappear from the marrow cells of older men as it does from male-derived neoplasms.

It is becoming apparent that some of these shared chromosome anomalies confer proliferative advantage to the tumour cell. Since every tumour must grow to survive, some common genetic anomalies are acquired to accelerate this process. On chromosome 7 are several genes critical to cell proliferation and chromatin functions, including the epidermal growth factor receptor (EGFR) gene (Kondo *et al.*, 1983). A close correlation between increased EGFR levels and an extra copy of chromosome 7 reported in malignant melanoma (Koprowski *et al.*, 1985) might provide additional growth advantage to already malignant melanoma cells in this late stage of progression. The HER-2/*neu* oncogene, a member of the tyrosine kinase family, is mapped on the long arm of chromosome 17 in the q21 region (Coussens *et al.*, 1985). The presence of multiple copies of chromosome 17 can amplify this gene and provide selective growth advantage. Amplification of this particular gene has been correlated in breast tumour cell lines derived from lymph node metastases (Slamon *et al.*, 1987).

Examples of shared chromosomes in mammalian species could be the X and Y chromosomes. In a number of mammalian species the X chromosome harbours the same gene loci, and the Y chromosome contains the gene or genes responsible for maleness (Ohno, 1983). Supernumerary and B chromosomes observed in a wide variety of organisms including several species of mammals are other examples of shared anomalies, and these could be compared with the homogeneously staining region (HSR) and double minute (DM) chromosomes representing certain gene amplifications in a variety of tumour specimens. (This will be discussed in greater detail in the next section.)

III. Constitutive Heterochromatin and Gene Amplification

The presence of heteromorphism of the C band (constitutive heterochromatin) in different mammalian (Pathak *et al.*, 1973a and b) and grasshopper species (John, 1981; John and King, 1982) is a well-known fact. There have been numerous speculations and hypotheses proposed regarding the possible function of heterochromatin (Yunis and Yasmineh, 1971; Walker, 1972; Hsu, 1975; and John and Miklos, 1979). The 'bodyguard' hypothesis, proposed by Hsu (1975) as a possible function of the heterochromatin, was based on the restricted distribution of such chromatin in the interphase nuclei. He demonstrated that the inner wall of the nuclear membrane is guarded by the heterochromatic material that protects the euchromatin present in the centre of the nucleus. Any clastogen entering the nucleus will strike the heterochromatin first. The euchromatin present in the middle of the nucleus might be protected from such insults. This glass-box hypothesis can be tested in the light of gene-amplification-induced *in vitro* systems (Biedler and Spengler, 1976; Bostock and Clark, 1980).

In 1973, my associates and I reported the presence of a disproportionate amount of heterochromatin in two species of the cactus mouse, *Peromyscus eremicus* and *P. crinitus* (Pathak *et al.*, 1973a). The G-banding pattern of the euchromatic segments in both species are almost identical. Compared to that of *P. eremicus* the genome of *P. crinitus* has approximately 36 per cent extra DNA in the form of heterochromatic segments (Figure 2). The antelope squirrel, *Ammospermophilus harrissi* has 70 per cent more DNA in the form of heterochromatin, compared with the related ground squirrel species (Mascarello and Mazrimas, 1977). With G-banding, the interstitially located heterochromatin of the antelope squirrel chromosomes

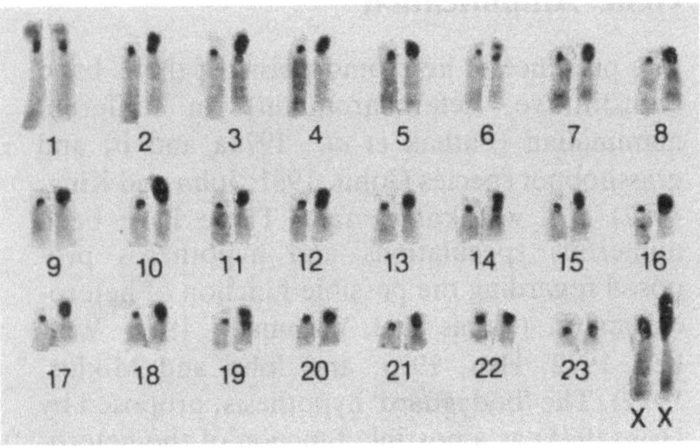

FIGURE 2: A C-banded composite karyotype of *Peromyscus crinitus* and *P. eremicus*. For each pair, the right-side chromosome is from *crinitus* and the left-side is from *eremicus*. Note the presence of excessive amounts of heterochromatin in the short arms of *eremicus* chromosomes. The long arms in these two species have identical G-banding patterns.

stains homogeneously very light, while the euchromatic segments show typical G-bands (Figure 3). A similar situation exists in the whale karyotype (Jalal *et al.*, 1974).

Supernumerary chromosomes have been reported in a number of organisms starting from insects (White, 1973) to man (Yip *et al.*, 1982). In most organisms the supernumerary chromosomes are heterochromatic. In most cases the origin and possible function of such extra heterochromatic elements are not known (John, 1973; Muntzing, 1974). The basic karyotype of *Rattus rattus* consists of a diploid number of 42. In some populations of this species the diploid numbers of 40 and 38 have also been reported. The reduction in the diploid number has been attributed to Robertsonian translocations. Studies of the karyotypes of some specimens of *R. rattus* I collected from Nepal in 1969 revealed that the chromosome numbers in these specimens could be as high as 2n = 49. However, specimens with a diploid number of 43, 44, 45, 47 and 48 chromosomes were also present (Pathak, 1971). The extra (supernumerary) chromosomes were predominantly small metacentric elements. The heterochromatic nature of such supernumerary B chromosomes in the genus *Rattus* has since been reported by others (Raman and Sharma, 1974; Yosida, 1977). The presence of C-band-positive B chromosomes has also been reported in Australian rodents *Uromys caudimaculatus* (Baverstock *et al.*, 1976), *Mus shortridgei* (Gropp *et al.*, 1973), *Perognathus baileyi* (Patton, 1977), and raccoon dogs, *Nyctereutus procyonoides viverieus* (Wurster-Hill *et al.*, 1986).

Minute to medium-sized supernumerary chromosomes have been reported in a wide variety of organisms (Muntzing, 1974; Jones, 1975). A metaphase spread of microchiropteran species *Pipistrellus mimus*, with micro-supernumerary chromosomes (MSC) is shown in Figure 4. Due to the small size and dark printing, these MSCs do not appear as paired entities. Such MSCs should not be confused with the microchromosomes of birds and reptiles that are regular elements of their genomes (Ray-Chaudhuri, 1967). The number of MSCs varies from specimen to specimen of different organisms and also from one cell to another of the same specimen (Ray-Chaudhuri *et al.*, 1967). Supernumerary heterochromatic segments in different autosomes of an Australian grasshopper, *Heteropternis obscurella*, have been reported (John and King, 1982). What is the origin of these extra chromosomal bodies and extended supernumerary segments? Do they represent amplification of certain genes? What advantage do they render to those organisms when present? There are no satisfactory answers to these and several other questions on supernumerary chromosomes.

The supernumerary chromosomes present in organisms caught in nature and the double minutes (DMs) and homogeneously staining regions (HSRs) induced and observed in cell

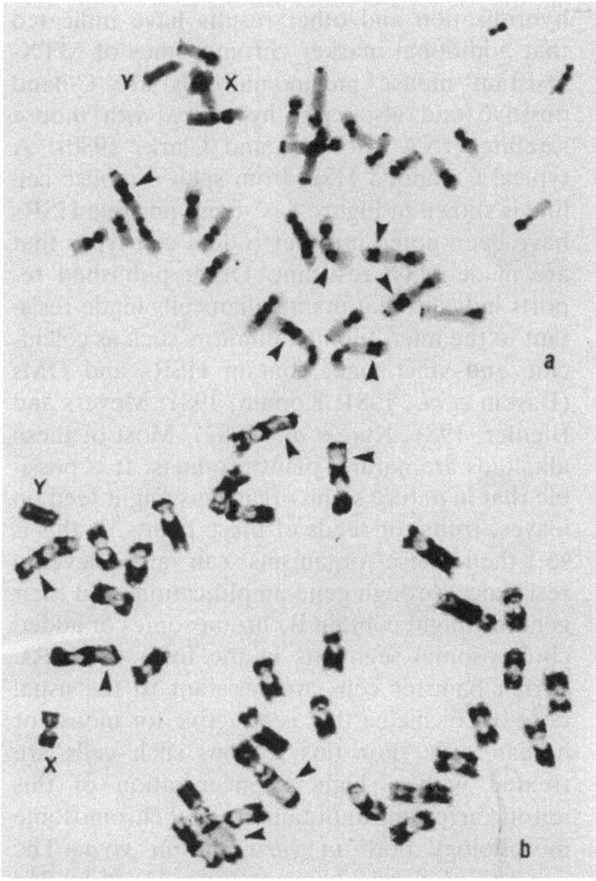

FIGURE 3: C-banded (A) and G-banded (B) metaphase plates from a male antelope squirrel, *Ammospermophilus harrisii*. Note the presence of intercalary heterochromatic segments in six autosomes (arrowheads). These chromosomal segments show homogeneously staining regions (HSRs) after G-banding induction (arrowheads).

culture systems share many characteristic features:

1. They all contain DNA.
2. They are reported in a wide variety of organisms and tumour types.
3. They may be present as extrachromosomal bodies or added segments in the regular chromosomes.
4. They are mostly C-band positive.
5. Their number varies from cell to cell, also from specimen to specimen and tumour to tumour.
6. They are not present in every cell, or specimen, collected.
7. DMs and MSCs may not have a functional centromere (possibly explaining why they are not distributed equally in every cell).
8. They all represent amplification of a gene or genes.
9. They all provide protection and growth advantage against toxic agents (bodyguard hypothesis).

The role or roles of amplified DNA sequences against multidrug resistance has been extensively reviewed (Cowell, 1982; Schimke, 1982; Stark and Wahl, 1984). DMs and HSRs have been reported in both treated and non-treated patients with a wide variety of tumours. In some tumour

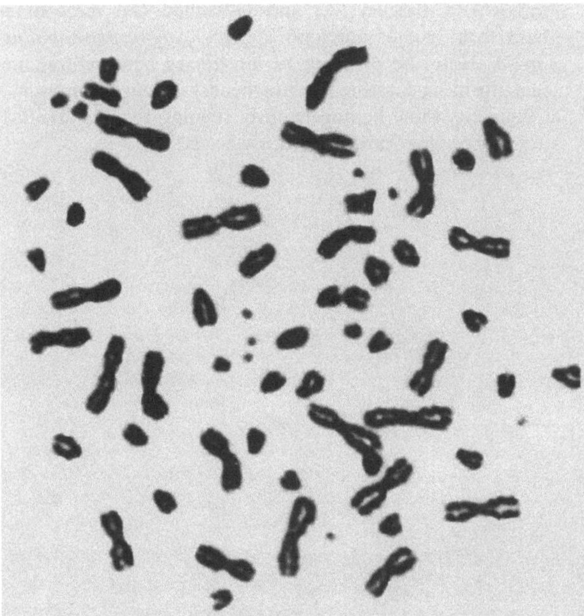

FIGURE 4: A conventionally stained metaphase spread of a microchiropteran species, *Pipistrellus mimus*, showing micro-supernumerary chromosomes (MSC). The number of these MSC varies from cell to cell.

cell lines, abnormally banded regions (ABR) and differentially staining regions (DSR) have also been observed as examples of amplification. As mentioned earlier, DMs and HSRs can be induced in cell cultures or *in vivo* cultures (body fluids) treated with cytotoxic chemotherapy. Numerous studies on both cytological and molecular levels have concentrated on understanding the mechanisms of production of these cytological novelties. From these studies it has become clear that DMs and HSRs may play a key role in neoplastic progression and the development of tumour resistance against chemotherapy.

Molecular and cytogenetic analysis of the methotrexate-(MTX) resistant Chinese hamster and mouse cell lines clearly demonstrated the amplification of dihydrofolate reductase (DHFR) genes and the presence of DMs and HSRs (Nunberg *et al.*, 1978; Kaufman *et al.*, 1979; Bostock *et al.* 1979). Moreover *in situ* hybridization and other results have indicated that additional marker chromosomes of MTX-resistant mouse melanoma cells are C-band positive and also cross-hybridize with mouse satellite DNA (Bostock and Clark, 1980). A typical C-banded HSR from such a mouse cell line is shown in Figure 5. C-band positive HSRs have been noted in other rodent cell types that are made MTX-resistant. Other published reports indicate that mammalian cells made resistant to the microtubule inhibitors such as colchicine and vincristine, contain HSRs and DMs (Baskin *et al.*, 1981; Kopnin, 1981; Meyers and Biedler, 1981; Kuo *et al.*, 1982). Most of these alkaloids are natural plant products. It is possible that in nature some organisms might feed on leaves, fruits, or seeds of these plants. If this is so, then these organisms can also develop resistance through gene amplification, and their genome might contain B chromosomes or added chromosomal segments in the form of HSRs. Syrian hamster cells are resistant to the usual dose of colchicine that is effective for mouse or human cells. For this reason, such cells are treated with a higher concentration of this mitotic arrestant to obtain optimal chromosome morphology both *in vitro* and *in vivo*. The genome of Syrian hamster ($2n = 44$) is loaded with an extra amount of chromosomal segments that are C-band positive (Hsu and Arrighi, 1971; Popescu and DiPaolo, 1979). Whether these C-band-positive arms represent gene amplification is not known. Most planthoppers and grasshoppers feed on certain types of plants, and their genomes have been characterized by supernumerary B-chromosomes. Do these additional chromosomes or chromosome segments represent gene amplification? The answer to such a question could be yes, because HSRs and DMs have been observed in the body fluid of some patients who have been treated with methotrexate (Pathak, 1980; Pathak *et al.*, 1979), representing gene amplification *in vivo*.

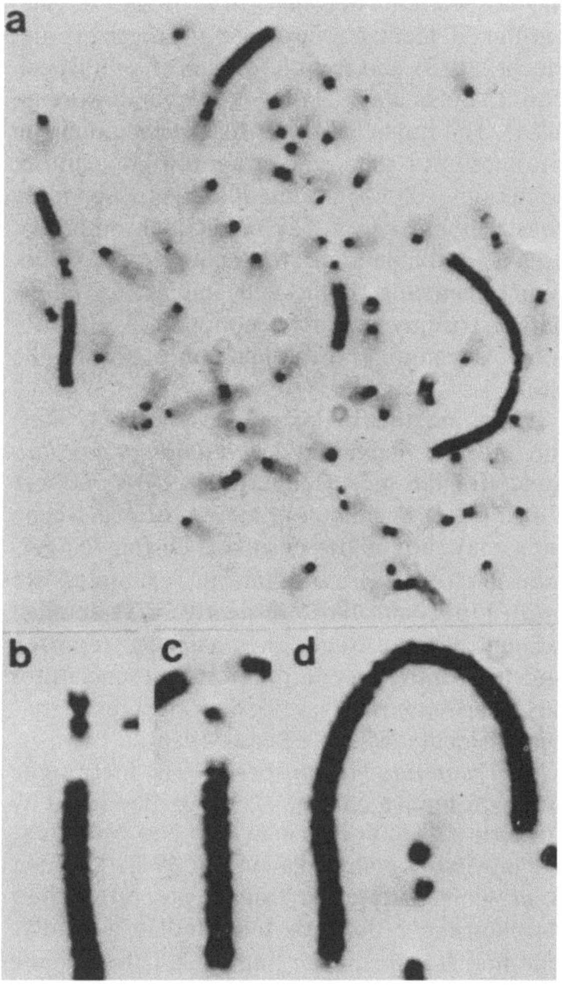

FIGURE 5: A C-banded metaphase spread from a mouse melanoma cell line (PG19T3: MTX_R10-4M) that was made resistant to methotrexate *in vitro*. Note the presence of heterochromatin in the centromeric regions of normal and altered chromosomes and dark staining of HSR of four marker chromosomes (a). Marker chromosomes containing HSR from other cells are also shown (b, c, and d). These C-band positive HSRs have high degree of homology with the satellite DNA of mouse genome (from Bostock and Clark, 1980; Courtesy of MIT Press).

DMs and HSRs may not be the properties of tumour cells only. Graves (1984) reported multiple copies of DMs in cell lines independently derived from the F_1 embryos produced by artificially inseminating *Mus musculus* females with spermatozoa of *M. Caroli*. These two *Mus* species do not hybridize in nature. The simplest explanation this author offered was that DMs were already present in cells of the *M. musculus* × *M. caroli* embryo before the cells were put in culture. If so, then that fact supports the hypthesis that gene amplification occurred in early embryogenesis. Cells have not been tested to determine whether they are tumourigenic in nude mice.

IV. Fragile Sites

The fragile sites (*fra*) on human chromosomes have attracted the attention of both clinicians and experimental oncologists lately, because of the association between the fragile X and sex-linked mental retardation (Sutherland and Hecht, 1985) and its role in cancer predisposition (Yunis, 1983, 1984; Yunis and Soreng, 1984). The fragile sites are the specific points on chromosomes that, under appropriate culture conditions, show chromatid gaps, secondary constrictions, chromatid breaks, chromosome breaks, and sometimes 'branching' of chromosomal segments. Some cells might also show partial trisomy and monosomy of a chromosome, depending on the location of the fragile site.

There are two major groups of fragile sites: the *rare or heritable* and the *common or constitutional* fragile sites (Sutherland, 1983; Glover *et al.*, 1984; Smeets *et al.*, 1986). Rare fragile sites may be folate-sensitive, Distamycin A-inducible, and BrdU-requiring; examples are 6p23, 17p12, and 10q25 respectively. A detailed account of these fragile sites has been reported and the controversy in different nomenclatures has been extensively reviewed (Berger *et al.*, 1985; Hecht, 1986; LeBeau, 1986).

(a) *Fra in non-human species*. The first fragile site on a human chromosome was described by Dekaban (1965) in a woman who had been treated previously with radiation. In 1968, Lejeune *et al.* demonstrated a fragile site on human chromosome 2 that was inherited in a family. The first fragile site on human X chromosome was demonstrated by Lubs (1969). Since then, a number of articles have been published on fragile sites (Sutherland and Hecht, 1985).

Very rarely a fragile site in non-human species has been reported. Only a few papers have described a *fra* in non-human mammalian species. It has been thought that chimpanzees and gorillas would be next after humans to show such lesions in their chromosomes (Yunis and Soreng, 1984); however, the first fragile site in any mammalian species, including human, was reported in Chinese hamster chromosome 1 (Hsu and Somers, 1961). While working on the chemical effect of BrdU on chromosome morphology of Chinese hamster cell lines, they observed that chromosome No. 1 showed a fragile site in one arm at a particular locus. This observation was recently confirmed on kidney cells of an inbred strain of Chinese hamsters (Lin *et al.*, 1984).

Since then, fragile sites specific for autosomes, X chromosomes, or both, have been reported in the human (Sutherland and Hecht, 1985), chimpanzee and gorilla (Yunis and Soreng, 1984), Chinese hamster (Lin *et al.*, 1984), Persian vole, *Ellobius lutescens* (Djalali *et al.*, 1985), cattle (Uchida *et al.*, 1986), Indian mole rat, *Nesokia indica* (Tewari *et al.*, 1987) and rats (Robinson and Elder, 1987). A fragile site on the X chromosome of the laboratory mouse, *Mus musculus*, has been seen in the form of a gap by a number of investigators including me. This gap in the X chromosome is very obvious in meiotic preparations of mouse testes. In future, fragile sites will be reported in other mammalian species as well.

(b) *Fra and cancer-specific chromosome defects*. Neither the mechanism of production of a *fra*, nor its role or roles in pathogenesis is clearly defined (Berger *et al.*, 1985). In 1983, Yunis was the first to report that at least three autosomal fragile sites on chromosomes 11q13, 12q13, and 16q22 coincide with chromosomal defects reported in human neoplasia. He further pointed out that certain *fra* might act as predisposing factors, and others might represent oncogenic sites. In a more recent article, Yunis and associates (1987) have shown experimentally that fragile sites are the targets of different mutagens and carcinogens in human genomes. Of the 110 fragile sites Yunis and associates induced with 16 different mutagens and carcinogens, 50 coincide with the breakpoints of 50 out of 75 specific chromosomal breakpoints and 21 with the location of 26 of 36 oncogenes (Yunis *et al.*, 1987).

These observations suggest that diverse mutagens and carcinogens produce a large number of chromosomal lesions at fragile sites and that these sites may be the targets of carcinogenic action. Also, in leukemia and lymphoma, a concordance between the fragile sites and the breakpoints that occur in the specific chromosome regions have been identified (LeBeau and Rowley, 1984). In spite of all these correlations between the cancer breakpoints and *fra*, the committee (Berger *et al.*, 1985) on chromosomal rearrangements in neoplasia and on fragile sites during the 8th International Human Gene Mapping Workshop in Helsinki concluded, 'There is no evidence that possession of a rare fragile site predisposes its carrier to cancer. All individuals are probably homozygous for the common fragile sites; thus these are unlikely to be important to any individual in terms of increased risks of cancer.' In acute myelomonocytic leukemia (AMMoL), the specific chromosome defect involved rearrangement of band 16q22. Some of these AMMoL patients who have shown 16q22 rearrangements in their leukemic cells also carry the rare fragile site at 16q22. By using *in situ* hybridization technique, the metallothioneine gene cluster was mapped at 16q22, and it was also discovered that in some AMMoL patients this cluster was split due to chromosomal rearrangement at 16q22 (LeBeau *et al.*, 1985). However, a recent report by Simmers *et al.* (1987) has contradicted this observation and has shown that the metallothioneine gene cluster is proximal to the rare fragile site (16q21) and to a common fragile site (16q22) in AMMoL patients, therefore, neither of these fragile sites was considered to be at the chromosomal breakpoint of those patients. These observations clearly indicate that a fragile site seen on chromosomes with light microscopy may be at a different location when probed with a specific gene probe aimed at a *fra* site. Demonstration of a causal relationship between fragile sites and cancer predisposition awaits further research.

V. Concluding Remarks

The purpose of this brief review has been to compare the cytogenetic process of speciation in nature with that of the neoplastic transformation and tumour growth within an organism. By providing examples of the restructuring of a genome to certain stimuli or challenges, a parallel has been drawn between chromosome alterations in speciation and neoplastic transformation. I consider every human or animal tumour to be a natural species, with its own genomic characteristics. Both have *unique* and *shared* types of chromosomes, as described earlier. The genome of a species in the wild is constantly exposed to natural forces in the form of cosmic radiation, changing ecology, food habits and preferences for particular types of plant leaves, roots and seeds, and natural and man-made chemicals. Similarly, a neoplasm is exposed to the body's natural immune system, chemicals produced within the organ or body, chemotherapy, radiation therapy, and other biological products within the host system. Exposure to these environmental factors brings about changes in the genome of the species and of the neoplasm.

Chromosome-banding data have shown that cells of a particular species can be identified only when attention is focused on their *unique* chromosomes. In other words, identification of a species can easily be made by identifying the unique chromosome or chromosomes in a given sample or in a somatic cell hybrid. Similarly, by identifying the unique chromosomal defect in the peripheral blood or fibroblast cultures, a person predisposed to a particular neoplasm can be identified. In this respect, *shared* chromosome anomalies are of less importance. However, their importance in tumour growth, heterogeneity, metastasis of the neoplasm, and survival of the species cannot be overlooked. In tumour cells, three types of primary chromosomal defects have been identified: (a) reciprocal translocation between two non-homologous chromosomes and inversion, (b) deletion involving a specific region of a chromosome, and (c) amplification of a particular gene. Translocations and amplifications have affected mainly the cellular oncogenes, while deletions might explain the existence of suppressor or antioncogenes whose complete deletion might result in tumourigenesis.

Translocations and inversions observed in leukemias and lymphomas can activate or affect the expression of oncogenes. For example, in Burkitt's lymphoma (B-cell neoplasm), a translocation between chromosomes 8 (*c-myc*) and 2 and chromosomes 14 and 22 (various immunoglobulin genes) can activate *c-myc* at the wrong times. In chronic myelogenous leukemia (CML), a translocation between chromosomes 9 (*c-abl*) and 22 (*bcr* = breakpoint cluster region) can transfer a portion of the *c–abl* oncogene to *bcr* and thereby affect the level of expression of a gene product. In certain solid tumours that have been extensively studied, recessive mutations have been shown to be responsible for neoplastic characteristics. Examples of such cases are a deletion 13q14 in retinoblastoma, a deletion 11q13 in Wilms' tumour, complete or partial monosomy of chromosome 22 in meningioma, and a breakpoint in 3p14 in renal cell carcinomas. In retinoblastoma and Wilms' tumour, loss of heterozygosity has been demonstrated in tumour cells (Cavenee *et al.*, 1983; Dao *et al.*, 1987). Whether a similar mechanism of tumourigenesis operates in most other solid tumours needs further investigation.

Gene amplification can be induced in drug-resistant cell lines in the form of double minutes, homogeneously staining regions, or both. In natural populations, a species could develop resistance against exposure to its environment and to its food habits in the form of extra chromosomal elements called supernumerary or B chromosomes. The size and morphology of such chromosomes vary from species to species. A species could amplify certain genes and restructure its genome in the form of added chromosomal segments. Most of these extra chromosomal segments and B chromosomes are heterochromatic in nature. A few examples of such natural amplifications are included to support my hypothesis.

Fragile sites are present not only on human chromosomes but also on chromosomes of other mammalian species. I am confident they will be reported in other groups of organisms also. Fragile sites can provide raw materials for establishing a particular translocation between two non-homologous chromosomes, can aid in inversion, or aid in deletion. I have presented examples of such chromosomal mechanisms operating in karyotypic evolution in certain groups of organisms and their possible roles in neoplastic transformation and tumour progression.

Finally, I have tried to establish a parallelism between chromosomal alterations in speciation and neoplastic transformation and tumour growth. I realize that my discussion is based purely on cytogenetic evidence. I hope to stimulate other investigators to provide molecular evidence in support of this hypothesis.

Acknowledgements

This work was supported in part by grants from the John S. Dunn Research Foundation, Kelsey-Seybold Foundation, Houston, Texas and by the M.D. Anderson Hospital Annual Campaign Fund. I thank Ms Ruby Kirkpatrick for secretarial assistance and Ms Susan Wilkinson for editorial corrections.

References

Aurias A, Rimbaut C, Buffe D, Zucker JM, Mazabraud A (1984). Translocation involving chromosome 22 in Ewing's Sarcoma. A cytogenetic study of four fresh tumors. Cancer Genet Cytogenet 12; 21–5.

Autio K, Turunen O, Penttila O, Eramao E, de la Chapelle A, Schroder J (1979). Human chronic lymphocytic leukemia: Karyotypes in different lymphocyte populations. Cancer Genet Cytogenet 1; 147–55.

Baskin F, Rosenberg RN, Dev V (1981). Correlation of double minute chromosomes with unstable multidrug cross-resistance in uptake mutants of neuroblastoma cells. Proc Natl Acad Sci USA 78; 3654–8.

Baverstock PR, Watts CHS, Hogarth JT (1976). Heterochromatin variation in the Australian rodent *Uromys caudimaculatus*. Chromosoma 57; 397–403.

Becher R, Gibase Z, Sandberg AA (1983). Chromosome 6 in malignant melanoma. Cancer Genet Cytogenet 9; 173–5.

Berger R, Bernheim A, Brouet JC, Daniel MT, Flandrin G (1979a). t(8;14) translocation in a Burkitt's type of lymphoblastic leukemia (L3). Br J Haematol 43; 87–90.

Berger R, Bernheim A, Weh H-J, Flandrin G, Daniel MT, Brouet J-C, Colbert N (1979b). A new translocation in Burkitt's tumor cells. Hum Genet 53; 111–2.

Berger R, Bloomfield CD, Sutherland GR (1985). Report of the committee on chromosome rearrangements in neoplasia and on fragile sites. Human Gene Mapping 9, Helsinki Conference. Cytogenet Cell Genet 40; 490–535.

Biedler JL, Spengler BA (1976). Metaphase chromosome anomaly: Association with drug resistance and cell-specific products. Science 191; 185–7.

Bloomfield CD, Lindquist LL, Brunning RD, Yunis JJ, Coccia PF (1978). The Philadelphia chromosome in acute leukemia. Virchows Arch 29; 81–91.

Bloomfield CD, Peterson LC, Yunis JJ, Brunning RD (1977). The Philadelphia chromosome (Ph^1) in adults presenting with acute leukemia: A comparison of Ph^1+and Ph^1–patients. Br J Haematol 36; 347–58.

Bolger GB, Stamberg J, Kirsch IR, Hollis GF, Schwarz DF, Thomas GH (1985). Chromosome translocation t(14;22) and oncogene (*c-sis*) variant in a pedigree with familial meningioma. New Engl J Med 312; 564–7.

Bostock CJ, Clark EM (1980). Satellite DNA in large marker chromosomes of methotrexate-resistant mouse cells. Cell 19; 709–15.

Bostock CJ, Clark EM, Harding NGL, Mounts PM, Tyler-Smith C, van Heyningen V, Walker PMB (1979). The development of resistance to methotrexate in a mouse melanoma cell line I. Characterization of the dihydrofolate reductases and chromosomes in sensitive and resistant cells. Chromosoma 74; 153–77.

Brodeur GM, Green AA, Hayes FA, Williams KJ, Williams DL, Tsiatis AA (1981). Cytogenetic features of human neuroblastomas and cell lines. Cancer Res 41; 4678–86.

Cavenee WK, Dryja TP, Phillips RA, Benedict WF, Godbout R, Gallie BL, Murphree AL, Strong LC, White RL (1983). Expression of recessive alleles by chromosomal mechanisms in retinoblastoma. Nature 305; 779–84.

Cohen AJ, Li FP, Berg S, Marchetto DJ, Tsai S, Jacobs SE, Brown RS (1979). Hereditary renal-cell carcinoma associated with a chromosomal translocation. New Engl J Med 301; 592–5.

Coussens L, Yang-Feng TL, Liao Y-C, Chen E, Gray A, McGrath J, Seeburg PH, Libermann TA, Schlessinger J, Francke U, *et al.* (1985). Tyrosine kinase receptor with extensive homology to EGF receptor shared chromosomal location with *neu* oncogene. Science 230; 1132–9.

Cowell JK (1982). Double minutes and homogeneously staining regions: Gene amplification in mammalian cells. Ann Rev Genet 16; 21–59.

Dao DD, Schroeder WT, Chao LY, Kikuchi H, Strong LC, Riccardi VM, Pathak S, Nichols WW, Lewis WH, Saunders GF (1987). Genetic mechanisms of tumor-specific loss of 11p DNA sequences in Wilms' tumor. Am J Hum Genet 41; 202–17.

Dekaban A (1965). Persisting clone of cells with an abnormal chromosome in a woman previously irradiated. J Nucl Med 6; 740–6.

Dhaliwal MK, Pathak S, Winn RJ (1986). Involvement of chromosome 12 in colorectal carcinoma: evidence of specific deletion of 12p12 (Abstract). 39th M.D. Anderson Symposium of Fundamental Cancer Research, 'Critical Molecular Determinants in Carcinogensis', 16–19 September.

Djalali M, Barbi G, Steinbach P (1985). Folic acid sensitive fragile sites are not limited to human karyotype: demonstration of non-random gaps and breaks in Persian vole *Ellobius lutesceus* Th. inducible by methotrexate, fluorodeoxyuridine and aphidicolin. Hum Genet 70; 183–5.

First International Workshop on Chromosomes of Solid Tumors (1986). Cancer Genet Cytogenet 19; 1–197.

Glover TW, Gerger C, Coyle J, Echo B (1984). DNA polymerase inhibition by aphidicolin induces gaps and breaks at common fragile sites in human chromosomes. Hum Genet 67; 136–42.

Graves JAM (1984). Gene amplification in a mouse embryo? Double minutes in cell lines independently derived from a *Mus musculus* × *M. caroli* fetus. Chromosoma 89; 138–42.

Gropp A, Marshall J, Markvong A (1973). Chromosomal findings in the spiny mice of Thailand (genus *Mus*) and occurrence of a complex intraspecific variation in *M. shortridgei*. Z Saugetierkunde 38; 159–68.

Hagemeizer A, Hahlen K, Sizoo W, Abels J (1982). Translocation (9;11) (p21;q23) in three cases of acute monoblastic leukemia. Cancer Genet Cytogenet 5; 95–105.

Hecht F (1986). Rare, polymorphic, and common fragile sites: A classification. Hum Genet 74; 207–8.

Hsu TC (1974). Longitudinal differentiation of chromosomes. Ann Rev Genet 7; 153–76.

Hsu TC (1975). A possible function of constitutive heterochromatin: The body guard hypothesis. Genetics 79: 137–50.

Hsu TC, Arrighi FE (1971). Distribution of constitutive heterochromatin in mammalian chromosomes. Chromosoma 34: 243–53.

Hsu TC, Somers CE (1961). Effect of 5-Bromodeoxyuridine on mammalian chromosomes. Proc Natl Acad Sci USA 47; 396–403.

Jalal SM, Clark RW, Hsu TC, Pathak S (1974). Cytological differentiation of constitutive heterochromatin. Chromosoma 48; 391–403.

John B (1973). The cytogenetic systems of grasshoppers and locust II. The origin and evolution of supernumerary segments. Chromosoma 44; 123–46.

John B (1981). Heterochromatin variation in natural populations. Chromosomes Today 7; 128–37.

John B, King M (1982). Meiotic effects of supernumerary heterochromatin in *Heteropternis obscurell*. Chromosoma 85; 39–65.

John B, Miklos GLG (1979). Functional aspects of satellite DNA and heterochromatin. Int Rev Cytol 58; 1–114.

Jones RN (1975). B-chromosome systems in flowering plants and animal species. Int Rev Cytol 40; 1–99.

Kaufman RJ, Brown PC, Schimke RT (1979). Amplified dihydrofolate reductase genes in unstably methotrexate-resistant cells are associated with double minute chromosomes. Proc Natl Acad Sci USA 76; 5669–73.

Kondo I, Shimizu N (1983). Mapping of the human gene for epidermal growth factor receptor (EGFR) on the p13q22 region of chromosome 7. Cytogenet Cell Genet 35; 9–14.

Kopnin BP (1981). Specific karyotypic alterations in colchicine-resistant cells. Cytogenet Cell Genet 30; 11–14.

Koprowski H, Herlyn M, Balaban G, Parmiter A, Ross A, Nowell P (1985). Expression of the receptor for epidermal growth factor correlates with increased dosage of chromosome 7 in malignant melanoma. Somatic Cell Mole Genet 11; 297–302.

Kuo T, Pathak S, Ramagli L, Rodriguez L, Hsu TC (1982). Vincristine-resistant Chinese hamster ovary cells. In Gene Amplification (ed. Schimke RJ). Cold Spring Harbor Laboratory; 53–7.

LeBeau MM (1986). Chromosomal fragile sites and cancer-specific rearrangements. Blood 67; 849–58.

LeBeau MM, Diaz MO, Marin M, Rowley JD (1985). Metallothionein gene cluster is split by chromosome 16 rearrangements in myelomonocytic leukemia. Nature 313; 709–11.

LeBeau MM, Rowley JD (1984). Heritable fragile sites in cancer. Nature 308; 607–8.

Lejeune J, Dutrillaux B, Lafourcade J, Berger R, Abonyi D, Rethone MO (1968). Endoreduplication selective du bras long du chromosome 2 chez une femme et sa fille. C R Acad Sci III 266; 24–6.

Limon J, Turc-Carel C, Dalcin P, Rao U, Sandberg AA (1986). Recurrent chromosome translocations in liposarcoma. Cancer Genet Cytogenet 22; 93–4.

Lin MS, Takabayashi T, Wilson MG, Marchese CA (1984). An in vitro and in vivo study of a BrdU-sensitive fragile sites in the chinese hamster. Cytogenet Cell Genet 38; 211–5.

Lubs HA (1969). A marker X chromosome. Am J Hum Genet 21; 231–44.

Mascarello JT, Mazrimas JA (1977). Chromosomes of antelope squirrels (Genus *Ammospermophilus):* A systematic banding analysis of four species with unusual constitutive heterochromatin. Chromosoma 64; 207–17.

Mascarello JT, Stock AD, Pathak S (1974). Conservatism in the arrangement of genetic material in rodents. J. Mammal 55; 695–704.

Mecucci C, Van Orshoven A, Tricot G, Michaux JL, Delannoy A, Van den Berghe H (1986). Trisomy 4 identifies a subset of acute nonlymphocytic leukemias. Blood 67; 1328–32.

Meyers MB, Biedler JL (1981). Increased synthesis of a low molecular weight protein in vincristine-resistant cells. Biochem Biophys Res Commun 99; 228–35.

Mitelman F (1985) Catalog of chromosome aberrations in cancer. In Progress and Topics in Cytogenetics, (ed. Sandberg AA). Vol. 5, Alan R. Liss, Inc., New York.

Miyoshi I, Hiraki S, Kimura K, Miyomoto K., Sato J (1979). 2/8 translocation in a Japanese Burkitt's lymphoma. Experientia 35; 742–3.

Muntzing A (1974). Accessory chromosomes. Ann Rev Genet 8; 243–68.

Nash WG, O'Brien SJ (1982). Conserved regions of homologous G-banded Chromosomes between orders in mammalian evolution: carnivores and primates. Proc Natl Acad Sci USA 79; 6631–5.

Nunberg JH, Kaufman RJ, Schimke RT, Urlaub G, Chasin LA (1978). Amplified dihydrofolate reductase genes are localized to a homogeneously staining region of a single chromosome in a methotrexate resistant Chinese hamster ovary cell line. Proc Natl Acad Sci USA 75; 5553–6.

Ohno S (1967). Sex Chromosomes and Sex-linked Genes. Springer-Verlag, Berlin/Heidelberg/New York.

Ohno S (1983). Phylogeny of the X-chromosome of man. In Cytogenetics of Mammalian X Chromosome, Part A, Basic Mechanisms of X Chromosome Behavior (ed. Sandberg AA). Alan R. Liss, Inc., New York; 1–19.

Pathak S (1971). A new type of intraspecies chromosome polymorphism in *rattus rattus berunnesculus* (Hodgson), collected from Nepal. Mamm Chrom Newsletter 12; 92–3.

Pathak S (1980). Cytogenetic analysis in human breast tumors. Cancer Genet Cytogenet 1; 281–9.

Pathak S (1983). Chromosome constitution of human solid tumors. Cancer Bull 35; 126–31.

Pathak S (1986). Constitutional translocations and predisposition to T– and B– cell malignancies. Cytogenet Cell Genet 43; 223–5.

Pathak S, Drwinga HL, Hsu TC (1983). Involvement of chromosome 6 in rearrangements in human malignant melanoma cell lines. Cytogenet Cell Genet 36; 573–9.

Pathak S, Goodacre A (1986). Specific chromosome anomalies and predisposition to human breast, renal cell and colorectal carcinoma. Cancer Genet Cytogenet 19; 29–36.

Pathak S, Hsu TC, Arrighi FE (1973a). Chromosomes of *Peromyscus* (Rodentia, Cricetidae). IV. The role of heterochromatin in karyotypic evolution. Cytogenet Cell Genet 12; 315–26.

Pathak S, Hsu TC, Shirley L, Helm JD (1973b). Chromosome homology in the climbing rats, genus *Tylomys* (Rodentia, Cricetidae). Chromosoma 42; 215–28.

Pathak S, Siciliano MJ, Cailleau R, Wiseman CL, Hsu TC (1979). A human breast adenocarcinoma with chromosome and isozyme markers similar to those of the HeLa line. JNCI 62; 263–71.

Pathak S, Stock AD (1974). The X chromosomes of mammals: karyological homology as revealed by banding techniques. Genetics 78; 703–14.

Pathak S, Strong LC, Ferrell RE, Trindade A (1982). Familial renal cell carcinoma with a 3; 11 chromosome translocation limited to tumor cells. Science 217; 939–41.

Patton JL (1977). B-chromosome systems in the pocket mouse *Perognathus baileyi:* Meiosis and C-band studies. Chromosoma 60; 1–14.

Popescu NC, DiPaolo JA (1979). Heterogeneity of constitutive heterochromatin in somatic Syrian hamster chromosomes. Cytogenet Cell Genet 24; 53–60.

Raman R. Sharma T (1974). DNA replication, G- and C-bands and meiotic behavior of supernumerary chromosomes of *Rattus rattus* (Linn.) Chromosoma 45; 111–9.

Ray-Chaudhuri R (1967). Mitotic and meiotic chromosomes of the koel, *Eudynanys scolopacea scolopacea.* The Nucleus 10; 179–89.

Ray-Chaudhuri SP, Kakati S, Pathak S, Sharma T (1967). Studies on the chromosomes of four species of Indian Homoptera (Auchenorrhyncha). Proc Zool Soc, Calcutta 20; 25–31.

Riccardi VM, Sujansky E, Smith AC, Francke U (1978). Chromosomal imbalance in the aniridia Wilms' tumor association: 11p interstitial deletion. Pediatrics 61; 604–10.

Robinson TJ, Elder FFB (1987). Chromosomal fragility in Sprague-Dawley and Fisher 344 rats. Abstract. 14th Annual Texas Genetics Society Meeting, College Station, 2–4 April 1987.

Rowley JD (1973a). A new constitent abnormality in chronic myelogenous leukemia identified by quinacrine fluorescence and Giemsa staining. Nature 243; 290–3.

Rowley JD (1973b). Identification of a translocation with quinacrine fluorescence in a patient with acute leukemia. Ann Genet (Paris) 16; 109–12.

Rowley JD, Golumb HM, Vardiman J, Fukuhara S, Dougherty C, Potter D (1977). Further evidence for a nonrandom chromosomal abnormality in acute promyelocytic leukemia. Int J Cancer 20; 869–72.

Sandberg AA (1980). The Chromosomes in Human Cancer and Leukemia. Elsevier North-Holland, Inc., New York.

Sandberg AA (1981). Chromosomal changes in the lymphomas. Hum Pathol 12; 531–40.

Schimke RT (1982). Gene Amplification. Cold Spring Harbor Laboratory, New York; 317–33.

Simmers RN, Sutherland GR, West A, Richards RI (1987). Fragile sites at 16q22 are not at the breakpoint of the chromosomal rearrangement in AMMoL. Science 236; 92–4.

Slamon DJ, Clark GM, Wong SG, Levin WJ, Ullrich A, McGuire WL (1987). Human breast cancer: correlation of relapse and survival with amplification of the HER-2/*neu* oncogene. Science 235; 177–82.

Smeets DFCM, Scheres JMJC, Hustinx TWJ (1986). The most common fragile site in man is 3p14. Hum Genet 72; 215–20.

Stark GR, Wahl GM (1984). Gene amplification. Ann Rev Biochem 53; 447–91.

Sumner AT (1982). The nature and mechanisms of chromosome banding. Cancer Genet Cytogenet 6; 59–87.

Sutherland GR (1983). The fragile X chromosome. Int Rev Cytol 81; 107–43.

Sutherland GR, Hecht F (1985). Fragile Sites on Human Chromosomes. Oxford University Press, New York.

Tewari R, Juyal RC, Thelma BK, Das BC, Rao SRV (1987). Folate-sensitive fragile sites on the X-chromosome heterochromatin of the Indian mole rat, *Nesokia Indica.* Cytogenet Cell Genet 44; 11–7.

Trent JM, Rosenfeld SB, Meyskens FL (1983). Chromosome 6q involvement in human malignant melanoma. Cancer Genet Cytogenet 9; 177–80.

Trujillo JM, Cork A, Ahearn MJ, Youness El, McCredie KB (1978). Hematologic and cytologic characterization of 8/21 translocation acute granulocytic leukemia. Blood 53; 695–706.

Turc-Carel C, Dalcin P, Limon J, Li F, Sandberg AA (1986a). Translocation X; 18 in synovial sarcoma. Cancer Genet Cytogenet 22; 93–4.
Turc Carel C, Lizard-Nacol S, Justrabo E, Favrot M, Philip T, Tabone E (1986b). Consistent chromosomal translocation in alveolar rhabdomyosarcoma. Cancer Genet Cytogenet 19; 361–2.
Uchida IA, Freeman VCP, Basrur PK (1986). The fragile X in cattle. Am J Med Genet 23; 557–62.
Van den Berghe H, David G, Orshoven AB, Louwagie A, Verwilghen R, Casteels-Van Daele M, Eggermont E, Eckels R (1979). A new chromosome anomaly in acute lymphoblastic leukemia. Hum Genet 46; 173–80.
Wake N, Hreshchyshyn MM, Piver SM, Matsui S, Sandberg AA (1980). Specific cytogenetic changes in ovarian cancer involving chromosome 6 and 14. Cancer Res 40; 4512–8.
Walker PMB (1972). Repetitive DNA in higher organisms. Progr Biophys Molec Biol 23; 67–101.
Whang-Peng J, Kao-Shan CS, Lee EC, Bunn PA, Carney DN, Gazdar AF, Minna JD (1982). Specific chromosome defect associated with human small-cell lung cancer: deletion 3p (14–23). Science 215; 181–2.
White MJD (1973). Animal Cytology and Evolution. 3rd Ed. Cambridge at the University Press, London; 312–33.
Wilson MG, Towner JW, Fujimoto A (1973). Retinoblastoma and D-chromosome deletion. Am J Hum Genet 25; 57–61.
Wurster-Hill DH, Ward OG, Kada H, Whittemore S (1986). Banded chromosome studies and B chromosomes in wild-caught raccoon dogs, *Nyctereutes procyonoides viverrinus*. Cytogenet Cell Genet 42; 85–93.
Yip MY, Mark J, Hulten M (1982). Supernumerary chromosomes in six patients. Clin Genet 21; 397–406.
Yosida TH (1977). Supernumerary chromosomes in the black rat (*Rattus rattus*) and their distribution in three geographic variants. Cytogenet Cell Genet 18; 149–59.
Yunis JJ (1983). The chromosomal basis of human neoplasia. Science 221; 227–36.
Yunis JJ (1984). Fragile sites and predisposition to leukemia and lymphoma. Cancer Genet Cytogenet 12; 85–8.
Yunis JJ, Bloomfield CD, Ensrud BS (1981). All patients with acute nonlymphocytic leukemia may have a chromosomal defect. N Engl J Med 305; 135–9.
Yunis JJ, Oken MM, Kaplan ME, Ensrud KM, Howe RR, Theologides A (1982). Distinctive chromosomal abnormalities in histologic subtypes of non-Hodgkin's lymphoma. N Engl J Med 307; 1231–6.
Yunis JJ, Soreng AL (1984). Constitutive fragile sites and cancer. Science 226; 1199–204.
Yunis JJ, Soreng AL, Bowe AE (1987). Fragile sites are targets of diverse mutagens and carcinogens. Oncogene 1; 59–69.
Yunis JJ, Yasmineh WG (1971). Heterochromatin, satellite DNA and cell function. Science 174; 1200–9.
Zankl H, Zang KD (1972). Cytological and cytogenetical studies on brain tumors. IV. Identification of the missing G chromosome in human meningiomas as No. 22 by fluorescence technique. Humangenetik 14; 167–9.
Zech L, Haglund U, Nilsson K, Klein G (1976). Characteristic chromosomal abnormalities in biopsies and lymphoid-cell lines from patients with Burkitt and non-Burkitt lymphomas. Int J Cancer 17; 47–56.

15 Flow-cytometry: A Tool in Cytogenetics

V.K. JAIN AND H.C. GOEL

Institute of Nuclear Medicine and Allied Science
Lucknow Road
Delhi 110007, India

1. Introduction

The conventional cytogenetic techniques to investigate the cellular DNA and chromosomes are laborious and time consuming. A number of attempts are being made, therefore, to develop rapid and sensitive automated methods. Computer aided scanning microphotometry allows automatic or semi-automatic analysis of whole cells and chromosomes in good metaphases deposited on solid carriers such as slides or tapes. The scanning process, however, remains slow, thereby limiting the number of cells which can be analysed in a sample within a reasonable period of time. The technique of flow-cytometry, in comparison, enables rapid and precise measurements to be made on whole cells, isolated nuclei or chromosomes in a monodisperse suspension. Flow-cytometry (FCM) is being extensively used in various fields of biology and medicine because of the relative ease, fast speed (>1000 particles/sec), precision of measurements and sorting capabilities of modern instruments. In this chapter, basic principles of flow-cytometry and typical applications in genetic research will be briefly described.

1.1. Basic principles of flow-cytometry

In flow-cytometry, cells or subcellular particles are made to flow at high speed through a sensor which generates signals corresponding to certain characteristic properties of objects under measurement. Most often physical properties such as light absorption, scattering, fluorescence emission and electrical conductivity are measured. Hydrodynamic methods are employed to force the particles to move through the sensor in almost identical trajectories at uniform speeds, well separated from each other. The signals are suitably amplified, collected, displayed and quantitatively analysed by the data-processing systems. For measurements involving light absorption and fluorescence emission, dyes which bind specifically to certain biomolecules of interest, can be also used for staining. Flow-cytometric measurements and analysis can thus provide useful information on various structural and functional parameters of the cell-populations and subcellular particles.

Depending upon the magnitude of the measured signals, particles can also be sorted out into different groups from a heterogeneous population. For this purpose, the liquid stream, after passing through the sensor, is made to form droplets which are imparted electrical charges proportional to the magnitude of the signals, by means of computer-controlled charging circuits. The droplets can be separated by deflection in an electric field. In this way specific subpopulations can be separated and used for further analysis or experiments.

1.2. Instrument design and development

Development of instruments suitable for measurements of particles in flow has been continuing for more than 50 years. Moldavan (1934) attempted to count individual cells flowing in a capillary tube using brightfield microscopy and a photoelectric sensor. Crossland-Taylor (1953) was the first to describe a method of hydrodynamic focusing using faster moving sheath flow to contain particles in a central narrow stream. Coulter fabricated an instrument to measure cell volume by means of changes in the electrical resistance as the cells pass through a small orifice in a conducting electrolyte. Instruments using absorption and scattering of light and fluorescence emission were developed and used in the late sixities (Kamentsky *et al.*, 1965; Van Dilla *et al.*, 1969; Dittrich and Gohde, 1969); fluorescence activated cell sorting (FACS) equipment was also developed about the same time (Hulett *et al.*, 1969). Subsequent to these pioneering developments, more sophisticated multiparameter systems with sorting capabilities have been built.

Present commercially available flow-through systems differ in basic designs depending upon the light source, the flow chamber and the light collection used.

1.2.1. *Laser-based instruments*. The particles are constrained to flow through a cylindrical chamber of small diameter (10–100 μm), upon which the laser beam is focused, thus intensively illuminating the particles (Figure 1). In this system the axis of illumination and collection of light are orthogonal to each other and to the direction of sample flow. The fluorescent light emitted is allowed to fall on one or more photomultipliers, which act as sensors. Besides the fluorescent light, scattered light can also be collected and measured independently and simultaneously, by using appropriate filters to distinguish between the scattered and fluorescent light. Lasers can be focused to very small areas and offer the advantage of high radiance, spectral purity and stability. Argon-ion lasers are most commonly used, followed by Krypton-ion He-Ne and dye lasers.

1.2.2. *Arc-lamp-based instruments*. These instruments use high pressure (mercury or xenon) arc lamps as the light source and the principle of epi-illumination with a fluorescence microscope is employed.

The cell suspension is made to flow parallel to the optical axis in a measuring chamber (Figure 2). The mercury high-pressure arc lamp (100 W) has enough energy in the short wavelength UV region down to 200 nm to excite the various fluorochromes; the wavelength of interest can be selected by means of appropriate filters.

As compared to laser-based instruments, high pressure mercury arc lamp systems are versatile, simpler and less expensive to operate and maintain. The resolutions of the two systems are comparable. Laser-based instruments are, however, more sensitive and can analyse even very small particles rapidly.

1.2.3. *Scanning flow-cytometers*. The instruments described above provide very little information about the morphological features of the particles. These are 'zero resolution' instruments in which the excitation or the measuring apertures are larger than the objects under

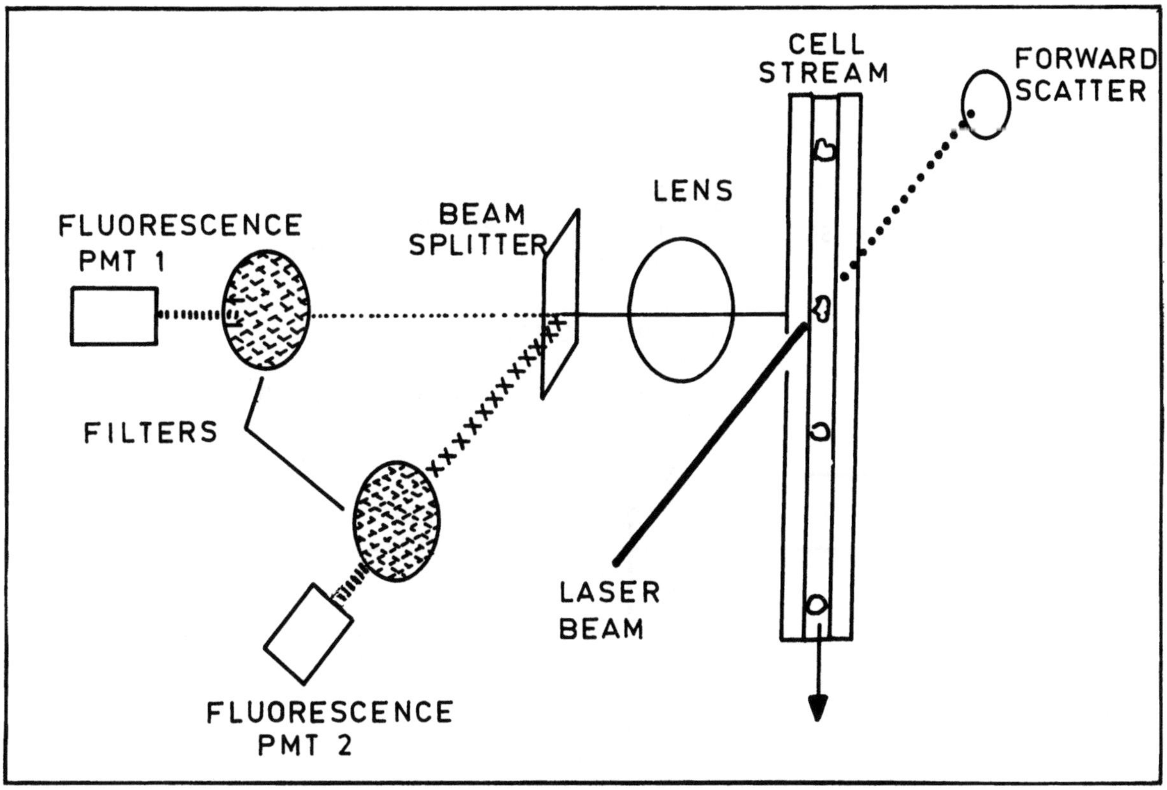

FIGURE 1: Schematic diagram of laser-based multi-parameter flow-cytometric optical system.

measurement. In the scanning flow-cytometer (SFCM), the excitation/measuring aperture is made smaller than the objects, for example, by using a very small and thin laser beam or by providing a narrow slit in the optical system (Wheeless and Patten, 1973). In slit scan flow cytometry (SSFCM), time of flight measurements yield very high resolution. Morphological information can be obtained from these instruments by analysing the shape of the signal as the object passes through the scanning region of the system. Using such a system, information about the shapes of chromosomes has been obtained and the relative position and number of centromeres have been determined (Gray *et al.*, 1984) as shown in Figure 3.

2. DNA Flow-cytometry

2.1 DNA fluorochromes and staining procedures

Information on the content, structure and function of the genetic material—viz. chromosomes and DNA—is of prime importance for genetic studies. Precise measurements of DNA contents of chromosomes, nuclei or whole cells can be made by FCM with the help of fluorochromes which bind specifically and stoichiometrically to DNA. Some of the fluorochromes can also be helpful in providing information about the base composition and conformation of DNA. Important DNA fluorochromes are listed in Table 1.

A variety of dyes with light absorption over a

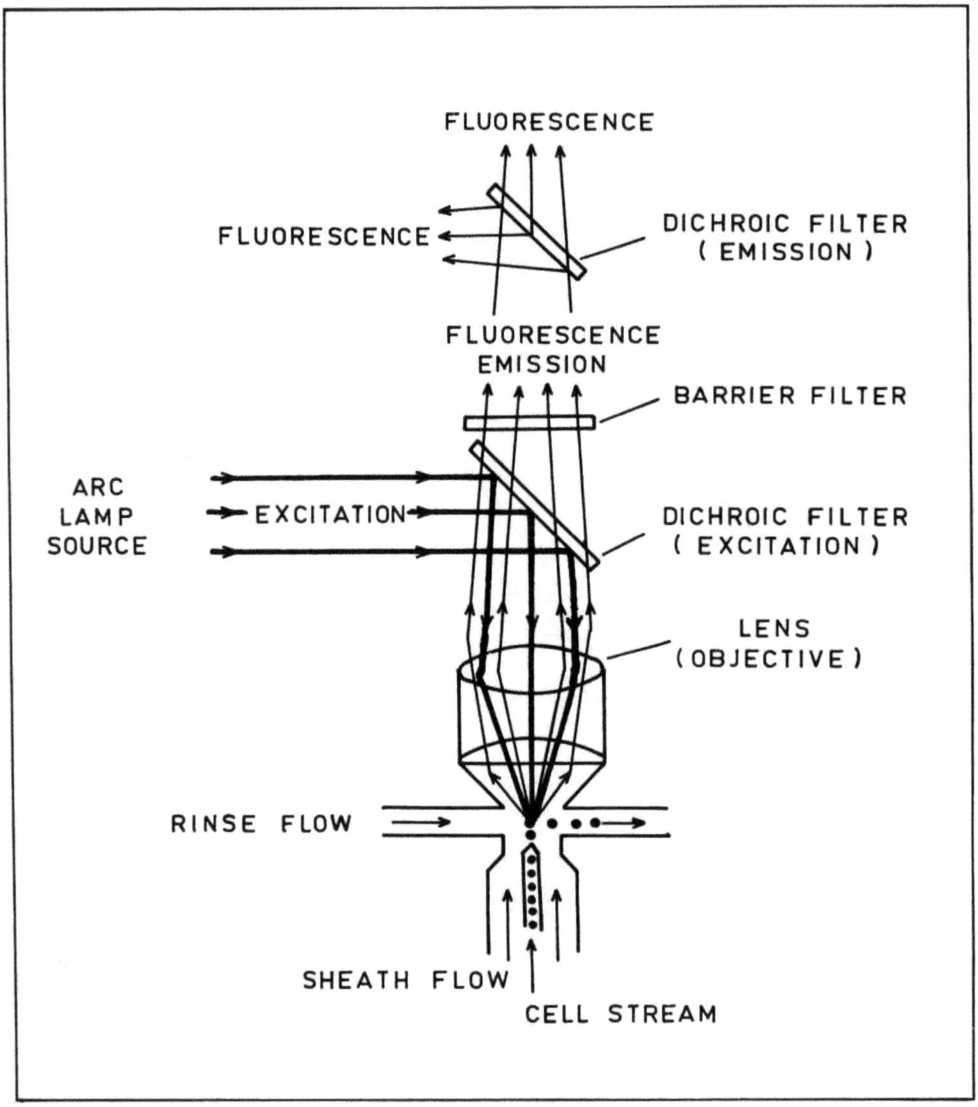

FIGURE 2: Schematic diagram of flow-cytometric optical system using high-pressure mercury arc lamp.

wide range, from the near UV to red light are available. These include intercalating dyes such as ethidium bromide (Le Pecq, 1971), propidium iodide (Krishan, 1975) and acridine orange (Darzynkiewicz *et al.*, 1977), which intercalate into double-stranded nucleic acids and do not show any preference for the base composition. However, these dyes bind also to double-stranded RNA, which needs to be removed by treatment with RNase. Acridine

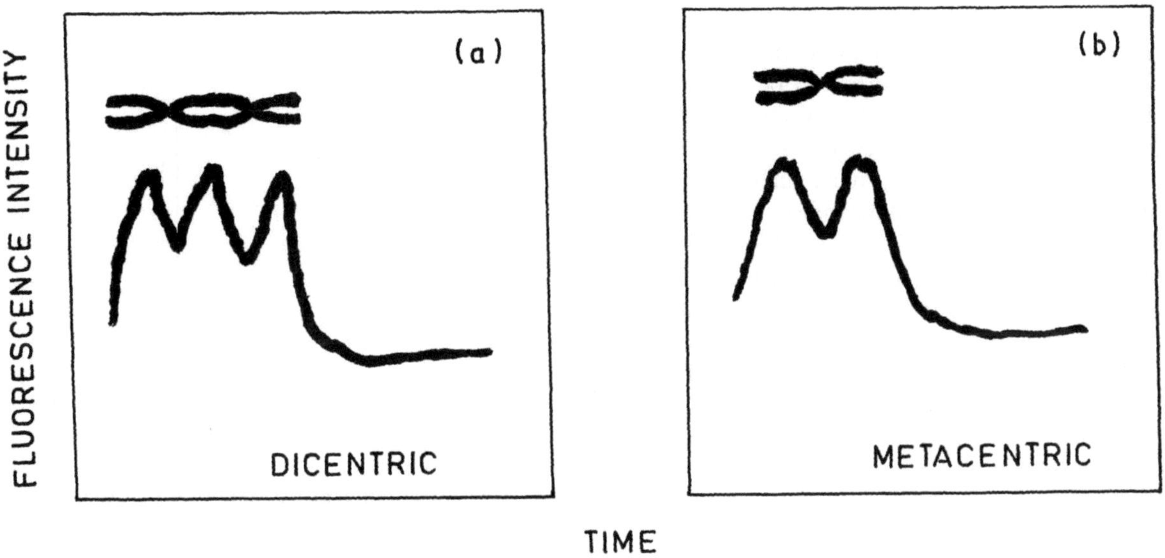

FIGURE 3: Slit-scan fluorescence profiles of a normal metacentric chromosome (b) and in abnormal dicentric chromosome (a). The chromosome silhouettes show the chromosomes expected to produce the two slit-scan profiles (Gray *et al.*, 1984).

TABLE 1

Important fluorochromes for DNA content measurements by flow cytometry

S. No.	Dye	Spectral characteristic of the dye-DNA complex absorption (nm)	Emission (nm)	Molecular specificity	Base specificity
1.	HO	360 (310–390)	450 (410–525)	DNA	A–T
2.	DAPI or DIPI	360 (310–400)	450 (410–530)	DNA	A–T
3.	MI	410 (350–470)	560 (470–660)	DNA	G–C
4.	CA_3	420 (350–460)	560 (480–660)	DNA	G–C
5.	AO	480 (420–520)	540 (500–600)	RNA/DNA	None
6.	EB	520 (440–580)	590 (590–690)	RNA/DNA	None
7.	PI	520 (450–585)	605 (580–690)	RNA/DNA	None
8.	OX-750	633 (580–710)	700 (670–760)	DNA	None
9.	LD 700	655 (560–690)	600 (630–760)	DNA	None
10.	LD 800	700 (600–720)	720 (680–760)	DNA	None

HO: Hoechst dye 33258 or 33342 (Bisbenzimidole); DAPI or DIPI: 4,6 Diamidino-2 Phenyl indole; MI: Mithramycin; CA_3: Chromomycin A_3; AO: Acridine orange: EB: Ethidium bromide; PI: Propidium iodide; OX-750: Oxazine-750; LD 700: Rhodamine-700; LD 800: Rhodamine-800.

orange also binds in a non-intercalative mode to single-stranded nucleic acids with resultant red fluorescence. This property can be exploited to simultaneously measure DNA and RNA contents of cells (Darzynkiewicz *et al.*, 1980).

The non-intercalating dyes, such as the antibiotics mithramycin and its analogue chromomycin A_3 (CA_3) (Crissman *et al.*, 1978), bisbenzimide (Hoechst 33258 or 33342) and 4, 6-diamidino-2-phenylindole (DAPI) show a marked dependence on DNA base composition and are DNA specific. Hoechst dyes and DAPI have similar DNA binding and fluorescent properties (A-T specific), while MI and CA_3 bind preferentially to G-C base pairs (Latt, 1979). The DNA-specific dyes minimize problems associated with cellular RNA.

Efforts to develop new dyes which can be excited with inexpensive and efficient light sources are being made. This may permit the development of simple and easy to maintain flow-cytometers. Recently Shapiro and Stephens (1986) have reported successful use of red excited (633 nm) laser (He-Ne) dyes such as oxazine 750, Rhodamine 700 and Rhodamine 800 for DNA estimation. These dyes have been shown to be specific and stoichiometric stains for DNA; their mode of binding to DNA remains, however, yet to be investigated.

The reliability, reproducibility and precision of FCM measurements of DNA contents depend to a large extent also on the methods of sample preparation in monodisperse suspensions and subsequent fixation and staining procedures. Most of the dyes are relatively large, highly charged molecules which do not easily penetrate viable cells. A number of fixation and staining procedures have been developed. The use of non-ionic detergents combined with hypotonic treatment has been found to be quite satisfactory in obtaining monodisperse nuclei from most of the cells and tissues (Krishan, 1975; Otto *et al.*, 1981; Vindelov *et al.*, 1983).

For investigations on chromosomes, the cells are arrested in mitosis, swollen hypotonically and disrupted mechanically in the presence of a detergent so as to release the chromosomal contingent into suspension. Subsequently, the chromosomes and cell fragements are treated with RNase and with DNA fluorochromes. A good chromosomal preparation should have uniformly stained chromosomes, minimum debris and clumps and the relative abundance of various chromosomes in suspension should correspond to chromosome numbers present in the cells. A number of factors critically influence the quality of chromosome preparation, which in turn affects the karyotype (van den Engh *et al.*, 1984).

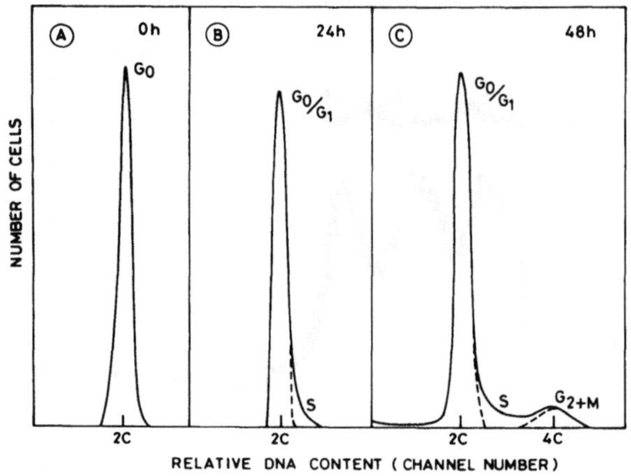

FIGURE 4; DNA histogram of human peripheral blood leukocytes measured at different time intervals after stimulation with PHA (Jain *et al.*, 1986).

2.2. Measurement of DNA content in whole cells and nuclei

DNA content measurements on intact cells or cell nuclei can be carried out with ease and great precision. Data obtained by FCM are displayed as frequency distributions (histograms). Significant changes occur in the cellular DNA content as the cells start proliferating. The corresponding changes are seen in the fluorescence intensity frequency distribution of the cell population as represented by the histogram. As an example, Figure 4 shows the changes observed in a human peripheral blood leukocyte culture stimulated to divide by PHA. From these DNA histograms, the average DNA contents of G_0/G_1 and G_{2+M} phase cells, their variability and the relative fractions of cells in the various phases of the cell-cycle can be calculated as briefly described below.

2.2.1. *Ploidy or DNA index.* DNA estimation by fluorochrome staining and flow cytometry is a relative measurement. Comparison with a reference standard having a known DNA content is, therefore, necessary for calibration of the method. One may use either an external, or internal standard. An internal standard which is stained and measured along with the sample should be preferred since errors due to instrumental drift or differences in staining could also be eliminated rendering the measurements more precise.

Using the internal reference standards, DNA content measurement can be carried out with a precision of 1–2 per cent. Small changes in the DNA content/nucleus can be quantitated and degree of aneuploidy or the DNA index estimated from the G_0/G_1 peaks of the DNA histograms by comparing with normal diploid cells. Thus DNA index (DI) can be determined as follows:-

$$DI = \frac{P_S}{P_R} - \frac{P_D}{P_R} \times 2 \tag{1}$$

Where P_S, P_R and P_D are the modal channel numbers corresponding to G_0/G_1 peaks of the sample, internal reference and standard diploid cells respectively.

2.2.2. *DNA content dispersion.* All cells of a particular species, while in the G_0/G_1 phase of the cell cycle, are expected to have identical DNA content. Similarly, the chromosomes of a particular type should have uniform DNA content. Reasons for the variability or dispersion in the DNA content distributions measured by flow cytometry have been discussed earlier (Otto *et al.*, 1984). In general, the observed dispersion could arise due to three main categories of factors as follows:

Observed dispersion = Intrinsic dispersion + Measurement dispersion + Induced dispersion

Taking variance (σ^2) as a quantitative measure of dispersion, and assuming that the deviations from a central value are small and follow a normal distribution, we can correspondingly write:

$$\sigma^2_{0b} = \sigma^2_0 + \sigma^2_m + \sigma^2_i \tag{2}$$

The intrinsic dispersion, σ^2_0, reflects the biological variability of the sample and may depend on its physiological state. Measurement dispersion σ^2_m, is introduced during various steps involved in the process of measurement; for example, random variations in the staining procedures, instability of the flow-rate, light intensity and electronic processing. Finally, induced dispersion σ^2_i, is due to genotoxic treatments to which cell populations might be exposed. Under optimal working conditions the intrinsic dispersion and dispersion due to measurement could be minimized and held constant within limits in untreated control population, so that:

$$\sigma^2_m + \sigma^2_0 = \sigma^2_c \tag{3}$$

The dispersion induced by treatments could then be given by:

$$\sigma^2_i = \sigma^2_0 - \sigma^2_c \tag{4}$$

For comparisons in the different cell populations it is convenient to use relative standard deviation or the coefficient of variation (CV) as an index of dispersion:

$$CV = \frac{\sigma}{m} \tag{5}$$

where σ is the standard deviation around the mean value m. In this case it can be seen that the

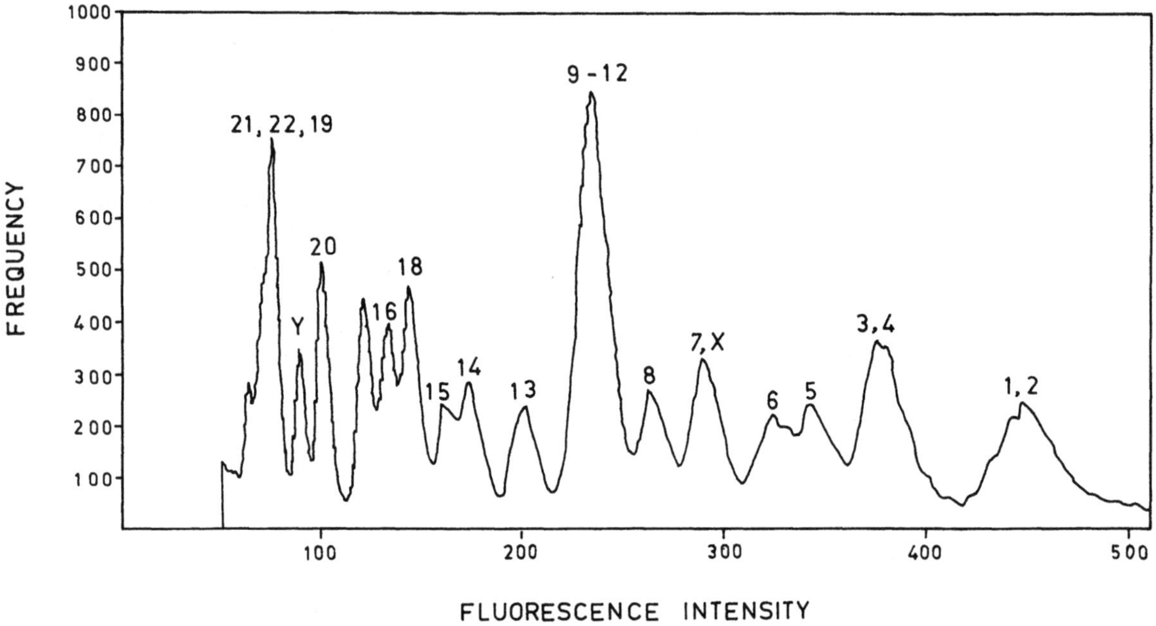

FIGURE 5: Hoechst 33258 fluorescence intensity distribution of chromosomes prepared from peripheral blood lymphocytes from a 46XY normal male donor. Chromosomes were sorted from the Y peak as marked (Fantes *et al.*, 1983).

increase in the dispersion due to treatment is given by the expression:

$$CV_i = (CV^2_{0b} - CV^2_c) \qquad (6)$$

The coefficient of variations CV_{ob} and CV_c can be calculated from corresponding DNA histograms of treated and control samples. DNA histograms can now be obtained with quite low values of CV. In untreated homogeneous cell suspension, CV_c of about 2 per cent can be easily obtained and values of less than 1 per cent have been reported under optimal conditions. The measurements are also highly reproducible and reliable.

2.2.3. *Cell-cycle analysis.* A number of graphical and numercial methods based on different models have been suggested and are being used to calculate the cell-cycle distributions from the DNA histograms of asynchronously growing and perturbed cell populations (Dean and Jett, 1974; Fried, 1977; Kim and Perry, 1977; Rich *et al.*, 1983). Graphical methods are relatively simple as compared to numercial methods, which employ mathematical techniques of various degrees of sophistication. Using numercial methods, on-line calculations can be conveniently carried out with the help of microprocessors. Comparing the relative merits of different methods it has been demonstrated, however, that the cell-cycle distributions cannot be determined with an accuracy of better than 10 per cent at present due to inherent sampling errors and staining variability (Dean *et al.*, 1982). Therefore, applications of very precise methods for calculations seem to be unnecessary.

2.3. Measurements on chromosomes: Flow-karyotyping

Application of fluorescence activated flow sorter system (FACS) and slit-scan flow cytometry in the study of karyogram, isolated chromosomes

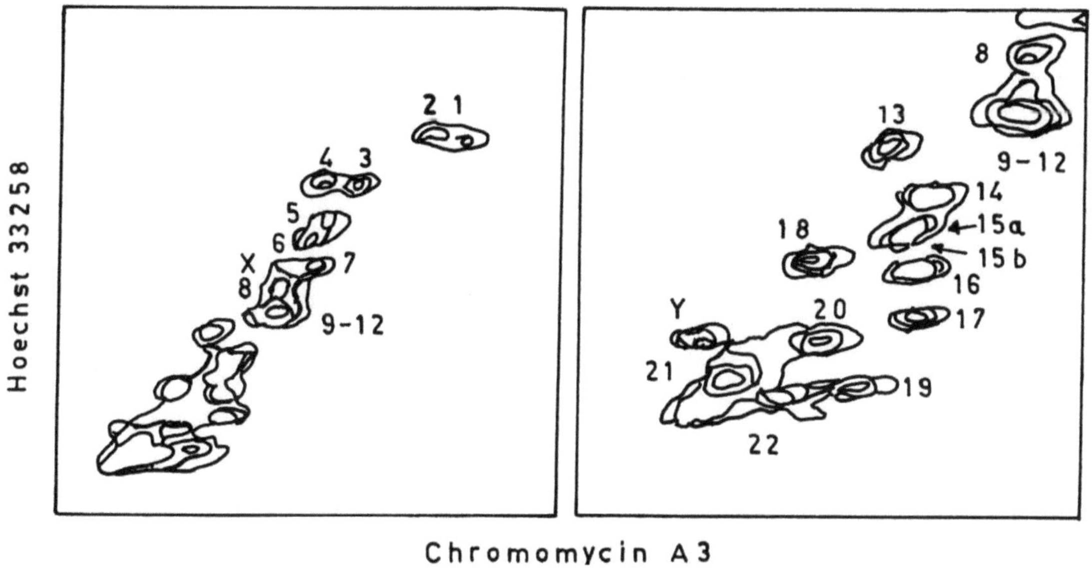

FIGURE 6: Bivariate flow karyotype of human chromosomes derived from peripheral lymphocytes from a normal donor. Left: whole karyotype. Right: expanded view of the smaller chromosome types from the same sample. The chromosomes producing each peak are indicated (Langlois *et al.*, 1982).

and parts of chromosomes, has been an important step forward in modern cytogenetics and has been reviewed recently (Gray and Langlois, 1986). In flow-karyotyping, the individual chromosomes could be identified on the basis of DNA content, base pair ratio, size, shape and centromeric index (Yu *et al.*, 1981; Lucas *et al.*, 1983). Abnormal chromosomes, even when present in a minority of the cells, can be identified and sorted for further details.

Typically a single parameter karyotype (a histogram of the DNA content of chromosomes) consists of a number of peaks which indicate the DNA content of the chromosomes (Figure 5). Thus the mean values of the peaks classify the chromosomes according to their DNA contents and area under each peak is proportional to the relative frequency of that chromosome type in the chromosome population. With the help of a sorter the chromosomes can be separated and purified according to their DNA; the purity depends to a large extent on the resolution with which the specific chromosomes can be resolved in the flow-karyotype. The uniformity of chromosome staining, variations in DNA content per chromosome and the measurement precision determine the width of the peak and the resolution of the measurements. Actual flow-karyotypes usually show broad distributions underlying the chromosome peaks. The background is caused by clumps of two or more chromosomes, chromosome fragments and other unidentified non-fluorescent particles. In fact the use of single fluorescent DNA dye permits only a limited use of FCM in karyotyping, specially due to the incapability of making

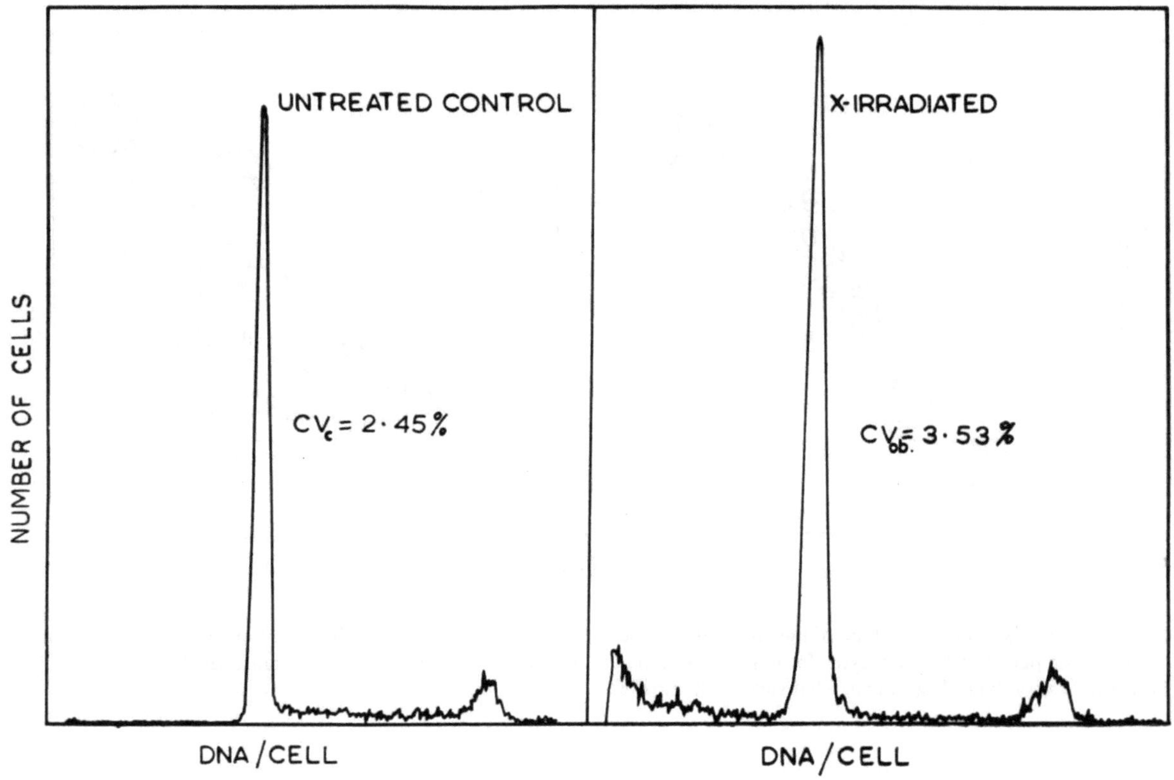

FIGURE 7: X-ray induced increase in DNA content dispersion in mouse marrow cells (D = 3.5 Gy) (Jain *et al.*, 1986).

distinction between chromosomes having very little difference in DNA contents, and the degree of unpredictable day-to-day variations.

Simultaneous staining of DNA by two or more fluorescent DNA dyes increases peak discrimination. Bivariate flow-karyotyping with high intensity lasers (5 to 19 W) utilizing the G-C specific chromomycin A_3 in addition to the A-T specific Hoechst 33258, allows resolution of all human chromosomes except 9–12 and 14–15 (Figure 6), yielding 95 to 99 per cent pure sorted chromosomes (Gray and Langlois, 1986). Due to different positions of the centromeres, slit-scanning flow-cytometry can partially resolve chromosomes 9 to 12 (Cram *et al.*, 1985; Lucas and Gray, 1987). Slit scan flow cytometry has been used to measure the frequency of dicentric chromosomes in X-irradiated Chinese hamster cells (Gray *et al.*, 1984) and a very good correlation with conventional visual scoring could be observed.

Chromosome banding patterns as revealed by certain fluorochromes could be enhanced or modified by counter-staining with base specific DNA binding agents (Schweizer, 1980, 1981). These techniques can also be adapted for flow studies. In human metaphase chromosomes, stained with the A-T specific fluorochromes DAPI, when counterstained with the fluorescent A-T specific DNA binding agent netropsin (NTR), the G-banding like pattern of DAPI fluorescence along the arm is depressed. How-

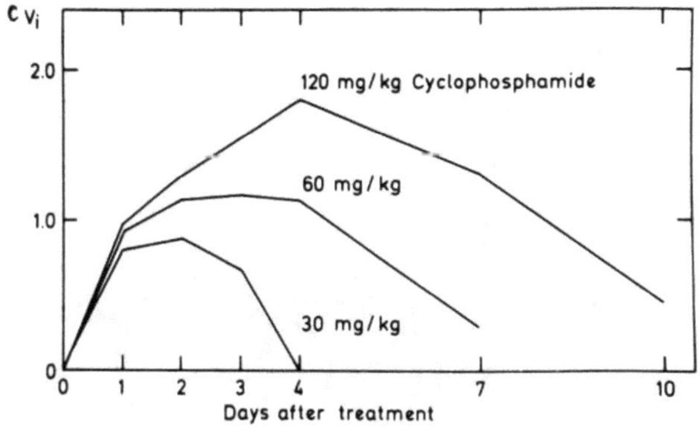

FIGURE 8: Time-dependence of DNA content dispersion induced by cyclophosphamide in bone marrow cells of mice (Otto *et al.*, 1984).

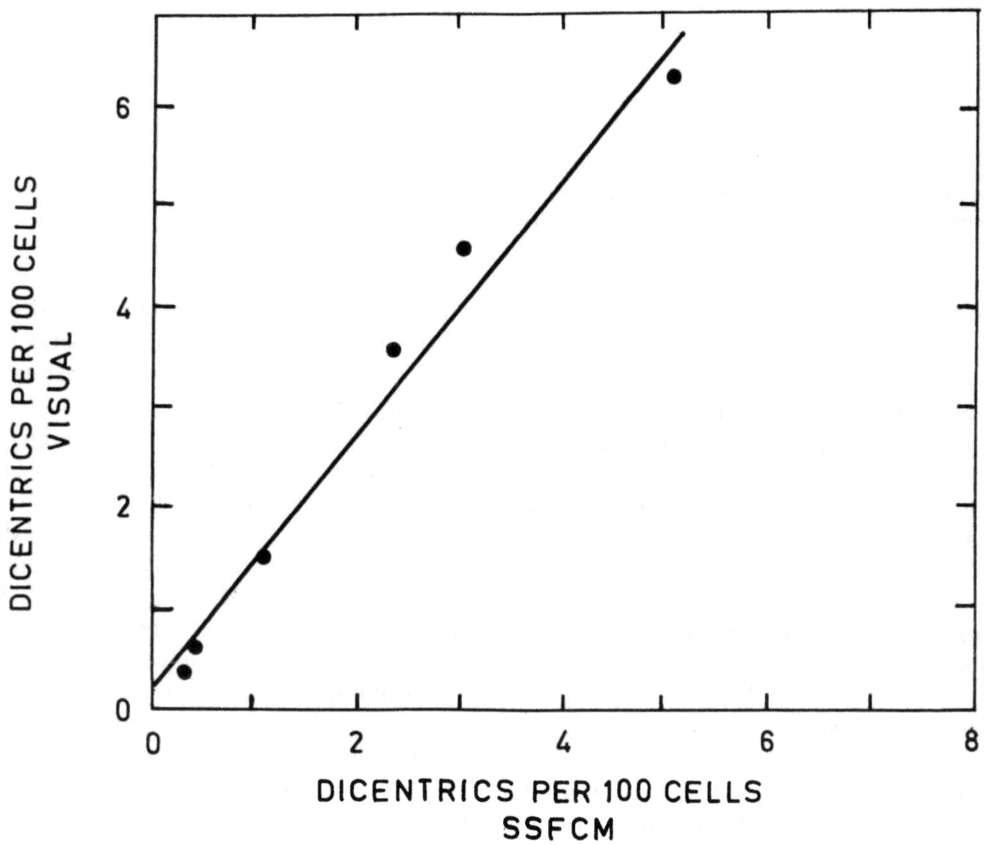

FIGURE 9: Dicentric chromosome frequencies measured for Chinese hamster cells exposed to 0–400 rad X-rays. Frequencies were estimated both by slit-scan flow cytometry and visual scoring of metaphase cells from the same samples (Gray *et al.*, 1984).

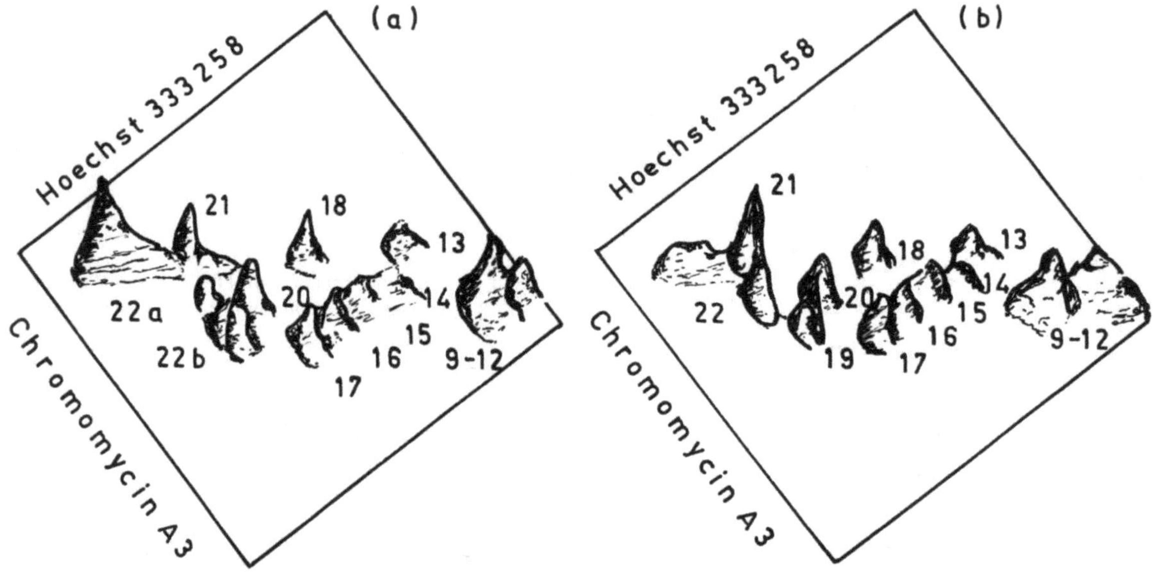

FIGURE 10: Isometric view of bivariate flow karyotypes measured for amniotic samples from a normal female (a) a female carrying three number 21 chromosomes (b). The increased volume of peak is corresponding to chromosome 21 (Gray *et al.*, 1986).

ever, the fluorescence of heterochromatic C-band regions of chromosomes 1,9,15,16 and Y is not depressed and remains bright. Use of NTR as a counterstain, with the G-C specific fluorochromes chromomycin A_3, increases contrast of the R-banding like pattern of A_3. These recent investigations demonstrate the immense potential of FCM in cytogenetics.

3. Application of Flow-cytometry in Genetics

3.1. Mutagen screening

Effects of environmental agents such as radiations and chemicals on DNA can be rapidly screened by measuring DNA content dispersion in whole cells or cell nuclei (Otto *et al.*, 1984; Jain *et al.*, 1986). Chromosomal damage and dysfunction of the mitotic apparatus induced by mutagenic agents lead to an unequal distribution of the DNA amount in the daughter cells and consequently to an increased variability of the cellular DNA content in the cell population. This effect can be observed in the DNA histograms as broadening of the (G_0+G_1)—and (G_{2+M})—peaks, as shown in Figure 7, and can be quantitated from the measured values of coefficient of variations of treated and control samples as described in section 2.2.2. Figure 8 shows an example of such measurements in bone marrow cells of mice. Detection of chromosome abnormalities, both numerical and structural can be accomplished by flow-karyotyping. Numerical alterations give rise to changes in the relative areas under the peaks corresponding to the chromosomes whereas structural alterations such as deletions or translocations result in appearance of new peaks in the flow-karyogram. Presence of damaged chromosomes also causes an increase in the background continuum and

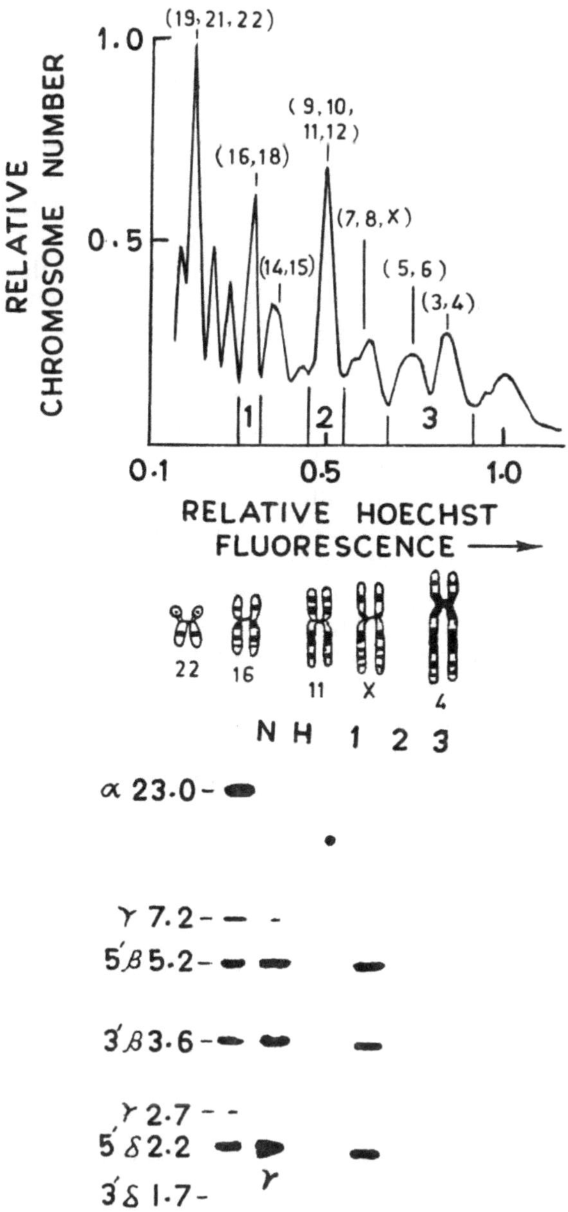

FIGURE 11: Flow-karyotype measured for human chromosomes (Top). Chromosomes were sorted from regions 1, 2 and 3. Below, the results of an EcoRI restriction enzyme analysis of DNA from a normal individual (N), from an individual lacking alpha-globin genes (H), and from chromosomes sorted from regions 1, 2 and 3 following hybridization with cDNA produced from globin m RNA (Lebo *et al.*, 1979).

leads to broadening of the peak widths, which can be used to quantitate the induced genetic damage by the environmental agents. Some specific types of chromosomal aberrations can be identified and quantitated by slit-scan FCM; for example, X-ray induced frequency of dicentrics has been accurately estimated using this technique (Gray *et al.*, 1984) as shown in Figure 9.

3.2. Population monitoring

Because of high speed of analysis by FCM, screening of a large number of members of a population becomes feasible. Thus, effects of carcinogens and mutagens can be evaluated in the exposed population by FCM by analysing DNA contents and DNA content dispersion yielding information about aneuploidy, polyploidy, etc. Flow-karyotyping of isolated chromosomes can help in investigating the incidence of a specific chromosomal disorder in a population, as illustrated in Figure 10. The dynamics of genes responsible for the disorder, could also be studied in context of a gene pool in the population. Such studies may be highly cumbersome conventionally but can be easily carried out by the FCM technique.

3.3. Oncogenetics

Based on DNA quantification, it has been possible to study the cell cycle, frequency of various cells in different phases of cell cycle like G_0, G_1, S and G_{2+M}, ploidy and the presence of specific tumour marker chromosomes. The frequency of cells in S-phase at a given time can help to evaluate tumour growth fraction. The role of ploidy in lymphomas, leukaemias, myelomas and solid tumours as studied by FCM has been reviewed by Barlogie *et al.* (1983). The correlation between ploidy and tumour progress, including objective grading of tumours and follow-up of the results of chemotherapy, has been quite reliable in some cases. Specific tumour marker chromosomes can be sorted out and studied further.

3.4. Genetic counselling and animal breeding

Labelling of lymphocytes from the blood of pregnant women with fluorescent antibodies to paternal histocompatability antigens permits the FACS to separate a very small population of lymphocytes bearing paternal antigens. These cells are in fact foetal cells in maternal circulation. Sex chromatin analysis of these cells by FCM has enabled successful prediction of the foetal sex and abnormalities. This could avoid cumbersome procedures of amniocentesis.

The isolation of spermatozoa based on X and Y chromosomes has been possible with the help of the FACS technique. Sperms of a particular type could be used for deciding the sex of progeny especially in context of test tube foetus technology. In animal breeding also, the isolation of desired spermatozoa based on sex or shape abnormalities, could be of immense utility (Pinkel *et al.*, 1979, 1982).

3.5. Checking pathological archives

The recently developed technique of recovering nuclei from paraffin embedded tissue blocks has opened pathological archives for retrospective DNA analysis by FCM (Hedley *et al.*, 1983, 1984). Direct comparison of fresh with paraffin embedded samples has shown good agreement (Camplejohn and Macartney, 1985; Coon *et al.*, 1986). This possibility can have important medicolegal applications.

3.6. Molecular cytogenetic research

Flow sorters have helped the molecular cytogeneticist to a great extent in gene mapping and in creating recombinant DNA libraries.

DNA from a purified chromosome fraction obtained after sorting, is digested using a restriction enzyme. The resulting DNA fragments are transferred to nitrocellulose paper after separation by electrophoresis and hybridized with a labelled nucleic acid probe coding for the gene under study (Lebo *et al.*, 1979). Using this approach, the β–, γ– and δ–, globin genes have been mapped to the distal portion of the short arm of human chromosome (Lebo *et al.*, 1979) (Figure 11). The methods can be further simplified to a considerable extent by collecting and hybridizing the sorted chromosomes directly on nitrocellulose filters (Lebo *et al.*, 1984).

For construction of recombinant DNA libraries, DNA from the sorted chromosomes is digested completely by restriction enzymes and cloned in vectors with high packaging efficiency such as lambda phage. For high quality library production high sorting purity is essential. Purification depends upon the resolution, which can be better achieved by dual beam chromosome sorter. Purification of specific chromosomes can be facilitated by using human × hamster hybrid cells which contain only a few human chromosomes (Cremer *et al.*, 1982) or by employing human cell lines in which chromosome polymorphisms (Harris *et al.*, 1985) or rearrangements (Lalande *et al.*, 1985) leave one chromosome type particularly well resolved. The Lawrence Livermore and the Los Almos National Laboratories in USA have produced a number of such libraries from human chromosomes.

4. Future Prospects

4.1. Chromosome isolation and staining procedures

The applications of flow-karyotyping in cytogenetics are likely to increase immensely if improved techniques for the isolation and staining of chromosomes from tissues and cell populations with low proliferation index can be developed. Presence of debris in the chromosomal preparations complicates quantitative estimations of chromosomal frequencies, therefore, better culture techniques, isolation procedures and more sophisticated algorithms for data analysis are needed. Use of immunochemical agents (Trask *et al.*, 1984) and/or specific recombinant DNA probes tagged with specific fluorochromes (Rappold *et al.*, 1984) may lead to improvements in flow-karyotyping and sorting. Monoclonal antibodies developed against proteins associated with chromosomal subregions like centromeres (Moroi *et al.*, 1980) will be useful for morphological classification of the chromosomes employing SSFCM. Many chromosome specific repetitive DNA probes have been recently constructed (Lau *et al.*, 1985; Willard, 1985) and a selective part of human Y chromosome has been successfully stained (Pinkel *et al.*, 1986). Development of specific antibodies against chromosomal proteins is also expected to provide useful information on the structure and function of chromatin. Further progress is rapidly taking place in this field.

4.2. Instrumentation

Attempts are being made to improve the spatial resolution ($<1 \mu m$) in SFCM and to extract meaningful morphological information. A new technique to scan chromosomes through a series of fringes produced by the constructive and destructive interference of two intersecting laser beams is being developed (Gray *et al.*, 1984). Increase in the speed of sorting combined with pre-enrichment techniques (such as velocity sedimentation) will allow faster purification of chromosomes, thus greatly facilitating experiments in molecular cytogenetics and gene mapping.

Improvements in optics, for example better efficiency of light collection, may permit use of low power lasers thus making the equipment less expensive and easier to maintain. Use of fluorochromes absorbing in the red region (Shapiro *et al.*, 1983), allowing employment of He-Ne laser or use of arc lamps, and economic microprocessors for dedicated application are likely to make the instruments cost-effective for routine clinical applications.

In conclusion, it may be stated that FCM has become a very useful and essential tool for advanced research in molecular genetics and is expected to be routinely used in future in clinical cytogenetics for detecting disease linked chromosome abnormalities.

References

Barlogie B, Raber MN, Schumann J, Johnson TS, Drewinko B, Swartzendruber DE, Gohde W, Andreef M, Freireich EJ (1983). Flow cytometry in clinical cancer research. Cancer Res 43; 3982–97.

Camplejohn RS, Macartney JC (1985). Comparison of DNA flow cytometry from fresh and paraffin embedded samples of non-Hodgkin's Lymphoma. J Clin Pathol 38; 1096–9.

Coon JS, Schwartz D, Summers JL (1986). Flow cytometric analysis of deparaffinized nuclei in urinary bladder carcinoma. Cancer 57; 1594–601.

Cremer C, Gray JW, Ropers H (1982). Flow cytometric characterization of a Chinese hamster × Man hybrid cell line retaining the human Y chromosome. Human Genet 60; 262–6.

Crissman HA, Stevenson AP, Orlicky DJ, Kissane RJ (1978). Detailed studies of the application of three fluorescent antibiotics for DNA staining in flow cytometry. Stain Technol 53; 321–30.

Crossland-Taylor PJ (1953). Device for counting small particles suspended in a fluid through a tube. Nature 171; 37–8.

Cram LS, Bartholdi MF, Wheeless LL, Gray JW (1985). Morphological analysis by scanning flow cytometry. In Flow cytometry: Instrumentation and data analysis (eds Van Dilla MA, Dean PN, Melamed M, Laerum O), Academic Press, New York; 163–94.

Darzynkiewicz Z, Tranganos F, Sharpless T, Melamed MR (1977). Recognition of cells in mitosis by flow cytofluorometry. J Histochem Cytochem 25; 875–80.

Darzynkiewicz Z, Tranganos F, Melamed MR (1980). New cell cycle compartments identified by multiparameter flow cytometry. Cytometry 1; 98–108.

Dean PN, Jett JH (1974). Mathematical analysis of DNA distributions derived from flow-microfluorometry. J Cell Biol 60; 523–7.

Dean PN, Gray JW, Dolbeare FA (1982). The analysis and interpretation of DNA distributions measured by flow-cytometry. Cytometry 3; 188–95.

Dittrich W, Gohde W (1969). Impulse fluorometry with single cells in suspension. Z. Naturforsch 24b; 360.

Fantes JA, Green DK, Cooke HJ (1983). Purifying human Y chromosomes by flow cytometry and sorting. Cytometry 4; 88–91.

Fried J (1977). Analysis of deoxyribonucleic acid histograms from flow cytometry. Estimation of the distribution of cells within S-Phase. J Histochem Cytochem 25; 942–51.

Gray JW, Langlois RG (1986). Chromosome classification and purification using flow cytometry and sorting. Ann Rev Biophys Biophys Chem 15; 195–235.

Gray JW, Lucas J, Yu LC, Langlois RG (1984). Flow cytometric detection of aberrant chromosomes. In Biological dosimetry, cytometric approaches to mammalian systems (eds Eisert WG, Mendelsohn ML), Springer Verlag, Berlin; 25–35.

Harris P, Boyd E, Ferguson-Smith MA (1985). Optimising human chromosome separation for the production of chromosome-specific DNA libraries by flow sorting. Human Genet 70; 59–65.

Hedley DW, Friedlander ML, Taylor IW (1983). Method for analysis of cellular DNA content of paraffin-embedded pathological material using flow cytometry. J Histochem Cytochem 31; 1333–5.

Hedley DW, Friedlander ML, Taylor IW (1984). Application of DNA flow cytometry to paraffin-embedded archival material for the study of aneuploidy and its clinical significance. Cytometry 6: 327–33.

Hulett HR, Bonner WA, Barrett J, Herzenberg LA (1969). Cell sorting automated separation of mammalian cells as a function of intracellular fluorescence. Science 166; 747–9.

Jain VK, Otto FJ, Gopinath PM (1986). Application of flow cytometry in genetic research. In Advances in biosciences, Vol. 56: Recent trends in medical genetics (eds Marimuthu KM, Gopinath PM), Pergamon Press, Oxford; 273–89.

Kamentsky LA, Melamed MR, Derman H (1956). Spectrophotometer: New instrument for ultrarapid cell analysis. Science 150; 630–1.

Kim M, Perry S (1977). Mathematical methods for determining cell DNA synthesis rate and age distribution utilizing flow microfluorometry. J Theoret Biol 68; 27–42.

Krishan A (1975). Rapid flow cytofluorometric analysis of mammalian cell cycle by propidium-iodide ·staining. J. Cell Biol 66; 188–93.

Lalande M, Schreck RR, Hoffman R, Latt SA (1985). Identification of inverted duplicated No. 15 chromosomes using bivariate flow-cytometric analysis. Cytometry 6; 1–6.

Langer-Safer P, Levine M, Ward D (1982). Immunological methods for mapping genes on *Drosophila* polytene chromosomes. Proc Natl Acad Sci USA 79; 4381–5.

Langlois RG, Yu LC, Gray JW, Carrano AV (1982). Quantitative karyotyping of human chromosomes by dual beam flow cytometry. Proc Natl Acad Sci USA 79; 7876–80.

Latt SA (1979). Fluorescent probes of DNA microstructure and synthesis. In Flow cytometry and sorting (eds Melamed MR, Mullaney PF, Mendelson ML), Wiley, New York; 263–84.

Latt SA, Marino M, Laldane M (1984). New fluorochromes, compatible with high wavelength excitation, for flow cytometric analysis of cellular nucleic acids. Cytometry 5; 339–47.

Lau LF, Ying KL, Donell GN (1985). Identification of a case of Y: 18 Translocation using a Y-specific repetitive DNA probe. Hum Genet 69: 102–5.

Lebo R, Carrano AV, Bur-Khart, Schultz K, Dozy AM, Yu LC, Khan YW (1979). Assignment of human β- γ- and-δ-globin genes to the short arm of chromosome by chromosome sorting and DNA restriction enzyme analysis. Proc Natl Acad Sci USA 76; 5804–8.

Lebo R, Gorin F, Fletterick RJ, Kao FT, Cheung MC, Bruce BD, Kan YW (1984). High resolution chromosome sorting and DNA spot-blot analysis assign Mcardie's syndrome to chromosome 11. Science 225; 57–9.

Le Pecq (1971). Use of ethidium bromide for separation and determination of nucleic acids of various conformational forms and measurement of their associated enzymes. Meth Biochem Analysis 20; 41–86.

Lucas JN, Gray JW (1987). Centromeric index versus DNA content flow karyotypes of human chromosomes measured by means of slit-scan flow cytometry. Cytometry 8; 273–9.

Lucas JN, Gray JW, Peters DC, Vandilla MA (1983). Centromeric index measurement by slit-scan flow cytometry. Cytometry 4; 109–16.

Meyne J, Bartholdi MF, Travis G, Cram LS (1984). Counter-staining human chromosomes for flow karyology. Cytometry 5; 580–3.

Moldavan A (1934). Scientific apparatus and laboratory methods. Photo-electric technique for the counting of microscopical cells. Science 80; 188.

Moroi Y, Peebles C, Fitzler M, Steigerwald F, Tan E (1980). Autoantibody to centromere (Kinetochore) in scleroderma sera. Proc Natl Acad Sci USA 77; 1627–31.

Otto FJ, Oldiges H, Jain VK, Gohde W (1981). Flow cytometric measurement of nuclear DNA content variation as a potential in-vivo mutagenecity. Cytometry 2; 189–91.

Otto FJ, Oldiges H, Jain VK (1984). Flow cytometric measurement of cellular DNA content dispersion induced by mutagenic treatment. In Biological Dosimetry (eds

Eisert WG, Mendalsohn ML), Springer Verlag, Berlin; 37–49.

Pinkel D, Dean P, Lake S, Peter D, Mendelsohn M, Gray J, Van Dilla M, Gledhill B (1979). Flow cytometry of mammalian sperm: Progress in DNA and modphology measurement. J. Histochem Cytochem 27; 353.

Pinkel D, Straume T, Gray JW (1986). Cytogenetic analysis using quantitative high sensitivity fluorescence hybridization. Proc Natl Acad Sci USA 83; 2934–38.

Pinkel D, Lake S, Gledhill BL, Van Dilla MA, Stephenson D, Watchmaker G (1982). High resolution DNA measurements of mammalian sperm. Cytometry 3; 1–9.

Rappold GA, Cremer T, Hager HD, Davies KE, Muller CR. Yan GT (1984). Sex chromosome positions in human interphase nuclei as studied by in situ hybridization with chromosome specific DNA probes. Hum Genet 67; 317–25.

Ritch PS, Shackney SE, Maccollum Smith CA (1983). A practical graphical method for estimating the fraction of cells in S in DNA histograms from clinical tumor samples containing aneuploid cell populations. Cytometry 4; 66–74.

Schweizer D (1980). Simultaneous fluorescent staining of R-bands and specific heterochromatic regions (DA-DAPI bands) in human chromosomes. Cytogenet Cell Genet 27; 190–3.

Schweizer D (1981). Counterstain-enhanced chromosome bending. Human Genet 57; 1–14.

Shapiro H, Glazer A, Christenson L, Williams J, Storm T (1983). Immunofluorescence measurement in a flow cytometer using low-power helium-neon laser excitation. Cytometry 4; 276–9.

Shapiro HM, Stephens S (1986). Flow cytometry of DNA content using oxazine 750 or related laser dyes with 633 nm excitation.Cytometry 7; 107–10.

Trask B, van den Engh G, Grey J, Vamder-Laan M, Turner B (1984). Immunofluorescent detection of histone 2B on metaphase chromosomes using flow cytometry. Chromosoma 90; 295–302.

Trask B, Van Den Engh G, Landengent J, Jansen I, de Wal N, Van Der Ploeg M (1985). Detection of DNA sequences in nuclei in suspension by in situ hybridisation and dual beam flow cytometry. Science 230; 1401–3.

van den Engh G, Trask B, Cram LS, Bartholdi M (1984). Preparation of chromosome suspensions for flow cytometry. Cytometry 5; 108–17.

Van Dilla MA, Trujillo TT, Mullaney PF, Coulter JR (1969). Cell Microfluormetry: A method for rapid fluorescence measurement, Science 163; 1213–4.

Vindelov LL, Christensen IJ, Nissen NI (1983). A detergent-trypsin method for the preparation of nuclei for flow cytometric DNA analysis. Cytometry 3; 323–7.

Wheeless LL, Patten SF (1973). Slit-scan cytofluorometry. Acta Cytol 17; 333–9.

Willard HF (1985). Chromosome-specific organisation of human alpha satellite DNA. Am J Hum Genet 37; 524–32.

Yu LC, Atten J, Gray JW, Carrano AV (1981). Human chromosome isolation from short-term lymphocyte culture for flow cytometry. Nature 293; 154–5.

16 Restriction Enzyme Banding of Metaphase Chromosomes

DOROTHY A. MILLER[1,2] AND ORLANDO J. MILLER [1,3,4]

[1]Department of Molecular Biology and Genetics
[2]Department of Pathology
[3]Department of Obstetrics and Gynecology
[4]The Center for Molecular Biology
Wayne State University
Detroit, Michigan 48201, USA

Introduction

Type 2 restriction endonucleases, or restriction enzymes, are derived from various bacteria where they serve to restrict the introduction of foreign DNA. Each enzyme recognizes a specific palindromic DNA sequence four or more base pairs (bp) in length and cuts DNA at this specific site to yield blunt ends or staggered ends, each with a short single-stranded region. DNA that has been methylated to form 5-methylcytosine or 6-methyladenosine within the recognition sequence is not cut, providing the bacterium with a mechanism for distinguishing self (specifically methylated) from foreign (unmethylated) DNA at the sites.

Each restriction enzyme has an abbreviated name indicating the bacterial strain from which it was derived. If more than one enzyme is derived from the same strain each enzyme is numbered sequentially using Roman numerals. For example, AluI is the first restriction enzyme derived from *Arthrobacter luteus*. The recognition sequences are listed in the 5′ to 3′ direction. AluI recognizes and cuts the sequence 5′ - AGCT-3′ in double-stranded DNA (Table 1). Because the sequence is palindromic, each strand of DNA will be cut between the same nucleotides.

Most restriction enzyme recognition sequences are available for cutting in chromatin (isolated nuclei), although some sites appear to be specifically blocked, e.g. HaeIII and HindIII sites in African green monkey satellite DNA, and EcoRI* sites in bovine satellite DNA, presumably buried within phased nucleosomes (Musich *et al.*, 1977). The recognition sequences are also available for cutting in metaphase chromosomes that have been fixed and air dried on a slide. Restriction enzyme digestion leads to removal of some DNA fragments, probably primarily those below a certain critical size; the remaining DNA can be stained to reveal the relative amount and location of the unextracted DNA. Enzymes that cut DNA into fragments whose average length is less than 1000 kb produce banding patterns on metaphase chromosomes (Table 1; also see below, Mechanism of restriction banding). Studies in a variety of species show that restriction enzyme treat-

TABLE 1

Characteristics of selected restriction enzymes

		Human DNA		
Enzyme	Recognition sequence	Av. fragment size (bp)[1]	% fragments <500 bp[2]	Human banding pattern
AluI	AGCT	180	94	C(G)[3]
DdeI	CTNAG	210	91	C(G)
HinfI	GANTC	300	81	negC
MboI	GATC	325	79	C(G)
HaeIII	GGCC	380	73	G
EcoRII	CC(A/T)GG	480	65	C(G)
RsaI	GTAC	515	62	C(G)
AvaII	GG(A/T)CC	820	45	G
MspI	CCGG	1650	24	none[4]
HindIII	AAGCTT	1750	25	none[5]
BalI	TGGCCA	2750	17	none
XbaI	TCTAGA	3050	15	none
PvuII	CAGCTG	3050	15	none
EcoRI	GAATTC	3200	14	none[5]
SstI	GAGCTC	4800	10	none
BamHI	GGATCC	5400	9	none
BstEII	GGTNACC	6950	7	none
KpnI	GGTACC	8500	6	none

[1]Drmanac *et al.*, (1986).
[2]Assuming an exponential distribution of sites, as described by Bishop *et al.* (1983)
[3]No enzyme left all C-bands intact. (G) = residual G – bands; negC = C – bands replaced by gaps.
[4]decrease in staining of nucleolus organizer region.
[5]banding reported by Bianchi *et al.* (1986) but not found by others.

ment can lead to G-banding or to specific modified C-band patterns, the latter reflecting extraction of major components of some or all C-bands. Some enzymes have no effect on chromosome staining.

Banding Patterns

Human

Restriction enzymes can be classified into three types based on their effect on human chromosomes when digestion is carried to completion: type one produces G-banding plus C-banding on some chromosomes, type two yields C-banding on some chromosomes, often with a small amount of residual G-banding, and type three produces no banding.

The chief example of the first type is HaeIII, which produces clear G-banding and leaves intact the C-bands on numbers 1, 9, 16, the Y, and a few acrocentric chromosomes (Mezzanotte *et al.*, 1983; Miller *et al.* 1983; M. Bianchi *et al.*, 1985; Babu and Verma, 1987a). The C-bands on all the other chromosomes are abolished by HaeIII, suggesting that the DNA in these regions has abundant and rather closely interspersed GGCC sequences, whose cutting by HaeIII produces fragments of DNA so small they are easily removed from the chromosomes.

There are several examples of the second type, enzymes which produce a C-band pattern on some, but not all, chromosomes. Because the chromosomes usually show faint (residual) G-bands, it seems reasonable to assume that good G-banding could be produced by partial enzyme digestion before the C-band stage is reached. This idea has not been tested with most enzymes. However, it has been shown for AluI by interrupting the enzyme incubation at an early stage (Ferrucci *et al.*, 1987). This group of enzymes, which includes AluI, DdeI, EcoRII, HinfI, MboI, and RsaI, is among the most useful for the study of human chromosomes because each enzyme produces a unique subset of C-band patterns (for example, see Table 2). MboI, for example, leaves intact the large C-band on no. 9, as well as smaller C-bands on a variety of Chromosomes, but reduces the C-bands on nos. 1 and 16 (Figure 1). HinfI is unique in that it abolishes the C-bands on every human chromosome, except nos. 3 and 4 in some individuals (Figure 2) (Miller *et al.*, 1983; M. Bianchi *et al.*, 1985).

Of less cytological interest are the enzymes that have little or no effect on the staining of

TABLE 2

Staining of C-bands after treatment by selected restriction enzymes

Enzyme	1	9	16	Y	15
AluI	+++	+++	+++	+++	++
RsaI	+++	+	+++	–	–
MboI	–	+++	+	+++	+
HinfI	–	–	–	–	—

fixed chromosomes when used alone. These enzymes are numerous and fall into two broad categories. Enzymes that include a CG doublet in their recognition sequence cut DNA infrequently, because the CG is underrepresented in mammalian genomes. Enzymes that have a 6 bp or longer recognition sequence also cut infrequently (Table 1), because the longer the recognition sequence the less often it will occur in the genome.

One exception may be MspI, whose recognition sequence, CCGG, includes the CG doublet. In our hands MspI does not produce a visible banding pattern throughout the genome but does reduce the staining of regions containing amplified rDNA (Miller *et al.*, 1983). In contrast, HpaII produces no effect on these regions. MspI and HpaII both recognize CCGG (i.e. they are isoschizomers), but HpaII does not cut this sequence if the internal C is methylated. rDNA contains many CCGG sequences and the internal C in many of these sequences is methylated in amplified, inactive copies of rDNA (Tantravahi *et al.*, 1981). The results with MspI-HpaII suggest that the amplified regions consist primarily of rDNA.

N. Bianchi *et al.* (1986) reported that they produced banding with MspI and that this enzyme extracted more than 40 per cent of the

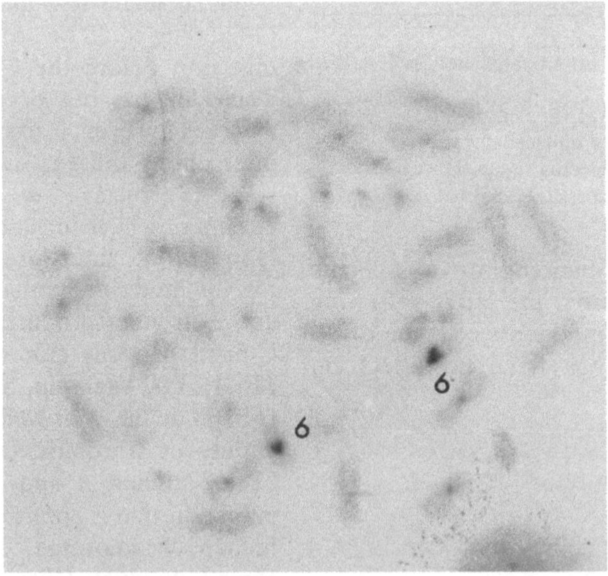

FIGURE 1: Human female metaphase spread stained with Giemsa after MboI treatment. The C-bands on no. 9 and various other chromosomes remain darkly stained; those on nos. 1 and 16 do not. Miller *et al* (1983). By permission of The American Association for the Advancement of Science.

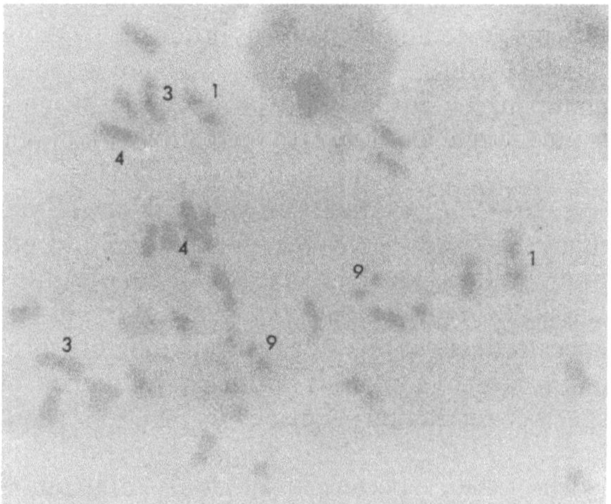

FIGURE 2: Human female metaphase spread stained with Giemsa after HinfI treatment. The C-bands on nos. 3 and 4 remain; all other C-bands are removed and the C-band regions on nos. 1, 9 and 16 appear as unstained gaps. Miller *et al.* (1983). By permission of The American Association for the Advancement of Science.

DNA from fixed human chromosomes on a slide. This is considerably higher than the theoretical fraction of fragments less than 500 bp long, though not out of line if fragments of 1000 bp were removed from the DNA. We have been unable to achieve banding with several different samples of MspI, and investigators studying various species (Ueda *et al.*, 1987; Lloyd and Thorgaard, 1988; Schmid and de Almeida, 1988) have reported results similar to ours. This discrepancy among laboratories cannot be explained at this time. It points up the desirability of testing each enzyme on DNA standards by gel electrophoresis to confirm the specificity and purity of the enzyme, or of using at least two batches of an enzyme.

Sequential application of two enzymes can produce interesting results. Some of the results are predictable from those obtained using each enzyme separately. HinfI followed by HaeIII removes virtually all staining, including the C-bands on nos. 3 and 4 which are resistant to cutting by HinfI but not HaeIII. Other results are difficult to predict from the action of the single enzymes. Neither HaeIII nor MspI removes the C-band on no. 9; however, HaeIII followed by MspI produces a reduction of part of the C-band on no. 9. HaeIII followed by HpaII has no such effect, indicating that many of the CCGG sequences present in this region are methylated (Miller *et al.*, 1983). Further studies using combinations of enzymes promise to reveal more detailed information about chromosome structure.

Human chromosome polymorphisms. Differences in the size of C-bands on homologous human chromosomes are observed often and represent common polymorphisms. The same differences can be reflected in the size of the regions that are stained after treatment with particular restriction enzymes (see for example, chromosomes 9 in Miller *et al.*, 1983). This makes restriction enzymes useful for detecting a variety of polymorphisms.

In addition, restriction enzyme banding occasionally reveals additional differences in the C-band regions of homologous chromosomes. Such differences are often revealed as a different pattern than that produced by classical C-banding. M. Bianchi *et al.* (1985) found three such examples in a sample of five individuals. In one case both no. 3's had equal sized classical C-bands, but only one of the pair showed a major C-band after treatment with AluI, DdeI,

HinfI, or MboI. In a second case, both no. 9's had equal C-bands. The same pattern was found after HaeIII treatment, but two different sized bands were stained after AluI and only one was stained after MboI. Finally two no. 18's had equal sized C-bands; after treatment with MboI only one was stained. N. Bianchi *et al.* (1986) reported an individual—whose no. 1's had equally large C-bands but showed different sized C-bands after MspI treatment. Babu and Verma (1986a) found a no. 4 which had a classical C-band that was cut by AluI. Such results indicate that quite long stretches of repetitive DNA either contain or do not contain specific short sequences.

Babu *et al.* (1987) carried out detailed studies on the effect of AluI treatment on chromosome 18. They reported differences in the size of the region that stained and on its location in either the short or long arm or both. It was not shown whether any of these reflected the size and location of classical C-bands. Five cases of trisomy 18 were also examined. In each case three patterns of C-band staining (all large) were observed after AluI treatment. No further analysis was possible because the parents were not available. However, this study indicates the potential use of restriction enzyme banding to determine parental origin of chromosomes in clinical studies.

Babu and Verma (1987b) made the interesting observation that the quinacrine-brilliant C-band regions on nos. 3 and 4 are cut by HaeIII but are resistant to AluI. It is unclear whether the HinfI–resistant C-bands on nos. 3 and 4 (Miller *et al.*, 1983) correspond to the Q-brilliant C-bands.

Primates

Restriction enzyme studies have been carried out on the chimpanzee, gorilla and orangutan in order to compare the banding patterns with those of humans. N. Bianchi *et al.* (1985) compared classical C-banding with the restriction enzyme banding patterns produced by AluI, HaeIII, HinfI, MboI, and RsaI in a single chimpanzee, gorilla, and orangutan. Chromosomes were identified by residual G-banding and the size of each C-band was scored as −, +, ++, or +++. Comparison with human showed many differences in the C-bands produced. Some of these were expected on the basis of the classical C-band studies, but many of the differences reflect subsets of constitutive heterochromatin which contain a different frequency of specific restriction enzyme recognition sequences.

The AluI patterns in four orangutans were studied by de Stefano *et al.* (1986). Identification of individual chromosomes was made by stopping the reaction at an early stage and photographing the G-bands before continuing the incubation to produce C-bands. This group also reported AluI banding patterns for a single chimpanzee and a single gorilla (Ferrucci *et al.*, 1987). The size of each C-band produced by AluI was also scored as −, +, ++, or +++.

The studies by the two groups gave similar, but not identical, results for AluI. For example, of the 24 types of chromosomes scored in the orangutan, the C-bands on 11 chromosomes were scored differently by plus or minus one gradation and C-bands on two chromosomes were scored differently by plus or minus two gradations. Such differences could result from technical differences or from differences between the animals studied.

N. Bianchi *et al.* (1985) included an extensive comparison of similarities and differences among the four higher primate species in order to shed light on the evolutionary relationships of the human and the great apes. The results must be interpreted cautiously. The sample size was small and the chromosomes were identified only by residual G-banding. In addition, most of the differences (23 of 30) depended on the presence or absence of telomeric bands on particular

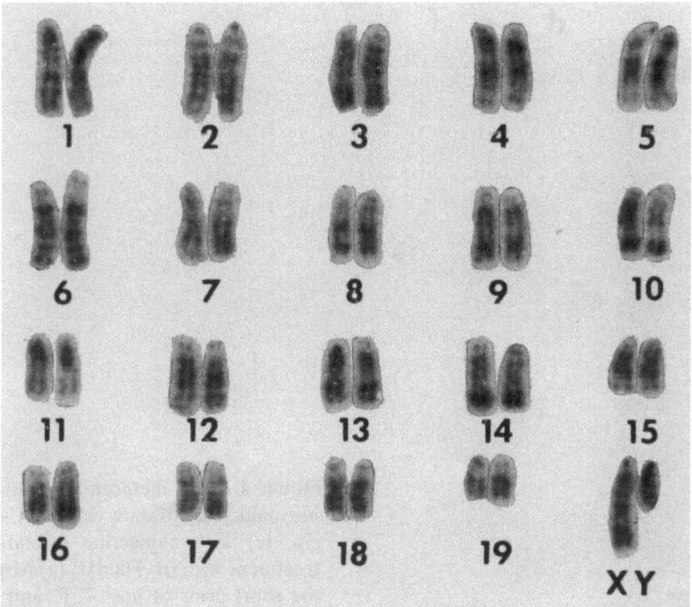

FIGURE 3: *Mus musculus* karyotype. G-band pattern produced by AvaII followed by Giemsa. Note the pale staining of the usual C-band region and the very small darkly stained region that remains near the centromere Kaelbling *et al.* (1984). By permission of Springer-Verlag.

chromosome arms, which were already known to be present in chimpanzees and gorillas but not in humans or orangutans (Miller *et al.*, 1974; Miller, 1977).

Mouse

In the laboratory mouse, *Mus musculus*, AluI, HinfI, or MboI treatment produces a C-band pattern, while EcoRII or HaeIII produces a G-band plus C-band pattern (Kaelbling *et al.*, 1984). AvaII and BstNI each produces a G-band pattern, but on most chromosomes a small segment of the C-band adjacent to the centromere remains (Figure 3). This is of particular interest because there is evidence (unpublished data) that this is the location of the minor satellite in *M. musculus*. That is, either the AvaII or BstNI enzyme cuts most of the major satellite into short monomers or dimers that are removed, leaving the intact minor satellite in place. It seems likely that similar findings will occur in other species (see Indian muntjac, below), i.e. more than one satellite is present and each is located in an adjacent or otherwise different chromosomal region. The minor satellite of *M. musculus* is the major satellite of *M. spretus* (Pietras *et al.*, 1983) so one would predict that in the latter species the C-bands would remain after treatment with AvaII or BstNI, unless mutation has led to the formation of the recognition sequences of these enzymes in the satellite in *M. spretus*.

Muntjac

Lima-de-Faria *et al.* (1980) showed that HaeIII produces G-banding in Indian muntjac chromosomes, but EcoRI or HpaII does not. Babu and Verma (1986b) described C-bands after treatment of Indian muntjac chromosomes with AluI, HaeIII, HinfI, and MboI. These results

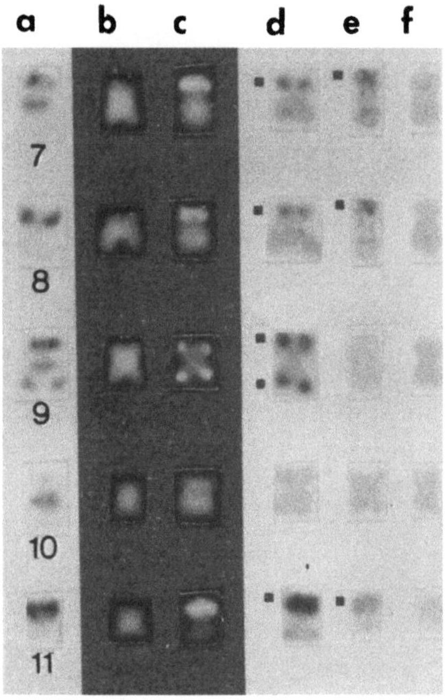

FIGURE 4: Small metacentric chromosomes of *Bufo americanus* stained (a) to show classical C-bands, (b) with mithramycin, (c) with quinacrine mustard, or with Giemsa after treatment with (d) HaeIII, (e) MspI, of (f) HpaII. Note that the short arms of nos. 7, 8, and 11 remain darkly stained after MspI but not after HpaII treatment. Schmid and de Almeida (1988). By permission of Springer-Verlag.

have been confirmed by Ueda *et al.* (1987), who also showed that the enzymes AluI, Sau3A, RsaI and EcoRII produce C-bands in muntjac chromosomes, and that a series of other enzymes produce no banding. These authors demonstrated homologue polymorphism using RsaI, and presented evidence that two repetitive DNAs are located in adjacent positions of a chromosome region, a finding that appears similar to that in *M. musculus*.

Birds

There is a single report of the use of restriction enzymes in the chicken, *Gallus domesticus*, in a paper dealing primarily with amphibia (Schmid and de Almeida, 1988). HaeIII treatment leads to faint G-banding in the chicken macrochromosomes and pronounced C-banding of the W sex chromosome found in the heterogametic female. The small pericentromeric C-bands on nos. 1–4, the large telomeric C-band on the Z chromosome, and the C-bands on some of the microchromosomes are removed.

Amphibia

In contrast to their effect in mammals, no enzyme has been found to produce G-bands in fish or amphibian chromosomes, reinforcing the idea that the DNA in chromosomes of these species is organized differently than that in mammals. Schmid and de Almeida (1988) studied the staining of metaphase chromosomes of two species of toad and one species of frog after treatment with HaeIII, MspI, HpaII or HindIII. Another species of frog and a newt were studied with HaeIII or MspI only. In no case was G-banding observed, and in almost every case only a subset of the C-bands was stained. In general the staining was not as distinct as after classical C-banding, but there were several striking findings. In *Bufo americanus* MspI did not remove the heterochromatin from the short

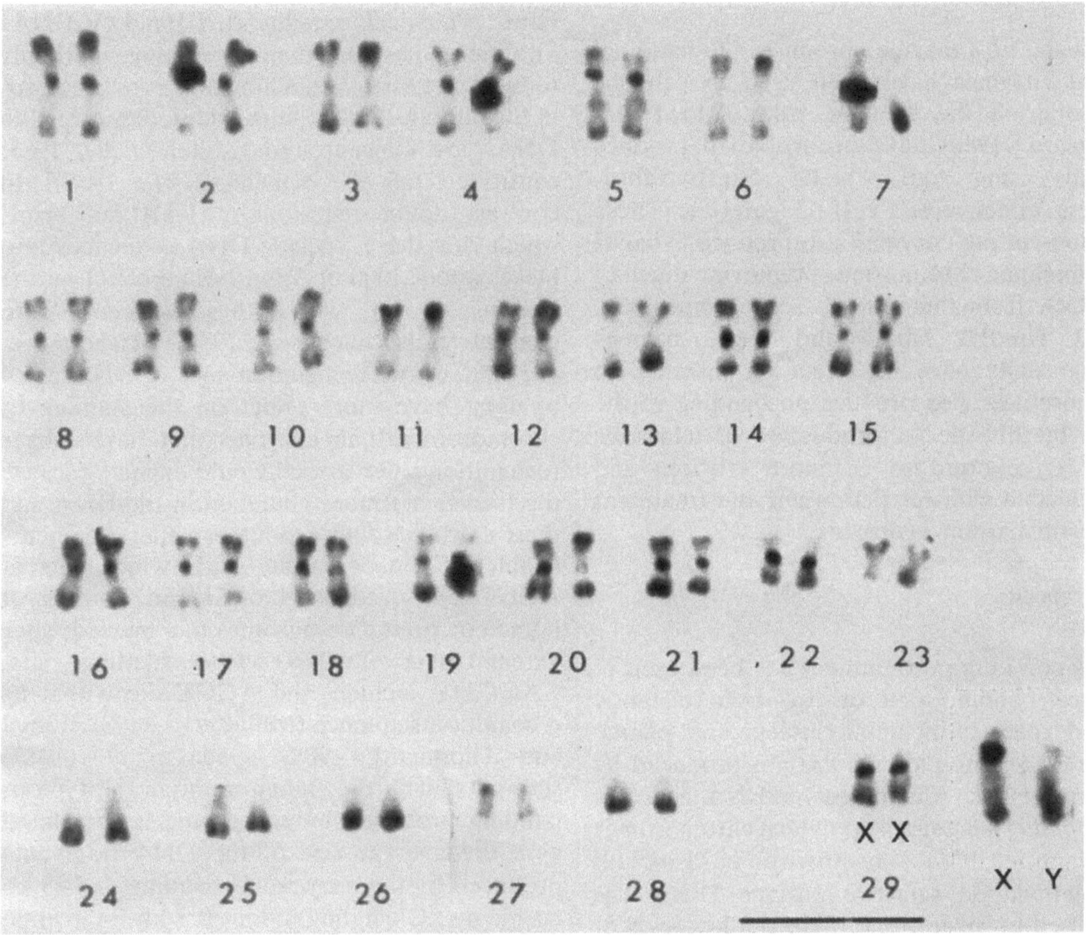

FIGURE 5: Karyotype of the rainbow trout. AluI treatment followed by Giemsa staining reveals centromeric and telomeric C-bands. Insert shows the different patterns of the X and Y chromosomes from a different fish. Lloyd and Thorgaard (1988). By permission of Springer-Verlag.

arms of three small chromosomes whereas HpaII did (Figure 4). This is the reverse of what would be expected if the internal C in the CCGG recognition sequence is methylated. It may indicate that, although methylation is found more commonly in the CG doublet, in this species the external C is methylated, because MspI cuts the mCCGG sequence less effectively. In *Rana erythrea* HaeIII produced very well stained bands at the centromere and/or telomere of several chromosomes. In *Gastrotheca riobambae* MspI or HaeIII treatment produced a gap only where the nucleolus organizer is located in the X, all other heterochromatic regions being resistant to these treatments.

Fish

Although no enzyme produces G-banding, several enzymes have been used to produce C-banding in the rainbow trout. Lloyd and Thorgaard (1988) have been able to produce C-bands using AluI, HaeIII, HinfI, MboI, or, less effectively, PvuII (Figure 5). These common-cutting enzymes also produce C-bands in mammalian chromosomes, removing much of the DNA from the rest of each chromosome. EcoRI, HindIII, HpaII, and MspI, enzymes which usually have no effect on mammalian chromosomes, also produce no banding in the trout. In this species production of telomeric bands is reported to be more efficient and reproducible with restriction enzyme treatment than with barium hydroxide.

Other species

Restriction enzyme treatment has been used to produce banding patterns in more distantly related species, including insects and plants. General banding patterns are not produced by such treatment. AluI detects 5SRNA genes in *Drosophila melanogaster* polytene chromosomes (Mezzanotte, 1986). A comparison of MspI with the methylation sensitive enzyme HpaII has been used to suggest that there is a low level of methylation of some satellite DNA in mosquito chromosomes (N. Bianchi *et al.*, 1986). In *Vicia faba* MboI or AluI produces 1–2 bands per chromosome in a pattern similar to C-banding (Frediani *et al.*, 1987).

Mechanism of Restriction Banding

Miller *et al.* (1983) proposed that the banding observed after chromosomes are treated with a restriction enzyme could be accounted for by removal of small pieces of DNA. This suggestion is supported by the following observations. HinfI, which cuts satellite II, III and IV DNAs into pieces no more than 80 bp long, markedly reduces the Giemsa staining and produces gaps in the chromosome regions where these satellite DNAs are concentrated (Miller *et al.*, 1983; confirmed by M. Bianchi *et al.*, 1985). In contrast, double digestion by HaeIII and MspI, which cuts the Y specific DNA sequences into pieces about 1000 bp long, has no effect on the staining of Yq, where these sequences are concentrated (Miller *et al.*, 1983). Furthermore, enzymes whose recognition sites in DNA are 4 bp long have more effect on the staining of chromosomes than enzymes that have longer recognition sites, as one would expect, because the former are more abundant in the DNA and their cutting would produce smaller fragments (Table 1). For example AluI, which cuts at AGCT sites, reduces the Giemsa staining of human or trout chromosomes to a much greater extent than PvuII, whose cutting site, CAGCTG, includes the AGCT site within its recognition sequence (Miller *et al.*, 1983; Lloyd and Thorgaard, 1988). Ueda *et al.* (1987) showed that in the Indian muntjac the type of banding produced by an enzyme is correlated with the average size of the DNA fragments produced by the enzyme: G-banding, 0.2–3 kb fragments; C-banding, about 0.5 kb fragments; and no banding, >3 kb fragments.

The relationship between the effect of a restriction enzyme on chromosome staining and the size of DNA fragment it produces is shown very clearly in Table 1, which includes most of the enzymes used in our original study (Miller *et al.*, 1983), as well as RsaI, which was not available at that time. The eight enzymes that have a significant effect on human chromosomes cut isolated human DNA into fragments whose average size varies from 180 bp (AluI) to 820 bp (AvaII). Enzymes that produce fragments of average size from 1650 bp (MspI) to 8900 bp (BglI) have no significant effect on chromosome staining, except for the specific abolition of staining of an HSR containing amplified rDNA by MspI.

These results make it clear that one should not choose a restriction enzyme at random to try to produce general chromosome banding. Instead, one must look for an enzyme (or combination) that cuts DNA into small fragments. In contrast, highly repetitive satellite DNAs in any species have, or do not have, a particular site within the repeating sequence, and the production of C-banding bears very little relationship to the overall average frequency of cutting. It is this specificity which makes restriction enzymes useful in evaluating C-band heteromorphisms. For example, human satellite 2, the major component of satellite II, contains abundant TaqI and HinfI sites but almost no other restriction enzyme sites; satellite 3 contains HinfI but virtually no other sites; and satellite 1 contains an RsaI site but no other sites (Prosser *et al.*, 1986). These simple satellite sequences are based on the ATTCC sequence or A + T-rich short sequences. Alphoid sequences, in contrast, are made up of 171 bp tandem repeats, some of them rich in most of the common 4 or 5 bp containing restriction enzymes (Jorgensen *et al.*, 1986).

Direct support for a mechanism involving loss of DNA has been presented. Miller *et al.* (1984) treated *M. musculus* chromosomes with BstNI, which cuts the major satellite primarily into 234 bp monomers and 468 bp dimers. This treatment led to a decrease in the amount of staining in the region where the satellite DNA is located and a simultaneous decrease in the amount of tritium-labelled cloned satellite DNA that could hybridize to the region. Thus BstNI treatment led to the loss of Z34 bp long monomers and, perhaps 468 bp dimers from the chromosomes.

M. Bianchi *et al.* (1985) labelled cells with tritium before making metaphase preparations, treated the slides with a restriction enzyme and measured the amount of tritium that could be washed off the slide after such treatment. In the case of AluI or HaeIII this amounted to 50–60 per cent of the counts. EcoRI and TaqI are enzymes whose recognition sequences are present less frequently in human DNA, and would be expected to produce longer DNA fragments. After treatment with these enzymes only about 13 per cent of the counts were removed.

Several groups have followed restriction enzyme treatment with DNA stains other than Giemsa. In almot every case the results using Giemsa were identical to those using other stains. The stains included Hoechst 33258 (Miller *et al.*, 1983), acridine orange (Mezzanotte and Ferrucci, 1983; Miller *et al.*, 1983), ethidium bromide (Ferrucci *et al.*, 1987; Frediani, 1987), and Feulgen (Ferrucci *et al.*, 1987). The results of such experiments are consistent with the conclusion that treatment of fixed metaphase chromosomes with specific restriction enzymes reduces the amount of stainable DNA remaining on the slide.

The fact that staining can be reduced in every region of every chromosome (Miller *et al.*, 1983) argues against chromosome conformation playing a major role in the differential staining resulting from the usual applications of these enzymes. In addition, M. Bianchi *et al.* (1985) have shown that restriction enzyme banding can be produced in G1 prematurely condensed chromosomes, which are longer and presumably are packaged somewhat differently than are metaphase chromosomes. Schmid and de Almeida (1988) noted that the restriction enzyme banding of prometaphase chromosomes is more efficient than that of shorter metaphase chromosomes. The primary evidence in support of a role for conformation was the report by Mezzanotte *et al.* (1983) that HindIII or EcoRI followed by Giemsa produced banding patterns but followed by ethidium bromide showed uniform staining. In most laboratories neither of these two enzymes has produced banding patterns, a result not surprising since each of these two enzymes is expected to produce fragments most of which are longer than 1000 bp, with no more than 5 per cent cut into fragments less than 200 bp long (Bishop *et al.*, 1983), and neither removes a significant amount of DNA from fixed chromosomes or nuclei (Mezzanotte *et al.*, 1985).

G-band patterns have been produced by treatment with a variety of restriction enzymes, but none have produced R-band patterns. This does not seem to be the result of the difference in base composition of the G- (more AT-rich) and R- (more GC-rich) bands. First these differences in base composition are very small, just a few per cent (Holmquist *et al.*, 1982). Second, many of the most effective enzymes show no obvious bias for GC-rich DNA, e.g. AluI, DdeI, HinfI, MboI, and RsaI. It is possible that chromosome organization, e.g. domain structure, is involved, with the R-band DNA more accessible to restriction enzymes as it is to DNase I.

Technical Factors

Restriction enzyme banding is a simple method which should find application in many laboratories because it is rapid, relatively inexpensive and requires only a standard light microscope. As with any staining procedure, various technical factors, some understood and others not, play a role. Slides must be relatively fresh, preferably one to a few days old. The amount of enzyme used depends on the activity of the particular sample of the enzyme and also on the length of incubation. The proper levels are generally achieved by trial and error. The level of glycerol is important because restriction enzymes are usually shipped and stored in 50 per cent glycerol and it has been reported that as little as 6–10 per cent glycerol in the incubation mixture may produce G-banding in the absence of enzyme (M. Bianchi *et al.*, 1985). The level of glycerol will rarely be a problem if concentrated enzymes are used, but a control slide with the same buffer and glycerol concentration as the test slides should be included in the experimental design.

One other note of caution is that not all batches of restriction enzymes have the same activity and they are also sensitive to degradation. If there is a problem it is preferable to check the level of activity by testing the enzyme on a sample of known DNA, such as lambda phage DNA, and examining the electrophoretic pattern to confirm the presence of the correct sizes of DNA fragments and the absence of nonspecific degradation.

Identification of individual chromosomes remains a problem. Babu and Verma (1987b) carried out quinacrine banding before treatment with restriction enzymes, but this has not been a successful method in other laboratories. Ferrucci *et al.* (1987), using an enzyme that produces C-banding, AluI, stopped the reaction after 2 hours, stained with Giemsa, and photographed the G-bands before adding more enzyme and continuing the reaction for a further 10 hours to produce C-bands. Most laboratories have depended on residual G-banding, although this is the least reliable method, except with HaeIII, which produces good G-banding.

A general method follows: Methanol-acetic acid fixed metaphase chromosomes are dropped onto slides and air dired. 30 units of enzyme in 25 μl buffer, as specified by the supplier, is dropped on the slide, which is coverslipped and incubated in a moist chamber for 2 hours at 37°C. The mixture is rinsed off with phosphate buffer, pH 6.8, and stained 5 minutes in 2 per cent Giemsa in the same buffer.

Nomenclature

There is as yet no agreed upon nomenclature for restriction enzyme banding. The abbreviation

CRe(AluI)G, *C*-banding produced by the *r*es-triction enzyme *AluI* followed by *G*iemsa staining would fit with the suggestions made by the International Committee, but this seems an awkward designation. The designation C(AluI) or C(AluI/MspI) seems preferable, with words used to indicate the stain used.

Summary

Treatment with restriction enzymes is an effective way to produce banding of fixed metaphase chromosomes. HaeIII treatment leads to G-banding of all mammalian species tested so far. Several enzymes that have a 4 or 5 base pair recognition sequence that does not include the CG doublet produce a subset of C-bands in a wide variety of species: mammals, birds, amphibia and plants. These enzymes have been used to demonstrate polymorphisms in the human and the existence of adjacent subsets of repetitive DNA in mouse or Indian muntjac. No enzyme has been reported to produce an R-band pattern in any species.

In 1983 we suggested several lines of investigation in which restriction enzyme digestion of fixed chromosomes might be useful. These included: (1) looking for chromosome banding sufficiently distinctive to permit identification of individual chromosomes in amphibian, fish or plant species lacking easily G- or R-banded chromosomes; (2) looking in various species for subclasses of satellite DNAs, and studying their location and evolution; and (3) looking for restriction enzymes that will reveal regions of specific gene amplification. Thus far, restriction enzymes have generally been useful in only the second of these lines of investigation, despite a few tantalizing leads in lines 1 and 3. Further progress may come from imaginative use of enzyme combinations and base-specific DNA ligands to block access of the enzymes to specific sequences (Schweizer, 1977). In addition restriction enzymes of different specificities are still being discovered, including some with 4 and 5 bp recognition sites (reviewed by Roberts, 1987), and these may reveal additional heterogeneity and polymorphism in repetitive sequence DNA in many species. Thus, continuing studies involving restriction enzymes may reveal unexpected facets of chromosome organization and evolution.

Acknowledgement

This work was supported in part by the National Foundation-March of Dimes.

References

Babu A, Verma RS (1986a). Heteromorphic variants of human chromosome 4. Cytogenet Cell Genet 41; 60–1.

Babu A, Verma RS (1986b). Expression of heterochromatin by restriction endonuclease treatment and distamycin A/DAPI staining of Indian muntjak (*Muntiacus muntjak*) chromosomes. Cytogenet Cell Genet 41; 96–100.

Babu A, Verma RS (1987a). Characterization of human chromosomal constitutive heterochromatin. Can J Genet Cytol 28; 631–44.

Babu A, Verma RS (1987b). Chromosome structure: euchromatin and heterochromatin. Int Rev Cytol 108; 1–60.

Babu A, Verma RS, Patil SR (1987). AluI-resistant chromatin of chromosome 18: Classification, frequencies and implications. Chromosoma 95; 163–6.

Bianchi MS, Bianchi NO, Pantelias GE, Wolff S (1985). The mechanism and pattern of banding induced by restriction endonucleases in human chromosomes. Chromosoma 91; 131–6.

Bianchi NO, Bianchi MS, Cleaver JE, Wolff S (1985). The pattern of restriction enzyme-induced banding in the chromosomes of chimpanzee, gorilla, and orangutan and its evolutionary significance. J Mol Evol 22; 323–33.

Bianchi NO, Vidal-Rioja L, Cleaver JE (1986). Direct visualization of the sites of DNA methylation in human and mosquito chromosomes. Chromosoma 94; 362–6.

Bishop DT, Williamson JA, Skolnick MH (1983). A model for restriction fragment length distributions. Amer J Hum Genet 35; 795–815.

de Stefano GF, Romano E, Ferrucci L (1986). The AluI-induced bands in metaphase chromosomes of orangutan (*Pongo pygmaeus*). Implications for the distribution pattern of highly repetitive DNA sequences. Hum Genet 72; 268–71.

Drmanac R, Petrovic N, Glisin V, Crkvenjakov, R. (1986). A calculation of fragment lengths obtainable from human DNA with 78 restriction enzymes: an aid for cloning and mapping. Nucl Acids Res 14; 4691–4.

Ferrucci L, Romano E, de Stefano GF (1987). The AluI-induced bands in great apes and man: implication for heterochromatin characterization and satellite DNA distribution. Cytogenet Cell Genet 44; 53–7.

Frediani M, Mezzanotte R, Vanni R, Pignone D, Cremonini R (1987). The biochemical and cytological characterization of *Vicia faba* DNA by means of MboI, AluI and BamHI restriction endonucleases. Theor Appl Genet 75; 46–50.

Holmquist G, Gray M, Porter T, Jordan J (1982). Characterization of Giemsa dark- and light-band DNA. Cell 31; 121–9.

Jorgensen AL, Bostock CJ, Bak AL (1986). Chromosome-specific subfamilies within human alphoid repetitive DNA. J Mol Biol 187; 185–96.

Kaelbling M, Miller DA, Miller OJ (1984). Restriction enzyme banding of mouse metaphase chromosomes. Chromosoma 90; 128–32.

Lima-de Faria A, Isaksson M, Olson E (1980). Action of restriction endonucleases on the DNA and chromosomes of *Muntiacus muntjak*. Hereditas 92; 267–73.

Lloyd M., Thorgaard GH (1988). Restriction endonuclease banding of rainbow trout chromosomes. Chromosoma 96; 171–7.

Mezzanotte R (1986). The selective digestion of polytene and mitotic chromosomes of *Drosophila melanogaster* by the AluI and HaeIII restriction enzymes. Chromosoma 93; 249–55.

Mezzanotte R, Ferrucci L (1983). The metaphase banding pattern produced in the chromosomes of *Culiseta longiareolata* (Diptera: Culicidae) by AluI restriction endonuclease or actinomycin D treatment followed by acridine orange staining. Genetica 62; 47–50.

Mezzanotte R, Bianchi U, Vanni R, Ferrucci L (1983). Chromatin organization and restriction endonuclease activity on human metaphase chromosomes. Cytogenet Cell Genet 36; 562–6.

Mezzanotte R, Ferrucci L, Vanni R, Sumner AT (1985). Some factors affecting the action of restriction endonucleases on human metaphase chromosomes. Exp Cell Res 161; 247–53.

Miller DA (1977). Evolution of primate chromosomes. Science 198; 1116–24.

Miller DA, Firschein IL, Dev VG, Tantravahi R, Miller OJ (1974). The gorilla karyotype. Chromosome lengths and polymorphisms. Cytogenet Cell Genet 13; 536–50.

Miller DA, Choi Y-C, Miller OJ (1983). Chromosome localization of highly repetitive human DNA's and amplified ribosomal DNA with restriction enzymes. Science 219; 395–7.

Miller DA, Gosden JR, Hastie ND, Evans HJ (1984). Mechanism of endonuclease banding of chromosomes. Exp Cell Res 155; 294–8.

Musich PR, Maio JJ, Brown FL (1977). Subunit structure of chromatin and the organization of eukaryotic highly repetitive DNA: indications of a phase relation between restriction sites and chromatin subunits in African green monkey and calf nuclei. J Mol Biol 117; 657–77.

Pietras DF, Bennett KL, Siracusa LD, Woodworth-Gutai M, Chapman VM, Gross KW, Kane-Haas C, Hastie ND (1983). Construction of a small *Mus musculus repetitive* DNA library: identification of a new satellite sequence in *Mus musculus*. Nucl Acids Res 11; 6965–83.

Prosser J, Frommer M, Paul C, Vincent PC (1986). Sequence relationships of three human satellite DNAs. J Mol Biol 187; 145–55.

Roberts RJ (1987). Restriction enzymes and their isoschizomers. Nucl Acids Res 15; r189–r215.

Schmid M, de Almeida CG (1988). Chromosome banding in amphibia. XII. Restriction endonuclease banding. Chromosoma 96; 283–90.

Schweizer D (1977). R-banding produced by DNase I digestion of chromomycin and DAPI. Chromosoma 64; 117–24.

Tantravahi U, Guntaka RV, Erlanger BF, Miller OJ (1981). Amplified ribosomal RNA genes in a rat hepatoma cell line are enriched in 5-methylcytosine. Proc Natl Acad Sci USA 78; 489–93.

Ueda T, Irie S, Kato Y (1987). Longitudinal differentiation of metaphase chromosomes of Indian muntjac as studied by restriction enzyme digestion, in situ hybridization with cloned DNA probes and distamycin A plus DAPI fluorescence staining. Chromosoma 95; 251–7.

17 DNase I-nick Translation Technique: A Cytogenetic Approach to Map Active Chromatin

K SPERLING

Institut für Humangenetik der Freien
Universität Berlin
Heubnerweg 6, D-1000, Berlin 19
Federal Republic of Germany

I. Introduction

Chromatin structure and functional activity are two closely related phenomena. Thus, RNA synthesis drops almost to zero when cells enter into mitosis and the interphase chromatin condenses into compact metaphase chromosomes. Even at interphase those heterochromatic regions which are heavily condensed and therefore intensely stained are genetically inactive and remain unlabelled after application of ^{3}H uridine (Hsu, 1962; Sieger *et al.*, 1970). Consequently, the chromosomal sites of RNA synthesis can be mapped only in those species in which interphase or meiotic prophase chromosomes are easily analysable.

Direct approaches for mapping the chromosomal sites of RNA synthesis

The most favourite objects in this respect are the polytene chromosomes found in the larval tissues of certain dipteran flies. Here, even the morphology of the chromsomes reflects to some extent their genetic activity. While the DNA of the most active genes is almost completely extended leading to large puffed regions, the Balbiani rings, the inactive genes, appear as compact bands. After ^{3}H uridine application the puffs are most heavily labelled but single discrete bands also regularly show silver grains indicating that active and inactive chromosomal regions cannot readily be discriminated at the level of the light microscope (Edström, 1976).

Large transcription units—greater than 1500 kb—can also be visualized in diplotene chromosomes of vertebrate oocytes, especially in amphibians and primary spermatocytes of *Drosophila.* Due to their loop-like structure these chromosomes were termed 'lampbrush' chromosomes. The lateral loops rapidly disappear when transcription has ceased. In *Drosophila* hyder five large lampbrush loops are formed by the constitutive heterochromatin of the Y chromosome (see Hennig, 1986). In Triturus, satellite DNA, a constituent of heterochromatin, is transcribed on lampbrush chromosomes (Varley *et al.*, 1980). This is clear evidence against the assumption of permanent inactivity of constitutive heterochromatin, still found in a number of textbooks of genetics.

Another direct approach to mapping the chromosomal sites of RNA synthesis—even in non-dividing cells—is based on the possibility of prematurely condensing interphase chromatin into discrete chromosomes. This can be achieved by fusion of interphase with mitotic cells (see Sperling, 1982 for review). The cytoplasm of the latter contains mitotic factors which initiate a prophase like process. This leads within 20–30 min to visible chromosomes which still continue to synthesize RNA until 40 min after fusion (Sperling and Marcus, 1984). It was thus feasible to determine the ^{3}H uridine labelling pattern of mammalian 'interphase' chromosomes. Using this technique it could be shown that in tissue culture cells of the Indian muntjac and the field vole *Microtus agrestis* the constitutive heterochromatin is transcribed (Sperling and Marcus, 1984; Sperling *et al.* 1987).

In principle, this approach can be applied to almost any cell type. However, it has considerable shortcomings: the resolving power is limited and due to the short half-life of the transcripts, they might become released from the chromatin and then bind unspecifically to the chromosomes. The number of analysable plates is therefore relatively low.

Here, a new approach is reviewed to identify indirectly genetically active and inactive sites on metaphase chromosomes: the DNAse I-nick translation technique.

II. Indirect Approach to Map Genetic Activity: DNase I-nick translation technique

A. Biological basis

The technique of DNase I-nick translation is based on the observation of Weintraub and Groudine (1976) that template active regions are more accessible to DNase I, an endonuclease with little sequence specificity, than non-transcribing regions. This finding was, perhaps, the first molecular evidence that active chromatin must be structurally distinct. Since this characteristic has now been found not only in nuclei of vertebrates and invertebrates but also in plants, it is obviously a general phenomenon (see Mathis *et al.*, 1980; Igo-Kemenes *et al.*, 1982; Weisbrod, 1982; Eissenberg *et al.*, 1985; Pederson *et al.*, 1986; Cartwright and Elgin, 1988; Fritton *et al.*, 1988 for reviews). However, DNase I sensitivity is only correlated with potential gene activity. It can precede gene expression as in precursor red cells of chick embryos in which the adult β-globin gene is DNase I sensitive though adult β-globin is not yet synthesized. It can also exist after gene activity has ceased as in chick erythrocytes in which the adult β-globin gene is still DNase I sensitive even though the cells have already stopped globin synthesis (Stalder *et al.*, 1980a, b; Weintraub and Groudine, 1976). Similarly, in Friend erythroleukemia cells the mouse β-globin gene is DNase I sensitive both before and after induction with DMSO (Miller *et al.*, 1978). In hormone stimulated oviduct tissue from hens the ovalbumin gene is preferentially digested by DNase, I also after hormone withdrawal (Lawson *et al.*, 1982). In all these cases the genes are insensitive in those tissues in which they are never expressed, e.g. in spleen, liver or nerve cells. In conclusion, the structural alteration of the chromatin fibre in the vicinity of a gene, leading to its DNase I sensitivity, is a necessary—although insufficient—condition for its transcription. Moreover, the region of DNase I sensitivity is not restricted to the gene itself but is extended for many kilobases up and downstream. Thus, the chicken and mouse β-globin gene clusters are located in a DNase I sensitive domain of about 30 kb, the ovalbumin gene is embedded in a sensitive domain of about 100 kb. These chromosomal domains are possibly analogous to the loops characterizing genetically active sites in lampbrush and polytene chromosomes, thus representing a general feature of chromatin organization in eukaryotes.

In this context it is important to distinguish between two kinds of DNase I sensitivity (Figure 1): the one which has been described above that extends over the entire domain and leads to an increase of about 3–10 fold as compared to inactive chromatin (intermediate or moderate DNase I sensitivity); and another which affects only single sites, that typically span 50–200 bp and that are about 100-fold more sensitive to digestion with DNase I (DNase I hypersensitivity, see Elgin, 1981 for review). These hypersensitive sites are usually near the 5′ end of genes transcribed RNA polymerase I and II but not by polymerase III (Figure 1). They are likely to be short nucleosome-free regions. In a number of cases they coincide closely with sequences which bind proteins and may be essential for transcription (i.e. enhancers and promoters). They have also been correlated with other functions such as initiation of replication and recombination.

DNase I hypersensitive sites (DHS) can be further divided into two broad categories: (1) constitutive DHS which are present regardless of transcriptional status of the gene and reflect potential gene activity, and (2) induced DHS which only appear upon activation of the gene and normally disappear upon inactivation. Both types can be associated with the same gene (Paul *et al.*, 1987).

The mechanism underlying the enhanced DNase I sensitivity is not yet understood. It must be a combination of various factors (e.g. methylation of DNA; binding of specific proteins and protein modification) which influence the topological state of the chromatin. One

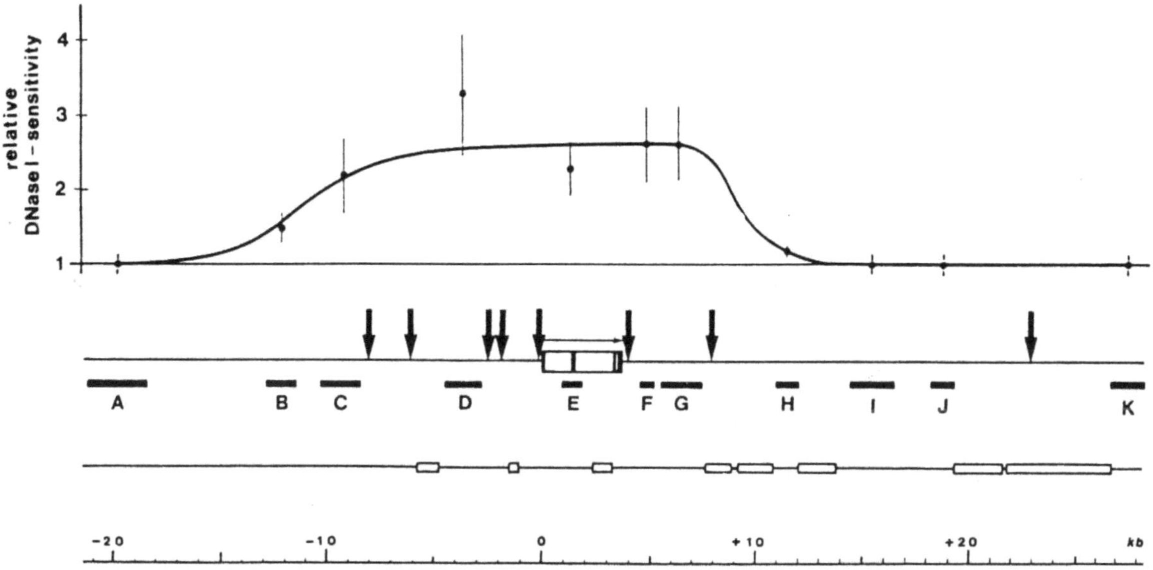

FIGURE 1: DNase I sensitivity around the lysozyme gene in viduct nuclei. The domain of moderate DNase I sensitvity of more than 20 kb is shown above, the DNase I hypersensitive sites are marked by the vertical arrows. The lysozyme gene is drawn as a box consisting of black exons and white introns. The horizontal arrow indicates start and direction of transcription. The bars labelled by capital letters represent the hybridization probes used. The open boxes in the third line indicate the position of repetitive sequences. A scale in Kb is shown in the lower part (from Fritton *et al.*, 1988).

factor is the presence of the nuclear proteins HMG 14 and HMG 17 which characterizes the DNase I hypersensitive sites but not those of moderate DNase I sensitivity.

The moderate DNase I sensitivity of potentially active genes is also maintained during metaphase. This has been demonstrated for the rRNA genes of the slime mold; Physarum, which remained DNase I sensitive during mitosis although rRNA synthesis was no longer detectable (Stalder *et al.*, 1978), and most clearly in mouse tissue culture cells containing an integrated, active leukemia virus. The viral genes retained their DNase I sensitivity at metaphase to the same degree as at interphase (Gazit *et al.*, 1982). This observation offered the possibility of mapping the DNase I sensitive, not the hypersensitive, sites on metaphase chromosomes. Both the theoretical basis and the experimental methodology have been most elegantly worked out by Howard Cedar and his group, and further refined by Menashe Marcus and his co-workers. The procedure depends on the induction of nicks in the DNA of fixed mitotic chromosomes followed by the replacement of the adjacent DNA with radio-labelled deoxynucleotide triphosphates by polymerase I and subsequent detection of the labelled sites by autoradiography. The resolution power can be significantly improved if biotinylated nucleotides are used, these can then be detected immuncytochemically (Figure 2).

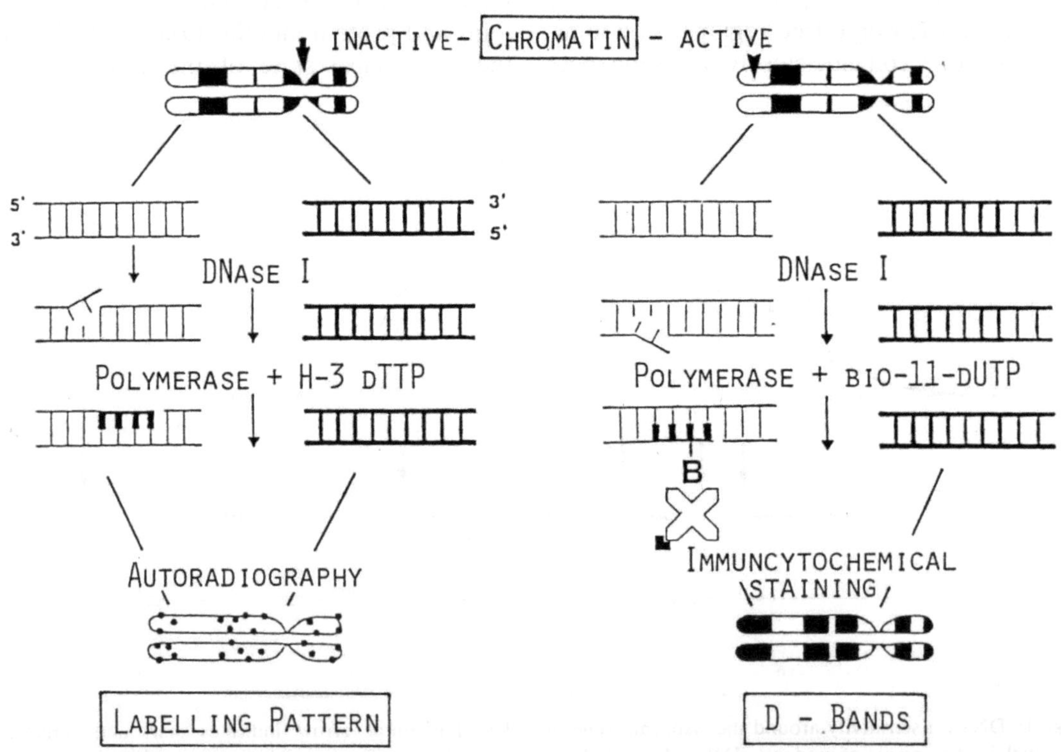

FIGURE 2: Schematic illustration of the DNase I-nick translation technique with radiolabelled (left) or biotinylated (B) probes. The light and dark chromosomal bands reflect the potentially active, DNase I sensitive and insensitive chromatin, respectively. For further details see text.

B. Methodological aspects

In the following section the experimental details of the DNase I nick translation technique are protocolled for both the radioactive and the non-radioactive approach (Figure 2). It should be noted that the technique can also be used to detect 'spontaneous' or mutagen induced nicks in various tissues if the DNase I treatment is omitted (Nose and Okamoto, 1983; Isaki, 1986).

1. Preparation of metaphase chromosomes.
   In order to obtain long chromosomes, cell cultures should be treated with a low dose of colcemid (0.01 to 0.05 μg/ml) for a relatively short time (5 to 60 min). The cells are then harvested as usual, treated with hypotonic KCl (0.075 M) for about 5 min at 37°C, and fixed with cold (−20°C) methanol/acetic acid (3:1). After three washes with the fixative the cell suspension is dropped on cold, wet slides and air dried. The metaphases should be free of cytoplasm. The duration of fixation (30 min or several hours) is not critical. The slides should be used either immediately or may be stored at −20°C to −70°C for several days before further processing.
2. *In situ* nick translation with radio-labelled dTTP (details of the various buffers and solutions are given in points 4 and 5 below).
   (a) 100 μl (20 μl) of the reaction mixture is added to the slide and covered with a 48×22 mm (24×18 mm) cover slip. The slides are incubated at room temperature for 5 min or at 15°C for 5 to 10 min.
   (b) Thereafter, the reaction is terminated by washing the slides three times for 5 min each with 2×SSC at room temperature.
   (c) The slides are dehydrated through an alcohol series (70%; 90%; 100%) and air-dried.
   (d) For autoradiography the dried slides are covered with photographic emulsion (e.g. Ilford K-5) and stored in the dark at 4°C. The first test slides should be developed after 4 days. The slides should be exposed until the metaphases show 100 to 200 grains.
3. *In situ* nick translation with biotinylated dUTP.
   The detection kits of either BRL or ENZO can be used. A number of different detection procedures have been described. In our experience the following protocol gave the most reliable results.
   (a) The nick-translation procedure is performed essentially as described above, except that the reaction mixture contains 4μM biotinyl-11-dUTP instead of radio-labelled dTTP.
   (b) After *in situ* nick translation the slides are rinsed twice in 2×SSC at 32°C for 5 min, and once in 0.1 % Triton X-100 in phosphate buffered saline (PBS; 0.13 M NaCl, 0.007 M Na_2HPO_4, 0.003 M NaH_2PO_4 pH 7.6) containing 2 mg/ml BSA for 5 min at room temperature.
   (c) Slides are then rinsed in PBS-BSA and incubated with streptavidin horse-radish peroxidase complex (diluted 1:200) at 37°C for 30 min in a humid chamber.
   (d) After three washes in PBS, the slides are stained with 0.5 mg/ml PBS 3.3' diaminobenzidine tetrahydrochloride (DAB)/ H_2O_2 (0.003 %) in the dark for about 10 min at room temperature which results in precipitation of an unsoluble brownish complex on the chromosomes. It is essential to prepare the DAB solution immediately before use.
   (e) Reaction is terminated by washing in PBS and air-drying of the slides.
   The detection sensitivity of the stain can be greatly increased by reflection contrast microscopy as described by Landegent *et al.* (1985).
   Note: the slides should not be allowed to dry during the procedure. Q-banding or DAPI staining of the chromosomes

should be performed after *in situ* nick translation but before the peroxidase reaction (Adolph and Hameister, 1985). Neither BrdU incorporation nor methotrexate or Hoechst 33258 treatment, prior to chromosome preparation, seems to affect the specificity of the DNase I-nick translation technique (Kerem *et al.*, 1983; 1984; Sperling *et al.*, 1985).

4. Preparation of buffers and stock solutions. The buffers and stock solutions described here should be stored as indicated to give reproducible results. Repeated freezing and thawing must be avoided to prevent loss of activity.
   (a) Nick translation buffer (10 × NTB) 500 mM Tris-HCl (pH 7.9)
   50 mM $MgCl_2$
   500 μg/ml nuclease-free bovine serum albumin (BSA)
   The substances should be dissolved in bidistilled water and aliquots stored at −20°C. After thawing and before use 2-β-mercaptoethanol is added to a final concentration of 100 mM.
   (b) Deoxynucleotide triphosphate solution (100 ×)
   0.4 mM dATP, 0.4 mM dGTP, and 0.4 mM dCTP are dissolved in 500 mM Tris-HCl (pH 7.9).
   Aliquots should be stored at −20°C.
   (c) ^{3}H dTTP (spec. activity about 80 Ci/mM). Aliquots of the original aqueous solution should be kept at −20°C or better at −80°C.
   The concentration is about 30 μM.
   (d) DNase I stock solution
   0.5 mg/ml DNase I in 0.1 M $MgCl_2$
   Aliquots should be stored at −20°C.
   For each experiment a fresh aliquot of the frozen stock should be used and further diluted to give a 5000-fold dilution: 10 μl of the stock solution are diluted in 990 μl 1 × NTB and from this solution 10 μl are further diluted with 490 μl 1 × NTB to give 100 ng DNase I/ml.
   (e) DNA polymerase I (endonuclease free!) in 50% glycerol is kept in the original vial (concentration 5000 U/ml) at −20°C. Take out the necessary volume for each experiment without warming up the stock solution.
   (f) 20 × SSC
   3.0 M NaCl; 0.3 M sodium citrate (pH 7.0).
   Store at room temperature.
   Prior to use 1 part of the stock solution should be added to nine parts of bidistilled water to give the 2 × SSC working solution.

5. Preparation of the reaction mixture (1000 μl).
   Add the reagents (on ice) in the following order:

(a) 10 × NTB buffer + 2-β mercaptoethanol	100 μl
(b) Bidistilled water	about 868 μl
(c) Deoxynucleotide triphosphate solution (final concentration 4 μM each)	10 μl
(d) ^{3}H dTTP (final concentration 0.3 μM)	about 10 μl
(e) DNase I (from the diluted stock) (final concentration 1 ng/ml)	10 μl
(f) DNA polymerase I (final concentration 10 U/ml)	2 μl
Total volume	1,000 μl

Note: each experiment should include a control without DNase I. The control should show only a few grains indicating that the label in the experimental series is strictly DNase I dependent.

III. DNase I Sensitivity of Mitotic Chromosomes

A. Facultative heterochromatin

During early embryonic development of female mammals one of the two X chromosomes becomes inactivated over most of its length, late replicating, and condensed during interphase, forming the sex chromatin body (facultative heterochromatin). The inactive state is stably inherited through mitotic divisions and *in vivo* reactivation occurs only at the entry into meiotic prophase.

As was first shown by Kerem *et al.* (1983) in the rodent, Gerbillus, the inactive chromatin conformation is also maintained in mitotic chromosomes, rendering the inactive X DNase I insensitive. Since Gerbillus has composite X chromosomes in which the 'original' X is flanked by two—genetically active—autosomal segments, these could serve as internal controls for the DNase I-nick translation procedure. In about 80 per cent of all metaphase plates both autosomal regions were heavily labelled while the 'original' X was almost free of grains except for a small, early replicating band. Perhaps this band represents a segment which escapes X inactivation like the pseudoautosomal region in the human X.

A similar observation was made in the European field vole, *Microtus agrestis* (Sperling *et al.*, 1985). This rodent has, perhaps, the largest mammalian sex chromosomes of which three-quarters (the whole long arm and the proximal part of the short arm (Figure 3) is composed of constitutive heterochromatin (see section 'B' below). In the cell line used for these studies the active and inactive X could be easily distinguished morphologically since the inactive X lacks part of the heterochromatin of the long arm (Figure 3). In about 60 per cent of all plates the euchromatic region of the inactive X was practically free of label except for an early replicating band at its distal end. The active X was uniformly labelled (Figure 3). Almost 40 per cent of all plates were considered 'overnicked' because they failed to show any difference between the X's which were both heavily labelled. Thus, due to slight preparational variation a considerable percentage of all metaphases was nonspecifically nicked, which underlines the importance of an internal control when applying this technique.

The molecular mechanism of X inactivation is not yet understood. It is, however, clear that methylation is involved in this process. Experimentally, hypomethylation can be induced by cultivation of cells in medium containing 5-azacytidine (5-aza C). As has been shown by Jablonka *et al.* (1985; 1987) in Gerbillus and Microtus cells this drug causes the late replicating, inactive X to replicate its DNA during early S-phase. This shift in DNA synthesis is correlated with a significant increase in DNase I sensitivity of the facultative heterochromatin. Upon removal of 5-aza C, the chromosomes revert to their original state. It is, however, unknown whether this change in DNase I sensitivity is also correlated with a corresponding change in gene activity.

Using a completely different approach, Kerem *et al.* (1988) analysed the DNase I sensitivity of the inactive Microtus X in which the gene for hypoxanthine guanine phosphoribosyltransferase (HPRT) has become spontaneously reactivated. The experiment was based on somatic cell hybrids obtained by fusion of female *Microtus agrestis* cells with mouse A 9 cells which are deficient for the X-linked enzyme HPRT. Immediately after isolation in HAT medium the hybrids contained both the active and the inactive Microtus X. By counter-selection in 6-thioguanine medium a cell line with only one, functionally inactive, Microtus X could be isolated. In order to select for cells

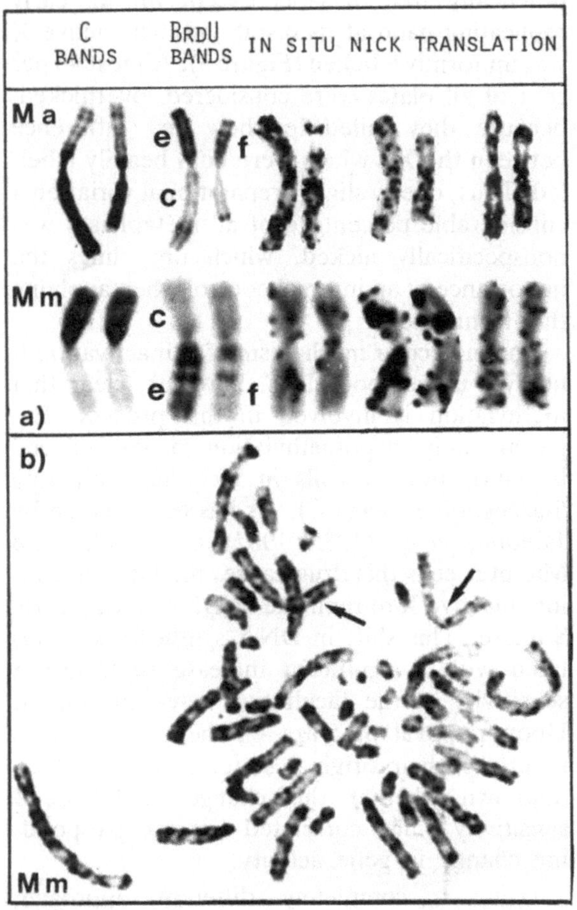

FIGURE 3: Labelling pattern of female *Microtus agrestis* (Ma) and *Millardia meltada* (Mm) chromosomes after *in situ* nick translation with ^{3}H dTTP (a) or biotinylated dUTP (b). The arrows point to the *Millardia meltada* X chromosomes. (a) Cut-outs of five pairs of Ma and Mm X chromosomes each. The first pair represents C-, the second replication banding. The other three sets are autoradiographs after *in situ* nick translation with ^{3}H dTTP.
Note that the facultative heterochromatin (f) is almost free of label, while the constitutive heterochromatin (c) and the euchromatin (e) is heavily labelled. The Ma photos are from Sperling *et al.* (1985), the Mm photos from I. Nanda (unpubl. observ.).

reactivated at the HPRT locus the hybrids were again cultivated in HAT medium. Thereafter, two cell lines containing a morphologically normal Microtus X could be isolated, both of which expressed the Microtus HPRT.

After DNase I-nick translation, no difference in the X labelling pattern was observed in the hybrids containing the inactive and the reactivated X. Both exhibited the same low DNase I sensitivity. Thus, reactivation of the HPRT locus is obviously only a local event, confined to this single gene and therefore below the resolution power of the DNase I-nick translation technique.

In this context it is noteworthy that neither Adolph and Hameister (1985) nor Murer-Orlando and Peterson (1985) observed a clear difference in the DNase I sensitivity of normal human or mouse active and inactive X, although in human interphase nuclei the sex chromatin body is completely unlabelled after DNase I-nick translation (Dyer *et al.*, 1985; Kerem, personal comm.). Perhaps the DNase I sensitivity differs only slightly between both human X's at metaphase.

In a case of X/autosome translocation, in which the X of the translocation chromosome was active in all cells of the mother (who had a balanced karyotype) but inactive in all cells of her daughter (who had an unbalanced karyotype with partial trisomy 1), reduced DNase I sensitivity was observed in the translocated autosomal material adjacent to the inactive X after quantitative autoradiography. This was correlated with a shift of DNA synthesis to late replication and interpreted as evidence that inactivation has spread from the X to the autosomal material (Wegner *et al.*, 1984).

B. Constitutive heterochromatin

In most species a significant amount of DNA consists of constitutive heterochromatin. In contrast to facultative heterochromatin it is a permanent feature of a chromosome, stains positive after C-banding, is usually late replicating, and preferentially located at the centromeres, but even whole chromosomes or chromosome arms may be constitutively heterochromatic.

Constitutive heterochromatin differs from euchromatin in its molecular organization. In the mouse, *Mus musculus*, the late replicating centromeric heterochromatin is enriched with highly repetitive sequences of about 234 bp which are almost 100 per cent methylated in somatic cells and not transcribed (Hsu, 1962). This centromeric heterochromatin is highly resistant to DNase I (Sperling *et al.*, 1985; Adolph, 1988) which can be seen most clearly in metaphases in which condensation of these segments is inhibited by the fluorochrome Hoechst 33258. Furthermore, at interphase the chromocentres formed by the centromeric heterochromatin are free of label (Figure 4). Even if the cells were grown in 5-aza C medium, in which the centromeric heterochromatin becomes hypomethylated and starts DNA replication in early S-phase, it remained DNase I insensitive (Selig *et al.*, 1988).

DNase I insensitivity has also been described for the constitutive heterochromatin of the human Y chromosome (Kerem *et al.*, 1984; Murer-Orlando and Peterson, 1985). Adolph and Hameister (1985), however, did not find any difference in the DNase I sensitivity between the euchromatin and the heterochromatin of the human Y, but identified their boundary as the most DNase I sensitive site. These conflicting results might be explained by the different cell lines used: while the first laboratories analysed metaphases from PHA stimulated lymphocytes, Adolph and Hameister performed their experiments on a transformed cell line (Burkitt lymphoma). In accord with this, Sperling and Marcus (1984) found that the heterochromatic Y chromosome of the Indian muntjac was DNase I sensitive in a spontaneously transformed cell line. Perhaps the non-centromeric heterochromatin has a more open conformation in rapidly growing cells.

The field vole, *Microtus agrestis,* has perhaps the largest blocks of non-centromeric heterochromatin in mammals (Figure 3). In nerve cells this heterochromatin is condensed into compact chromocentres which are genetically inactive and also DNase I insensitive (Figure 4). However, the boundary between this heterochromatin and the euchromatin is a preferential site of DNase I attack (Figure 4). This is in line with the findings of Hutchison and Weintraub (1985) in nucleated erythrocytes of newt and chicken in which the border of the condensed chromatin also showed the highest DNase I sensitivity.

In rapidly growing tissue culture cells of *M. agrestis* the constitutive heterochromatin is completely decondensed at interphase and not different in its DNase I sensitivity from euchromatin (Sperling *et al.*, 1985). Later it was shown that this does not reflect potential but, rather, actual transcriptional activity (Sperling *et al.*, 1987). In accord with this, the large gonosomal heterochromatin of the rodent, *Millardia, meltada,* is DNase I sensitive in tissue culture cells (Nanda unpubl. results, Figure 4), as well as the interstitial

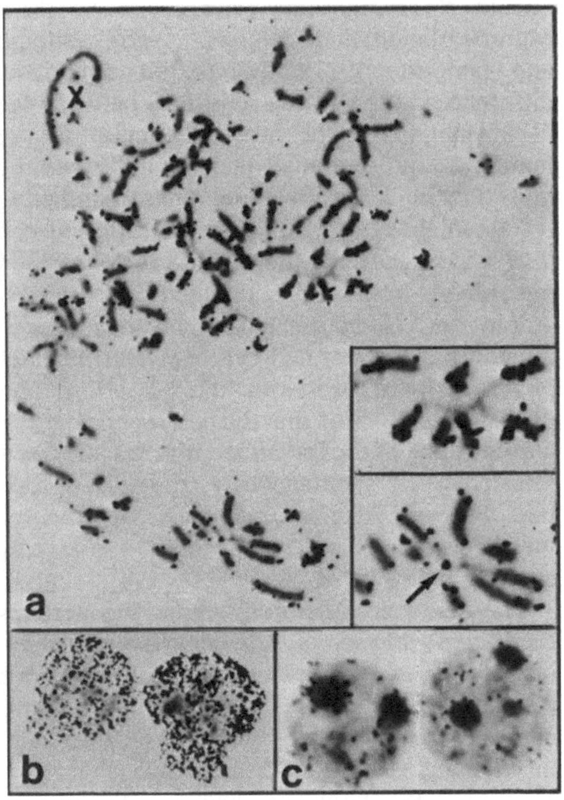

FIGURE 4 (a) *In situ* nick translated mitotic chromosomes of a mouse × microtus cell hybrid containing one inactive microtus X chromosome (X). The centromeric heterochromatin of the mouse chromosome is undercondensed since the cells were grown in medium containing the fluorochrome Hoechst 33258.
Note that the facultative heterochromatin of the X and the centromeric heterochromatin of the mouse chromosomes are DNase I insensitive. Occasionally the extreme tip (telomere) of the mouse heterochromatin is labelled (arrow).
(b), (c) Autoradiographs of *in situ* nick translated mouse (b) and *Microtus agrestis* brain (c) interphase nuclei. Note that the heteropycnotic regions are almost free of label and that the boundary of constitutive heterochromatin and euchromatin in the nerve cells is a preferential site of DNase I attack.

heterochromatin of the mouse chromosome no. 1 (Adolph, 1988). Generally, these findings indicate that the assumption of permanent inactivity of constitutive heterochromatin in mammals must be rejected. The biological relevance of this observation is, however, not yet understood.

C. Euchromatin

Differential Giemsa staining of mammalian chromosomes reveals an alternating pattern of dark G-bands (Giemsa bands) and light R-bands (reversed Giemsa bands). These bands correspond to units of replication and have also been considered to represent functional units (see Holmquist, 1988 for review).

After DNase I treatment of muntjac chromosomes, the labelling intensity closely correlated with chromosome length, and chromosomal segments rich in either G- or R-bands did not show great differences in their sensitivity (Sperling and Marcus, 1984). This confirms previous experiments in which the accessibility of G- and R-band chromatin was measured after staphylococcal nuclease treatment (Munroe and Latt, 1977) or DNase I treatment (Holmquist *et al.*, 1982) and was not found to be significantly different. Apart from this, the telomere ends of the muntjac chromosomes exhibited the highest, and the centromeres the lowest, DNase I sensitivity which was also reported for human and mouse chromosomes (Adolph and Hameister, 1985). It cannot be ruled out that due to preparational artifacts the chromosomal ends become more accessible to DNase I. However, after *in situ* hybridization with radio-labelled RNA from human or Chinese hamster cells the ends of several chromosomes were also more heavily labelled which might be indicative of—but does not prove—genetic activity (Kuo and Saunders, 1977; Pierpont and Yunis, 1977). Interestingly, the extreme tip of the mouse centromeric heterochromatin—which clearly is genetically inactive—is also labelled after DNase I-nick translation (Selig *et al.*, 1988; Figure 4 our unpublished observations). This might be explained by the unique structural properties of the real telomeric DNA which perhaps already contains single strand breaks and is not covalently closed (see Blackburn and Szostak, 1984, for review) and therefore more accessible to the nick translation procedure. The nucleous organizing regions are DNase I sensitive (Murer-Orlando and Peterson, 1985) and the active NORs can be distinguished from the inactive by their different sensitivity (Ferraro and Prantera, 1988). Also, the fragile site Xq27.3 of the human X appears DNase I sensitive (Nolin *et al.*, 1987).

The resolution power of the DNase I-nick translation technique, based on the autoradiographic detection of radio-labelled precursors, is limited and cannot discriminate between single G- and R-bands. This problem has been overcome by the use of biotinylated dUTP which can be detected immunocytochemically (Kerem *et al.*, 1984; Adolph and Hameister, 1985; Murer-Orlando and Peterson, 1985).

This non-radioactive approach led to a distinct banding pattern in which the sites of higher DNase I sensitivity were more darkly stained and were termed D-bands (Kerem *et al.*, 1984). The absolute number of bands was in the same order of magnitude as after Giemsa-banding. Kerem *et al.*, (1984) observed a good correspondence between R- and D-bands which fits with the concept that the early replicating R-bands harbour the active housekeeping genes while the G-band chromatin is more condensed and, perhaps, almost inactive (Holmquist, 1988). Other investigators, however, did not find such a close correlation. Obviously, the difference in DNase I sensitivity between G- and R-bands is less than that between euchromatin and facultative heterochromatin, or centromeric and non-centromeric constitutive heterochromatin and therefore more difficult to assess.

IV. DNase I Sensitivity of Meiotic Chromosomes

In mammalian male meiosis the XY bivalent appears condensed within the darkly stained sex vesicle and shows almost no RNA synthesis (Monesi, 1967; Tres, 1975). The DNase I sensitivity of the XY pair at pachytene and metaphase I in mouse and man is less than that of the autosomes (Sperling and Marcus, 1984; Richler *et al.*, 1987; Chandley and McBeath, 1987). In contrast to this, Murer-Orlando and Peterson (1985) did not find any obvious difference in labelling intensity. However, they applied higher DNase I concentrations and a longer treatment, so that this result could simply be explained by 'overnicking' especially since meiotic chromosomes in general seem to be more accessible to DNase I than mitotic chromosomes (Chandley and McBeath, 1987).

DNase I sensitivity along the sex chromosomes is non-random. Distinct spots of label have been found over the pairing region and the free terminal X and Y segments, as well as on a prominent early replicating band of the X in the mouse (Richler *et al.*, 1987; Raman *et al.*, 1988) and humans (Chandley and McBeath, 1987). Also, in pachytene cells of the shrew, the X-Y pairing region exhibits a higher DNase I sensitivity (cit. in Raman and Nanda, 1986). In a later investigation the DNase I sensitivity of the pairing region in mouse sex chromosomes could not be confirmed (Separovic and Chandley, 1987). Since grain localization after autoradiography is not very precise, application of the non-radioactive DNase I-nick translation-technique might clarify this point.

V. Conclusions

The DNase I-nick translation technique is a powerful approach to mapping genetically active or, at least, potentially active segments on mitotic and meiotic chromosomes. It is based on the observation that (1) chromosomal domains harbouring transcribed genes are preferentially digested by pancreatic DNase I and (2) this differential DNase I sensitivity is also maintained in fixed mitotic chromosomes.

The DNase I sensitive sites can be detected by *in situ* nick translation using either radiolabelled deoxynucleotides in combination with autoradiography or biotinylated dUTP with immunocytochemical staining. In the first case detection sensitivity is high and is quantifiable; in the second case the resolving power is significantly increased leading to distinct DNase I sensitive dark D-bands.

Since the difference in DNase I sensitivity is only gradual, both approaches are prone to various artifacts. The specificity of the DNase I-nick translation technique should therefore be checked for each metaphase analysed. Most suitable as internal controls are particular DNase I insensitive regions such as, for example, the centromeric heterochromatin or the facultative heterochromatin in female mammalian cells.

Generally, this new technique has led to a further subdivision of euchromatin and constitutive heterochromatin into DNase I sensitive and insensitive fractions. It offers the possibility of detecting clusters of tissue specific genes and following the kinetics of gene activation and inactivation during embryogenesis. In clinical cytogenetics it will improve our understanding of the relationship between chromosomal imbalances and their phenotypic expression and might even be of predictive value in cases of prenatal detection of supernumerary marker chromosomes of unknown origin.

Acknowledgements

This work was supported in part by a grant from the Deutsche Forschungsgemeinschaft and the GSF in the framework of German-Indian cooperation.

I am grateful to Mrs Ada Wilcke for her secretarial help and to Dr M. Digweed for critical reading of the manuscript.

References

Adolph S (1988). In situ nick translation distinguishes between C-band positive regions on mouse chromosomes. Chromosoma 96; 102–6.

Adolph S, Hameister, H (1985). In situ nick translation of metaphase chromosome with biotin-labeled d-UTP. Hum Genet 69; 117–21.

Blackburn EH, Szostak JW (1984). The molecular structure of centromeres and telomeres. Ann Rev Biochem 53; 163–94.

Cartwright IL, Elgin SCR (1988). Chromatin of active and inactive genes. In Architecture of eukaryotic genes (ed. Kahl G), VCH Verlagsgesellschaft, Weinheim; 283–300.

Chandley AC, McBeath S (1987). DNase I hypersensitive sites along the XY bivalent at meiosis in man include the Xp Yp pairing region. Cytogenet Cell Genet 44; 22–31.

Dyer KA, Riley D, Gartler SM (1985). Analysis of inactive X chromosome structure by in situ nick translation. Chromosoma 92; 209–13.

Edström J-E (1976). Meiotic versus somatic transcription with special reference to Diptera. Life Sci Res Rep 4; 301–16.

Eissenberg JC, Cartwright IL, Thomas GH, Elgin SCR (1985). Selected topics in chromatin structure. Ann Rev Genet 19; 485–536.

Elgin SCR (1981). DNase I hypersensitive sites of chromatin. Cell 27; 413–5.

Ferraro M, Prantera G (1988). Human NORs show correlation between transcriptional activity, DNase I sensitivity, and hypomethylation. Cytogenet Cell Genet 47; 58–61.

Fritton HP, Jantzen K, Igo-Kemenes T, Nowock J, Strech-Jurk U, Theisen M, Sippel AE (1988). Chromatin domains and gene expression—different chromatin conformations characterize the various functional states of the chicken lysozyme gene. In Architecture of eukaryotic genes (ed. Kahl G), VCH Verlagsgesellschaft, Weinheim; 333–54.

Gazit B, Cedar H, Lerer J, Voss R (1982). Active genes are sensitive to deoxyribonuclease I during metaphase. Science 217; 648–50.

Hennig W (1986). Heterochromatin and Germ L'ine-restricted DNA. In Results and Problems in Cell Differentiation, Vol. 13, Germline-Soma Differentiation (ed. Hennig W), Springer, Berlin, Heidelberg; 175–92.

Holmquist G, Gray M, Porter T, Jordan J (1982). Characterization of Giemsa dark- and light-band DNA. Cell 31; 121–9.

Holmquist G (1988). DNA sequences in G-bands and R-bands. In Chromosome and Chromatin Structure, Vol. II, (ed. Adolph KW), CRC Press; 75–121.

Hsu TC (1962). Differential rate in RNA synthesis between euchromatin and heterochromatin. Exp Cell Res 27; 332–4.

Hutchison N, Weintraub H (1985). Localization of DNase I sensitive sequences to specific regions of interphase nuclei. Cell 43; 471–82.

Igo-Kemenes T, Hörz W, Zachau HG (1982). Chromatin. Ann Rev Biochem 51; 89–121.

Iseki S (1986). DNA strand breaks in rat tissues as detected by in situ nick translation. Exp Cell Res; 167; 311–26.

Jablonka E, Goitein R, Marcus M, Cedar H (1985). DNA hypomethylation causes an increase in DNase-I sensitivity and an advance in the time of replication of the entire inactive X chromosome. Chromosoma 93; 152–6.

Jablonka E, Goitein R, Sperling K, Cedar H, Marcus M (1987). 5-aza-C-induced changes in the time of replication of the X chromosomes of Microtus agrestis are followed by non-random reversion to a late pattern of replication. Chromosoma 95; 81–3.

Kerem B. Goitein R, Richler C, Marcus M, Cedar H (1983). In situ nick-translation distinguishes between active and inactive X chromosomes. Nature 304; 88–90.

Kerem B, Goitein R, Diamond G, Cedar H, Marcus M (1984). Mapping DNase I sensitive regions on mitotic chromosomes. Cell 38; 493–9.

Kerem B, Kottusch-Geiseler V, Kalscheuer V, Goitein R, Sperling K, Marcus M (1988). DNase I sensitivity of Microtus agrestis active, inactive and reactivated X chromosomes in mouse-Microtus cell hybrids. Chromosoma 96; 227–30.

Kuo MT, Saunders GF (1977). Location of messenger specifying sequences in mammalian chromosomes. Chromosoma 63; 241–52.

Landegent JE, Jansen in de Wal N, Ploem JS, van der Ploeg M (1985). Sensitive detection of hybridicytochemical results by means of reflection contrast microscopy. J Histochem Cytochem 33; 1241–6.

Lawson GM, Knoll BJ, March CJ, Woo SLC, Tsai M-J, O'Malley BJ (1982). Definition of 5' and 3' structural boundaries of the chromatin domain containing the ovalbumin multigene family. J Biol Chem 257; 1501–7.

Mathis D, Oudet P, Chambon P (1980). Structure of transcribing chromatin. Progr Nucl Acid Res Mol Biol 24; 1–55.

Miller DM, Turner, P, Nienhuis AW, Axelrod DE, Gopalakrishnan TV (1978). Active conformation of the globin genes in uninduced and induced mouse Erythroleukemia cells. Cell 14; 511–21.

Monesi V (1965). Synthetic activities during spermatogenesis in the mouse. RNA and Protein. Exp Cell Res 39; 197–224.

Munroe SH, Latt SA (1977). Comparison of the subunit organization of early and late replicating chromatin. Exp Cell Res 110; 299–313.

Murer-Orlando ML, Peterson AC (1985). In situ nick translation of human and mouse chromosomes detected with a biotinylated nucleotide. Exp Cell Res 157; 322–34.

Nolin SL, Jenkins EC, Brown WT, Dobkin CS (1987). Sensitivity of fra(X) (q27.3) to DNase I. Am J Hum Genet 41; A 136.

Nose K, Okamoto H (1983). Biochem Biophys Res Commun 111; 383.

Paul A-L, Vasil V, Vasil IK, Ferl RJ (1987). Constitutive and anaerobically induced DNase-I-hypersensitive sites in the 5' region of the maize Adhl gene. Proc Natl Acad Sci USA 84; 799–803.

Pederson DS, Thoma F, Simpson RT (1986). Core particle, fibre, and transcriptionally active chromatin structure. Ann Rev Cell Biol 2; 117–47.

Pierpont ME, Yunis JJ (1977). Localization of chromosomal RNA in human G-banded metaphase chromosomes. Exp Cell Res 106; 303–8.

Raman R, Nanda I (1986). Mammalian sex chromosomes. I. Cytological changes in the chiasmatic sex chromosomes of the male musk-shrew Suncus murinus. Chromosoma 93; 367–74.

Raman R, Singh AP, Nanda I (1988). DNase-I nick translation in situ on meiotic chromosomes of the mouse, *Mus musculus* J Cell Sci 90; 629–34.

Richler C, Uliel E, Kerem B, Wahrman J (1987). Regions of active chromatin conformation in 'inactive' male meiotic sex chromosomes of the mouse. Chromosoma 95; 167–70.

Selig S, Ariel M, Goitein R, Marcus M, Cedar H (1988). Regulation of mouse satellite DNA replication time. EMBO J 7 no. 2; 419–26.

Separovic ER, Chandley AC (1987). Lack of evidence that

the XqYq pairing tips at meiosis in the mouse show hypersensitivity to DNase I. Chromosoma 95; 290–4.

Sieger M, Pera F, Schwarzacher HG (1970). Genetic inactivity of heterochromatin and heteropycnosis in Microtus agrestis. Chromosoma 29; 349–64.

Sperling K (1982). Cell cycle and chromosome cycle: morphological and functional aspects. In Premature chromosome condensation (eds. Rao PN, Johnson RT, Sperling K). Academic Press, New York; 43–78.

Sperling K, Marcus M (1984). Mapping of genetic activity on mammalian chromosomes. Chromosomes Today 8; 169–78.

Sperling K, Kerem BS, Goitein R, Kottusch V, Cedar H, Marcus M (1985). DNase I sensitivity in facultative and constitutive heterochromatin. Chromosoma 93; 38–42.

Sperling K, Kalscheuer V, Neitzel H (1987). Transcriptional activity of constitutive heterochromatin in the mammal Microtus agrestis (Rodentia, Cricetidae). Exp Cell Res 173; 463–72.

Stalder J, Seebeck T, Braun R (1978). Eur J Biochem 90; 452–63.

Stalder J, Groudine M, Dodgson JB, Engel JD, Weintraub H (1980a). Hb switching in chickens. Cell 19; 973–80.

Stalder J, Larsen A, Engel JD, Dolan M, Groudine M, Weintraub H (1980b). Tissue-specific DNA cleavages in the globin chromatin domain introduced by DNase I. Cell 20; 451–60.

Tres L (1975). Nucleolar RNA synthesis of meiotic prophase spermatocytes in the human testis. Chromosoma 53; 141–51.

Varley JM, Macgregor HC, Erba HP (1980). Satellite DNA is transcribed on lampbrush chromosomes. Nature 283; 686–9.

Wegner R-D, Lüdtke E-L, Sperling K (1984). Position effect in an X-1 translocation analyzed by means of replicational behaviour and DNase I sensitivity. In Chromosomes Today 8 (eds. Bennett MD, Gropp A, Wolf U), George Allen and Unwin Ltd, London; 352.

Weintraub H, Groudine M (1976). Chromosomal subunits in active genes have an altered conformation. Science 193; 848–56.

Weisbrod S (1982). Active chromatin. Nature 297; 289–95.

18 Chromosomal Speciation: Constitutive Heterochromatin and Evolutionary Differentiation of the Indian Pygmy Field-Mice

T. SHARMA, A.S. BALAJEE AND N. CHEONG*

Cytogenetics Laboratory
Centre of Advanced Study in Zoology
Banaras Hindu University
Varanasi-221 005, India

Introduction

During the last several decades many models of speciation have been put forward and it is obvious that there can be different paths to speciation. The genetic processes involved may vary in different organisms or groups of organisms. However, the idea that chromosomal changes have a role in speciation process gained credence only in the sixties and seventies. It was mainly M.J.D. White (1968, 1969, 1973, 1974, 1978a and b, and 1982) who emphasized the importance of different chromosomal changes in the establishment of genetic discontinuity among different populations leading to their speciation. The chromosomal mechanism of speciation seems to be especially prevalent in organisms of low vagility, whose populations consist of numerous local demes that persist in the same area for many generations, like wingless grasshoppers, stick insects, littoral isopods, ladybird beetles, lizards, small mammals, etc. (see White 1978a and references therein). It is, however, likely that not in all cases do chromosomal changes have a causal role in speciation though karyotypic differences may exist in closely related taxa.

The noted evolutionist Ernst Mayr had practically ignored the role of chromosomal changes in his discussions of the speciation process until as late as 1969 when in his article 'Species, speciation and chromosomes' he recognized the importance of chromosomal isolating mechanisms which, according to him, would especially occur in sedentary animals that tend to exist in small colonies (Mayr, 1969). He further pointed out that the ideal location of this type of speciation is the periphery of the species range, where species populations are often broken into small, widely separated and heavily inbreeding populations. According to the stasipatric mode of speciation originally proposed in 1967 by White *et al.* (1967) on the basis of cytogenetic investigation of wingless morabine grasshoppers, and subsequently further consolidated by White (1968, 1969, 1978a and b) by considering examples from other animals including mammals, speciation results when a chromosomal

*Present Address: Department of Medical Oncology, M.D. Anderson Hospital and Tumor Institute
Houston, Texas 77030, USA.

rearrangement becomes established and spreads in a small local population thereby converting an essentially continuous population into various contiguous taxa with different karyotypes. The stasipatric model essentially envisages (White 1978a, p. 177) 'a widespread species generating within its range daughter species characterized by chromosomal rearrangements that play a primary role in speciation because of the diminished fecundity or viability of the heterozygotes. The daughter species are assumed to gradually extend their range at the expense of the parent species, maintaining a narrow parapatric zone of overlap at the periphery of their distribution within which hybridization leads to the production of genetically inferior individuals (usually inferior because of irregularities at meiosis). The adaptive role of structural homozygote was initially assumed to have a specific effect on the phenotype but subsequently the 'protective' role of chromosomal rearrangements against introgression has been advocated in accordance with the idea put forward by Hall (1973; see White, 1978b). It is assumed that a locally adapted genetic population is protected by chromosomal rearrangements of certain types from the disruptive influence of introgression.

In the chromosomal mechanism of speciation structural rearrangements of chromosomes such as Robertsonian fusion or fission, inversion and reciprocal translocation have been considered important as they result in reduction in fecundity because of irregularities in meiosis in heterozygotes which may serve as a reproduction barrier. Since our study is on rodents, we discuss in this article examples mainly from mammals.

Chromosomal Rearrangements and the Cases of Active Speciation in Mammals

How a chromosomal change gets rapidly fixed in homozygous condition, which eventually gives rise to chromosomal race or potentially incipient species, can be critically examined in rapidly speciating organisms. In-depth studies are being carried out in a few cases of mammals which are in active phase of speciation. One of the most interesting cases has been the common shrew *Sorex araneus* of Europe known for long to exhibit chromosomal polymorphism and polytype. In *S. araneus* variations due to Robertsonian fusions exist extensively in different populations of Europe; the centric fusions involved vary in various populations or chromosome races which possess several arm combinations. In the Oxford meeting held in 1987 on 'The Population and Evolutionary Cytogenetics of *Sorex araneus*' it emerged that somewhere between 20 and 40 karyotypic races had already been described and there could be at least 100 karyotypic races over the entire range of the species (see Searle, 1988a).

It the West European phylogenetic group, so far 13 basic karyotypes have been identified, of which 3 are from Great Britain. These are described as the Oxford, Hermitage and Aberdeen chromosome races. In the zone of contact between the Oxford and Hermitage races where an ancestral acrocentric state also exists, Searle (1984) observed the highest frequency of acrocentrics in the narrow zone of occurrence of Oxford-Hermitage hybrids. According to him, this cannot be so readily explained on the basis of the allopatric model, whereas it can be on the basis of the stasipatric model. He has tried to explain the situation prevailing in the hybrid zone by proposing selection against the hybrids. The race-specific metacentrics occur at high frequency in areas away from the hybrid zone, which he presumes may be the consequence of selection, perhaps due to unequal transmission from heterozygotes (Searle, 1988b).

Hausser *et al.* (1985) have also opined that the formation of chromosome races in *S. araneus* should be considered as a paradigm of the speciation process in the European group. They have further said that the genetic proximity excludes every mechanism based on a long geographical isolation. They have suggested that the chromosomal changes occasion a stasipatric speciation. They take into account the breeding behaviour, territoriality and dispersal of these shrews and presume that neither small demes nor high breeding are likely in them, and that the centric fusions spread out by meiotic drive alone. They cited in support the observation of Fedyk (1980) who detected in *S. araneus* of Poland the difference in frequencies of karyotypes over a few years in favour of metacentric homozygotes. Hausser and his associates have assumed that chromosomal speciation occurs when two incompatible metacentrics meet and the situation in such a zone depends on the efficiency of the meiotic drive. They have argued that if the meiotic drive is strong the elimination of corresponding acrocentrics will be pursued, which will lead to a narrow sterile hybrid zone without genic introgression. And presumably based on olfaction (see Hawes, 1976), further ethological barriers against hybridization may possibly develop in a short period. On the contrary, if the meiotic drive is poorly efficient, selection against the hybrids for heterobrachial metacentric will favour the corresponding acrocentrics. Under such a condition a tension zone of polymorphism will occur, permitting genic exchanges between the two races. In support of their argument Hausser *et al.* cited Searle's observation in the area of contact between the Oxford and Hermitage races and the balanced polymorphism reported by Ford and Hamerton (1970).

In the mole rat *Spalax ehrenbergi*, Wahrman *et al.* (1969a and b) discovered the existence of four chromosomal forms with $2n=52$, 54, 58 and 60 in Israel and its neighbourhood; the difference in the karyotypes being mainly due to centric fissions. It was noticed that the four chromosomal forms were distributed clinally corresponding to biogeographic regions of increasing aridity from north to south, the regions being not separated by any obvious physical barriers. They came across narrow hybrid zones, too, with very few natural hybrids having 53 or 59 chromosomes. One of the authors, Nevo (1969), conducted experiments on mating behaviour. He observed much more pronounced aggression in the agonistic stage when matings were between the heterochromosomal rather than the homochromosomal forms.

The four chromosomal forms are morphologically uniform but Wahrman *et al.* (1969b) proposed that they should be considered as sibling species since they had achieved homozygosity in fixation of the karyotypes and they rarely formed natural hybrids. Furthermore, besides cytological barriers, they could possibly be ethologically isloated. Nevo (1969) suggested that ethological isolation might have played an important role in speciation concomitantly with chromosomal mechanism. Wahrman *et al.* (1969a) inferred from morabine grasshopper and mammalian examples which showed cytogenetic mechanisms leading to the creation of sibling species, that the common denominator in them was their relative restricted mobility and narrow environmental specialization. They suggested that the speciation processes in such groups characterized by limited mobility and narrow environmental specialization may proceed, in general, in the absence of geographical barriers. In a subsequent paper Wahrman *et al.* (1969b) postulated that ancestral populations of *Spalax* inhabited an unknown centre, which could have been somewhere in Asia Minor or southeastern Europe. In the process of expansion from this centre, including towards increasingly arid areas southward, new and selectively superior homozygous karyotypes became fixed.

Nevo and his associates subsequently conducted further studies on genic variation, ethological isolation, population structure and evolution (Nevo and Shaw, 1972; Nevo *et al.*, 1975;

Nevo and Cleve, 1978; Nevo, 1979, 1982; Heth and Nevo, 1981; Nevo *et al.*, 1982). The genic similarity between the chromosomal taxa was found very high and the level of heterozygosity low. The high genic similarity was taken to suggest speciation without 'genic revolution' and the low heterozygosity for minimal variation in the relatively constant subterranean niche of *Spalax*. Their study indicated that ethological isolation might have accumulated incidentally, at least in part, through adaptive divergence during geographic speciation and building up of ethological isolation seemed to be a very slow process. They observed that in the main range of distribution, populations of mole rats are largely continuous whereas in the ecologically marginal peripheries, populations become isolated among themselves and also from the main range. Initial reproductive isolation by chromosomal mutations may have arisen peripatrically by chance in small peripherally isolated populations followed by adaptation and the gradual accumulation of ethological reproductive isolation.

Intensive studies undertaken by Nevo and his associates (see Nevo, 1986), both at the molecular and organismal levels, to elucidate mechanisms of adaptive speciation in these subterranean mole rats, have led to the conclusion that the *S. ehrenbergi* superspecies complex represents an active case of ecological speciation; the four chromosomal species display progressive stages of late chromosomal speciation. Thus, in the *S. ehrenbergi* chromosome differentiation appears to have been causal to the speciation process.

Ever since Gropp *et al.* (1970) reported $2n = 26$ with seven pairs of metacentric autosomes in the 'Tobacco-mouse' from Poschiavo valley of eastern Switzerland, the wild house mouse populations of Europe have attracted a great deal of attention with a view to investigating the role of centric fusions in speciation. A number of studies have been carried out on the populations from the Orkney Islands to the Antarctic Convergence and from Greece to Spain (see Capanna, 1985 and references therein; Winking, 1986), but mainly the populations of mountainous regions of southern Switzerland and central and northern Italy have been found to exhibit high degrees of chromosome rearrangements. In all instances centric fusions have occurred involving almost all the autosomes of the complement except one of the smallest pairs and the sex chromosomes. In the Italian and Swiss populations 13 different chromosomal races have been found with 22 different fusions resulting in reduction of the diploid numbers to 22, 24, 26 and 28 from the original 40 which still persists in the surrounding populations (Capanna, 1985; Corti *et al.*, 1986).

Since these mice can be fairly easily collected from contact zones and hybrids can also be obtained between different chromosomal races as well as between laboratory bred strains and wild house mice, a wealth of information is available on the cytogenetic consequences at meiosis in both sexes of the heterozygotes (Ford and Evans, 1973; Capanna, 1976; Winking and Gropp, 1976; Gropp and Winking, 1981). Upto 70 per cent of aneuploid gametes could be produced in heterozygosity for seven or nine fusions. The reproductive fitness of mice heterozygous for one fusion is about 40 per cent, for seven fusions about 15 per cent and for nine fusions about 11 per cent. These estimates are the probability of formation of euploid offspring by matings between identical heterozygotes. This strongly suggests that reproductive isolation can be achieved to a substantial degree by reduction in fecundity in such heterozygotes.

That the centric fusions in *M. musculus domesticus* (some describe them as *M. domesticus*) are of very recent origin is strongly indicated by the fact that though these rearrangements act as strong isolating mechanisms due to reduction in fecundity of the natural hybrids, no significant allozyme differences exist between the 40-, 24- and 22-chromosomal races studied (Britton-Davidian *et al.*, 1980; Capanna *et al.*, 1985). According to Capanna (1982) the process of chromosomal evolution began around 6000 years ago in these mice. Since in these parapatric chromosomal races gene flow was inhibited by the post-zygotic isolation mechanism, White (1978b and 1982) took them to be a potential incipient species and consented that they might have evolved by the stasipatric mode of speciation. Capanna and Corti (1982) found the two 24- and 26-chromosomal races living sympatrically in Upper Valtellina in Italy but failed to obtain their hybrids. Further studies on intermale aggression between the two races indicated that possibly full speciation had already been accomplished (Capanna *et al.*, 1984). It has since been suggested that at such a stage pre-mating isolation mechanisms are rapidly selected in cases of sympatric and parapatric contacts, and behavioural diversification may play a decisive role in instituting pre-mating isolation (Capanna, 1985). In this particular instance since the 26-chromosome males dominated the 24-chromosome males, Capanna (loc. cit.) speculated that it is likely that under prolonged sympatric conditions, the 26-chromosome race will suppress its 24-chromosome counterpart. If that is indeed the situation, he remarks, 'the karyotypes are not in a situation of equilibrium and the creation of a viable population of 2n = 24 individuals was a fortuitous consequence of geographical isolation. It could even be the case that the 2n = 26 chromosomal race is contributing to the extinction of the 2n = 24 form not only by competing more successfully for demes but by carrying out sterile matings with 2n = 24 females.'

The foregoing clearly demonstrates the role of structural rearrangements of chromosomes in speciation notwithstanding difference of opinion about the modes of speciation involved.

Constitutive Heterochromatin/Highly Repeated DNA Variation and Evolutionary Significance

With the advent of chromosome banding techniques, a new and entirely different mechanism of karyotype evolution was brought into focus when constitutive heterochromatin could be specifically identified in the metaphase chromosomes. The first few observations were reported in the rodents *Peromyscus* (Duffey, 1972; Pathak *et al.*, 1973), *Bandicota* (Sharma and Raman, 1973) and *Mus* (Sharma and Garg, 1975; Markvong *et al.*, 1975), in which addition or deletion of totally heterochromatic whole arms of chromosomes had occurred. It was speculated that such conditions might have played important roles in the evolutionary divergence of the taxa. This mode of karyotypic change, which results in alternation of the genome size, has since been reported in various groups of rodents (see Patton and Sherwood, 1982; Gamperl *et al.*, 1982) and other animals (see White, 1978a and 1982; John and Miklos, 1979; Brutlag, 1980; Peacock *et al.*, 1981; Arnold and Shaw, 1985).

The complexity of eukaryotic genomes due to the presence of highly repeated DNA sequences had become apparent from DNA reassociation studies by Britten and Kohne (1968). Schildkraut and Maio (1969) had observed enrichment of highly repeated DNA in nucleolar associated heterochromatin isolated from the nuclear fraction. It was Pardue and Gall (1970) and Jones (1970) who demonstrated by *in situ* hybridization that the satellite DNA in *M. musculus* was located in the centromeric heterochromatin. Soon after, positive correlation was shown in many organisms between the amount of highly repeated DNA and the amount of heterochromatin present (Arrighi *et al.*, 1970; Jones and Robertson, 1970; Botchan *et al.*, 1971; Gall *et al.*, 1971; Yunis and Yasmineh, 1971). Not only that, heterochromatin exhibited compositional heterogeneity too.

While the initial studies (Hennig and Walker, 1970; Southern, 1970) were suggestive of rigid species specificity, *de novo* appearance and fast turnover rate of satellite DNA, later investigations revealed the occurrence of identical repeated sequences in related species, e.g. in rodents (Sutton and McCallum, 1972; Rice and Straus, 1973; Fry and Salser, 1977; Brown and Dover, 1980; Sen and Sharma, 1980) and in *Drosophila* (Gall and Atherton, 1974; Peacock *et al.*, 1977; Yamamoto and Miklos, 1978).

Higher eukaryotes possess a variable and often substantial amount of heterochromatin. But the functional significance of this component of the genome has not been established. One of the characteristics of heterochromatin, i.e. of the highly repeated DNA sequences comprising them is interspecific and intraspecific variations. Besides variation in amount and composition between and within species, lack of sequences coding for proteins and almost neutral effect of varying amounts of heterochromatin on the phenotype created a general impression that this component of the genome is redundant. None the less, many hypotheses have been put forward for its probable functional roles even in the absence of any hard evidence.

In an exhaustive review on the functional aspects of satellite DNA and heterochromatin, John and Miklos (1979) have emphasized that, except for the effects of heterochromatin on the frequency and distribution of chiasmata at meiosis, the other assumptions are hardly tenable. Since heterochromatin is mainly centromeric in placement, many workers, including those studying molecular biology of satellite DNA, postulated that causal relationship may exist between pairing of homologous chromosomes at meiosis and heterochromatin or satellite DNA of centromeric regions. Miklos *et al.* (1980) have argued by citing many experimental evidences from *Drosophila* and also from the mouse that pairing is not affected when quantity or position of the centromeric heterochromatin is altered and also when their homology is disturbed. How critical is initiation in the pairing phenomenon is difficult to say, but it has been always noticed that segment or segments of chromosomes, which need not necessarily be large, when translocated to different positions, regularly pair at meiosis. Every region has pairing affinity with its homologous region irrespective of whether they are situated on the same chromosome or different chromosomes. It therefore appears that heterochromatin or satellite DNA of centromeric regions is not crucial for the pairing of homologous chromosomes.

Many authors suggested a role which depended on the amount of heterochromatin or highly repeated DNA present in the genomes and the occurrence of chromosome rearrangements in their karyotypic evolution. Miklos *et al.* (1980) critically evaluated this aspect also and made it convincingly clear that to facilitate chromosome rearrangements it is not essential to have large amounts of heterochromatin or satellite DNA.

The *Thomomys bottae* complex, well known for very high degrees of intra-and interspecific variations of heterochromatin, is perhaps the only actively speciating system studied fairly intensively among mammals for the evolutionary role of heterochromatin (reviewed by Patton, 1981). On observing minimal structural rearrangements (Robertsonian or non-Robertsonian type) of chromosomes in the *bottae*-group of pocket gophers possessing highly variable and enormous amount of heterochromatin, in contrast to the *talpoides* (a group with very small amounts of heterochromatin and exhibiting high levels of chromosome rearrangements), Patton and Sherwood (1982) concluded that they agreed with Miklos *et al.* (loc. cit.) that there was lack of substantive data implicating heterochromatin as an agent of speciation. However, on noticing compositional heterogeneity in the heterochromatin of these gophers, Barros and Patton (1985) have remarked that it remains to be determined if changes in the balance between two or more repetitive sequences in heterochromatin may have had more crucial evolutionary consequences than simple increases or decreases of a homogeneous repetitive sequence as has been suggested by Sen and Sharma (1980) for the mice from the Indian subcontinent.

The case of *bottae*-group of *Thomomys* is rather complicated. Besides heterochromatin variations between the races and species, chromosomal rearrangements exist, though not always. The heterochromatin variation, too, is very extensive. It therefore becomes difficult to analyse the influence of heterochromatin variation using different parameters and to arrive at any conclusion regarding the functional role of the heterochromatin. The *Mus dunni* chromosome polytypes, on the other hand, have variations involving only heterochromatin. Also, the extent of variation is not high, and that too, apparently has occurred in a stepwise manner. The system is suitable to investigate the role, if any, of the heterochromatin in speciation.

Mus dunni complex—an Interesting System for Studying the Role of Constitutive Heterochromatin/Highly Repeated DNA in Speciation

The Indian pygmy field-mice are among the most common rodents of the subcontinent which heavily infest cultivated paddy and wheat fields. They live in burrows and never invade human dwellings. They include two morphologically extremely similar species *Mus booduga* and *Mus dunni* which were considered conspecific until recently. They are closely allied to the aboriginal mice *Mus musculus* and are distinguished from each other only on the basis of average characters (Ellerman, 1961). After studying evolutionary trends of the African and Indo-Australian murids, Misonne (1969) suggested that the genus *Mus* probably originated in tropical Asia. *Mus auctor*, known from isolated teeth and one dentary fragment of more than 5 million years age found in Siwalik deposits in Pakistan, is suggested by Jacobs (1978) to represent the basic stock of the genus *Mus*.

The diploid number in all the three species *M. musculus*, *M. booduga* and *M. dunni* is 40, but while the karyotypes of *M. musculus* and *M. booduga* with all acrocentric chromosomes are identical, that of *M. dunni* is distinct due to the invariable presence of large submetacentric X and acrocentric Y sex chromosomes. In contrast to the fixed karyotype displayed by *M. booduga* collected from different places of India, the sympatric *M. dunni* had been found polymorphic from different localities (Matthey and Petter, 1968; Sharma and Garg, 1975; Markvong *et al.*, 1975; Manjunatha and Aswathanarayana, 1979). Our study over the years has shown that *M. dunni* is in active phase of evolutionary differentiation, in which constitutive heterochromatin seems to have played a predominant role (Sen and Sharma, 1980 and 1983; Sharma *et al.*, 1986).

The 3 chromosomal forms or types of *M. dunni* (Figure 1) so far obtained (Sharma *et al.*, 1986) are apparently parapatric in distribution, and while conducting experiment on mate-preference behaviour between them it has been found that they exhibit some degree of ethological isolation. The populations studied from widely separated places like Varanasi, Khajuraho and Jhansi in northern India, Pune in western, and Alipurdaur in eastern India, are characterized by all autosomes as acrocentric displaying minute but discernible C-band posi-

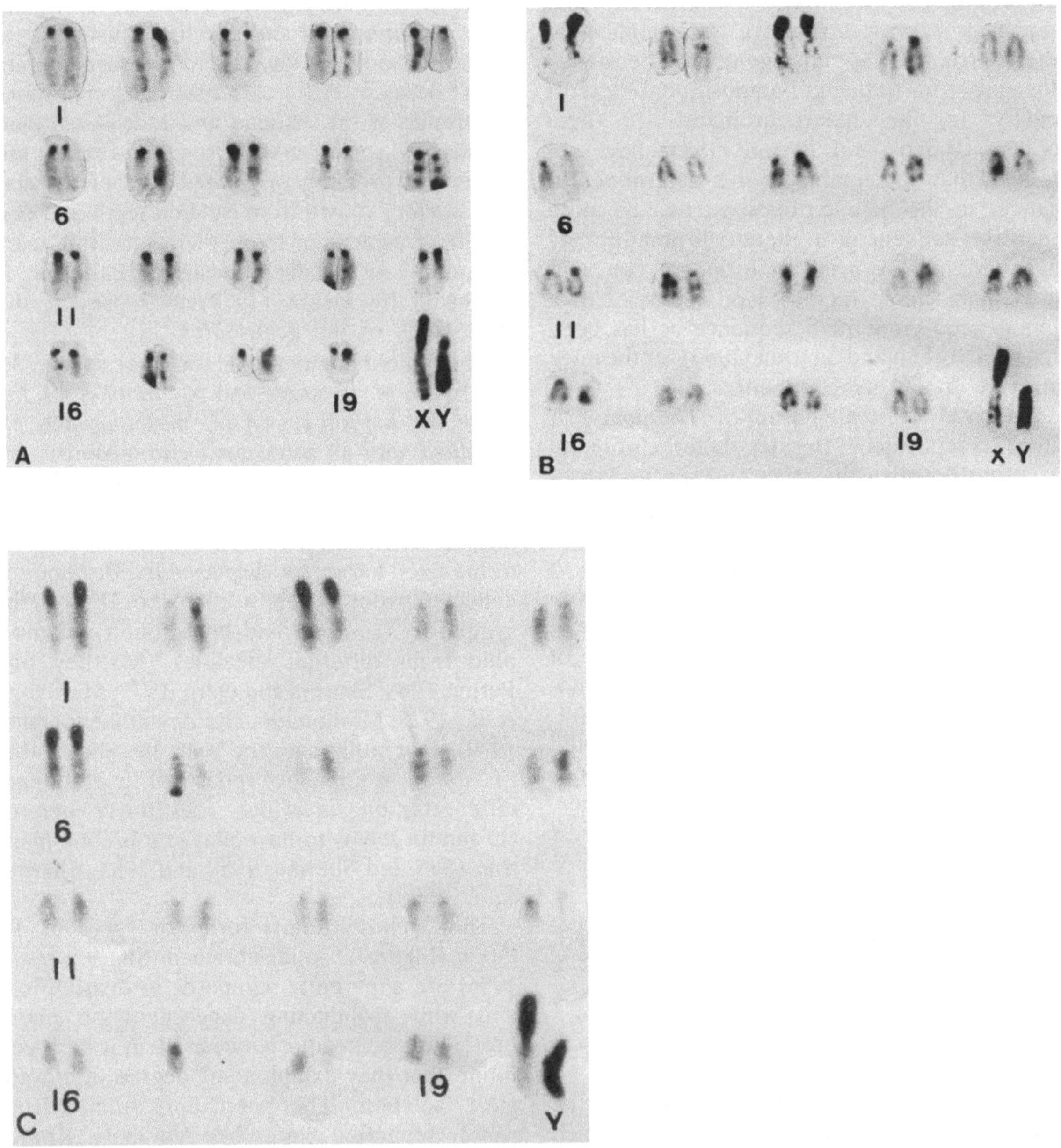

Figure 1: C-band male karyotypes of *M. dunni* chromosome types. Composite X and large Y chromosomes are present in all the threee chromosome types. (A) All autosomes of the type I are acrocentric possessing minute heterochromatic short arms; (B) with 2 pairs (1 and 3) of biarmed autosomes having prominent heterochromatic short arms in the type II; and (C) with 3 pairs (1, 3 and 6) of biarmed autosomes having prominent heterochromatic short arms in the type III.

tive short arms. This chromosomal form has been designated as chromosome type I. The populations of Mysore and Erode of south India, designated as chromosome type II, possess consistently two biarmed autosomal pairs, 1 and 3, with C-band positive short arms but other autosomes of the complement being acrocentric or telocentric. *M. dunni* individuals collected from Madras, Tirupati, Pondicherry and Madurai, all in south India, are of chromosome type III which invariably have the presence of one more autosomal pair (no. 6) with prominent heterochromatic short arms besides pairs 1 and 3. Nomenclature of the chromosomes is as per the G-band karyotype published by the Committee on Standardized Genetic Nomenclature for Mice (1972).

It is interesting to note that out of about 300 *M. booduga* studied karyologically from different places in north and south India, none showed any structural or numerical polymorphism. On the other hand, out of around 550 individuals of *M. dunni* studied from the same localities in the north and south showed a high degree of polymorphism. This polymorphism is besides the stable difference in the presence of heterochromatic short arms that distinguish the three chromosome types in homozygous condition (polytype). Out of more than 200 type III individuals studied so far, only in 4 or 5 did one of the homologues of pairs 1 and 6 had smaller C-band positive short arm. The variable numbers of biarmed autosomes of 4 to 6 in the chromosome type II and of 6 to 11 in the chromosome type III observed were due to polymorphism of pairs other than 1, 3 and 6, for which pericentric inversions in homozygous or heterozygous condition were responsible. The karyotypes analysed by G-banding had revealed that pairs 2 and 5 were involved in pericentric inversion. Occasionally the autosome pair 8 showed the presence or absence of distinct heterochromatic short arm. A total of 9 individuals belonging to all the chromosome types have been found so far to possess 39 chromosomes resulting from centric fusions of different chromosomes. Thus, in *M. dunni* the rate of visible chromosome rearrangements appears to be very high as compared with 1 in 500 individuals normally observed in such diverse groups as man and grasshoppers (White, 1973). At any rate, it is clear from the above that centric fusions and pericentric inversions occur in *M. dunni* very frequently but they always exist only in the polymorphic state without getting fixed in the population.

Comparative G-band analyses published by us at the level of 150 to 200 bands per haploid set (Sen and Sharma, 1983; Sharma *et al.*, 1986) revealed a striking conservation of euchromatic segments of the genomes of *M. musculus, M. booduga* and all the 3 chromosome types of *M. dunni*, which meant that no major structural changes involving euchromatin took place during the evolutionary divergence of these taxa. This has been further confirmed by comparison at higher resolutions of 260 and 370 bands (Figure 2 and Cheong and Sharma, unpublished). The relative values of chromosomal DNA contents measured with Leitz MPV3 cytofluorimeter (Sharma *et al.*, 1986) have been corroborated by measuring nuclear DNA values after staining with ethidium bromide; the mean values being 0.86 ± 0.01 in *M. booduga* and 0.89 ± 0.01, 1.22 ± 0.33 and 1.03 ± 0.02 in *M. dunni* chromosome types I, II and III, respectively, relative to a value 1.0 for *M. musculus*. Thus it appears the change in the genomic size is essentially due to the variation of constitutive heterochromatin.

The centromeric regions of all the chromosomes except the Y in both *M. musculus* and *M. booduga* are C-band positive. The Y in optimally stained plates looks totally heterochromatic. In the three chromosome types of *M. dunni*, the centromeric regions of acrocentric and the entire short arms of the biarmed autosomes are C-band positive. The whole short arm of the X chromo-

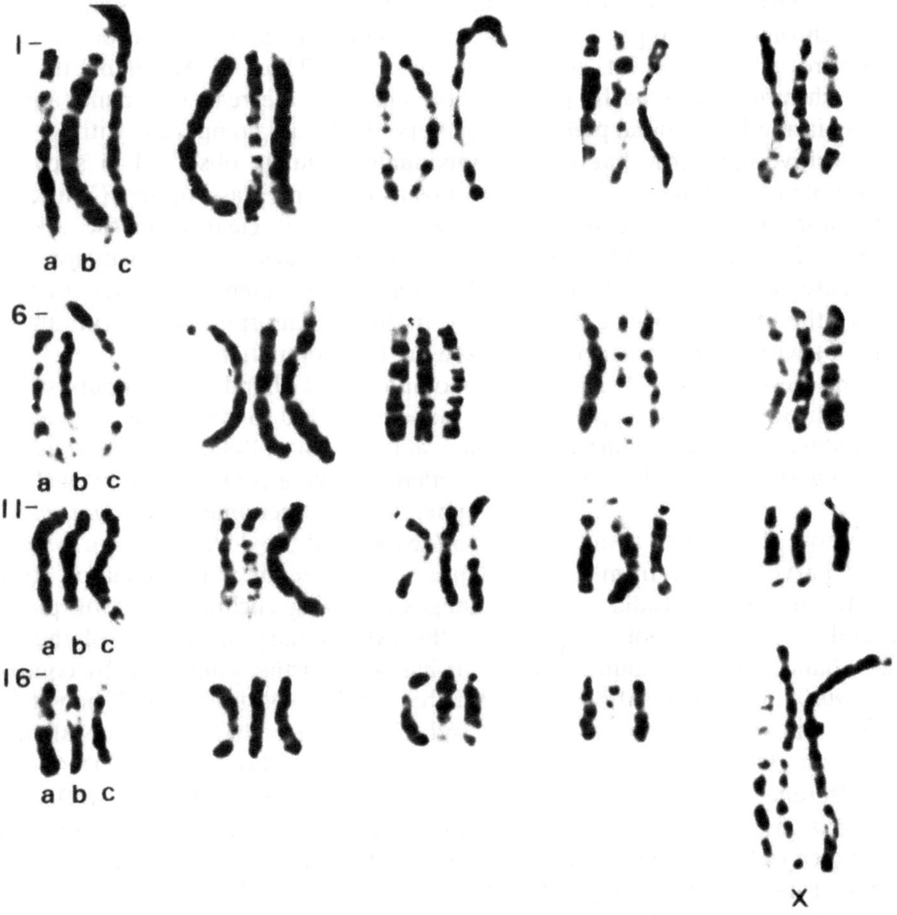

FIGURE 2: Composite G-band karyotype of 370 band stage showing homology of euchromatic portions of (a) *M booduga;* (b and c) *M. dunni* chromosome types I and III, respectively.

some as well as the terminal region of its long arm and the Y in its entirety are heterochromatic. On staining with different AT- or GC-specific fluorochromes the heterochromatin of these various taxa showed compositional heterogeneity (Sen and Sharma, 1983). The centromeric heterochromatin of all the chromosomes other than that of the Y in *M. musculus* and *M. booduga* fluoresced differentially brighter suggesting AT-rich nature of the heterochromatin. In contrast, in *M. dunni* brightly fluorescin with Hoechst 33258 '*M. musculus*-like' heterochromatin is confined as small segment(s) in the short arm of the X chromosome. This highly reduced '*M. musculus*-like' heterochromatin restricted to the X chromosomes in *M. dunni* exhibits polymorphism, both in quantity and location within the X, in individuals of all the three chromosome types (Sharma *et al.*, 1989). In case of *M. booduga*, although it displays a pattern similar to that of *M. musculus* with respect to Hoechst dye fluorescence, the intensity of fluorescence at the centromeric regions in *M. booduga* is much less. Thus

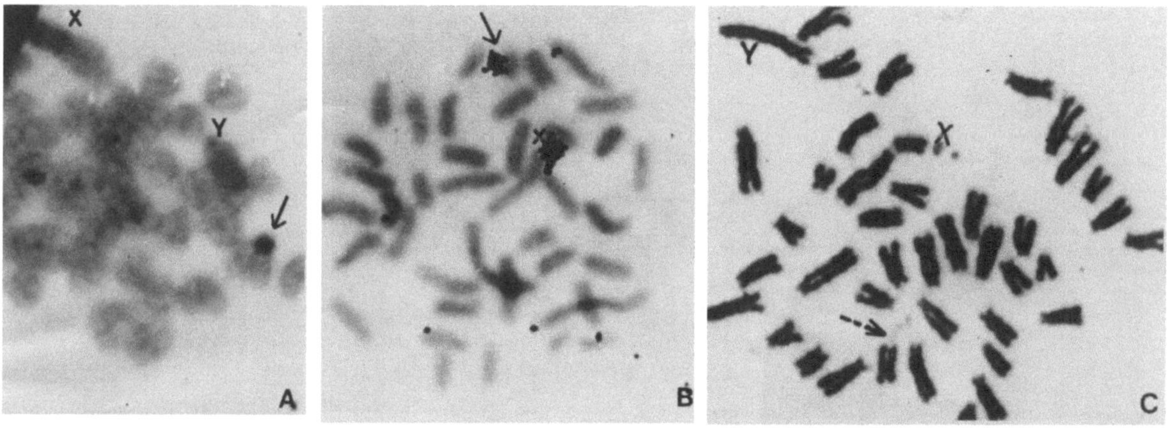

FIGURE 3: *M. dunni* metaphase plates (A) single autosome with '*M. musculus*–like' centromeric heterochromatin (arrow) and two small segments in the short arm of the X chromosome darkly stained with Giemsa after digestion with Alu I; (B) *in situ* hybridization with *M. musculus* satellite DNA as probe showing grains over the centromeric heterochromatin of the single autosome (arrow) and the small segments of the short arm of the X; and (C) condensation inhibition of the same regions with Hoechst 33258.

Hoechst 33258 fluorescence shows quantitative variation of '*M. musculus*-like' constitutive heterochromatin in *M. booduga* and *M. dunni.* Our investigations on condensation inhibition with Hoechst 33258 and *in situ* hybridization with *M. musculus* uncloned and cloned total satellite DNA probes (Sen and Sharma, 1980 and 1983; Balajee and Sharma, unpublished) have corroborated this. An interesting and unique feature has been observed in *M. dunni:* the presence of a complete block of *M. musculus*-like' centromeric heterochromatin (Hoechst fluorescing) in one (and on one occasion in two) autosomes obtained in 8 individuals belonging to different chromosome types (Sen and Sharma, 1983; Sharma *et al.*, 1986). The composition of this heterochromatin was confirmed by *in situ* hybridization with *M. musculus* satellite DNA probe as well as by digestion with Alu I restriction enzyme (Figure 3). What is remarkable, however, is that the qualitative shift in the composition of the centromeric heterochromatin appears to have occurred in an individual chromosome-specific manner; wholesale genomic revisions did not apparently take place.

Alu I restriction endonuclease digestion analysis on metaphase preparations has further corroborated our earlier observation. The efficiency of cleavage by restriction endonucleases both on fixed and unfixed chromatin has been fairly well established. The fact that many of these enzymes produce bands comparable to G-and C-bands on the metaphase chromosomes has already been looked into in a wide variety of organisms (see Miller and Miller, this volume). Alu I, whose recognition site is 5 'AG ↓ CT3 ↓ ', leaves centromeric heterochromatin intact in *M. musculus* , and when stained with Giemsa it resembles C-banding (Miller *et al.*, 1984). On digestion with Alu I, we have observed, *M. booduga* shows a comparable result to that of *M. musculus*, but the bands produced by Alu I in *M. booduga* are less pronounced (Figure 4). A few of its chromosome pairs display faintness or absence of the bands in the centromeric regions, which is in accordance with our earlier observation of condensation inhibtion with Hoechst

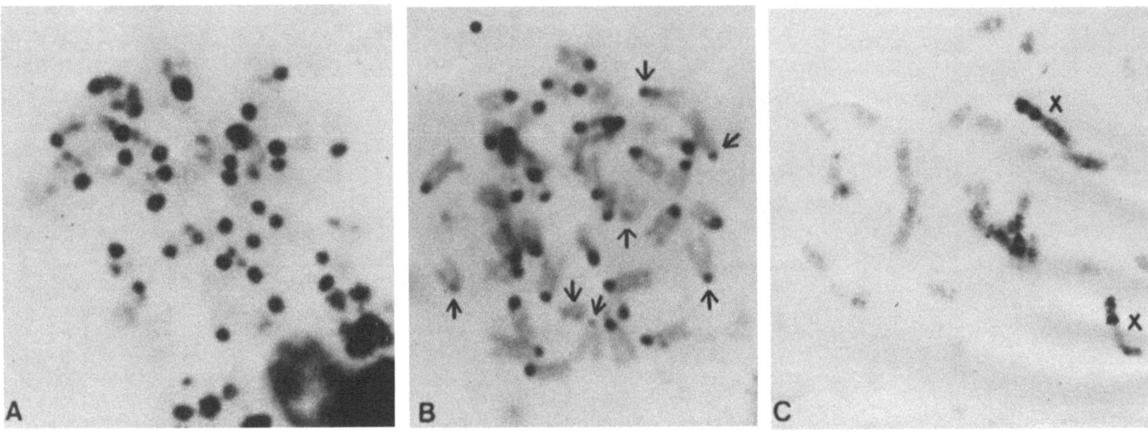

FIGURE 4: Metaphase chromosome preparations stained with Giemsa after digestion with Alu I. (A) *M. musculus*; (B) *M. booduga*; and (C) *M. dunni*.

33258 and *in situ* hybridization with *M. musculus* satellite DNA. In *M. dunni*, as expected, darkly stained regions after Alu I digestion are confined to small segmental regions mainly in the short arm of the X chromosome (Figure 4c). From the highly variable location of this heterochromatin observed in the X chromosome of different individuals of all the three chromosome types of *M. dunni* it seems, as if quite a chunk of it could get 'transposed' rather easily. It will be highly interesting to look into the sequence organization of this satellite DNA which might elucidate the probable mechanism involved in its frequent shift in position. The different approaches, i.e., condensation inhibition with Hoechst 33258, *in situ* hybridization of satellite DNA and Alu I digestion, have confirmed the presence of varying amounts of *M. musculus* related AT-rich satellite DNA in *M. booduga* and *M. dunni*. Obviously, in their genomes, besides varying amounts of this satellite DNA, other highly repeated sequences of their own are present and some may be even common to them. This remains to be investigated.

Hörz and Zachau (1977) had shown that the AT-rich mouse satellite DNA is cut into 'type A' and 'type B' patterns by different restriction enzymes. In the type A pattern essentially all the satellite sequences are cut into a regular series of fragments based on a monomeric unit of 234 bp (established subsequently by Hörz and Altenburger in 1981) by restriction enzymes like Eco RII, Ava II and Sau 96 I. The type B pattern is produced when only part of the total satellite DNA is digested into fragments of the same periodicity but with a size distribution that is skewed towards higher multiples by Eco RI, Alu I, Hinf I, Taq I, Hae III, etc. Southern hybridization experiments done by us earlier (Sharma *et al.*, 1986) using ^{32}P-labelled *M. musculus* uncloned total satellite DNA as probe, had revealed that in *M. booduga* with type A pattern producing enzyme Eco RII the homologous DNA was digested to monomer, dimer, trimer, etc., but as compared with *M. musculus* there was a distinct reduction of type A segments. In *M. dunni* the type A pattern was not detected at all. When digested with enzymes producing type B pattern, the proportion of sequences in all the three species cut by Hinf I were comparable, but the proportion of Alu I sites varied showing distinctly lower than Hinf I sites both in *M. booduga* and *M. dunni* and with difference in higher and lower multiples of monomers and in relative abundance of multimer repeat sizes.

Southern hybridization experiments done subsequently with cloned total satellite DNA using type A pattern producing Ava II and Sau 96 I enzymes and type B pattern producing Taq I, Eco RI, Hae III, Pst I, Bgl II, etc., enzymes (Balajee and Sharma, unpublished), have revealed comparable type B segments in *M. booduga* and all the three chromosome types of *M. dunni* with only slight differences here and there in the proportion of enzyme target sites and/or in the relative abundance of monomer and multiple repeats. When digested with Taq I there are hardly any monomer or dimer fragments, and relatively less abundant higher multiple fragments in *M. booduga* in comparison with *M. dunni* (Figure 5). In striking contrast to our earlier observation of absence of type A segments in *M. dunni* when digested with Eco RII, there are monomers and ascending series of multimers when digested with both Ava II and Sau 96I in the genomes of all the three chromosome types of *M. dunni* (Figure 5). The segment patterns are identical after digesting with Ava II and Sau 96I and with both the enzymes it is noticed that the type A segments are less in amount in *M. booduga* than in *M. musculus* and still far less in all the chromosome types of *M. dunni*. The number of subrepeats is different from *M. musculus* but is almost identical in *M. booduga* and *M. dunni*, which signifies that though there is a difference in the quantity of type A segments in *M. booduga* and *M.*

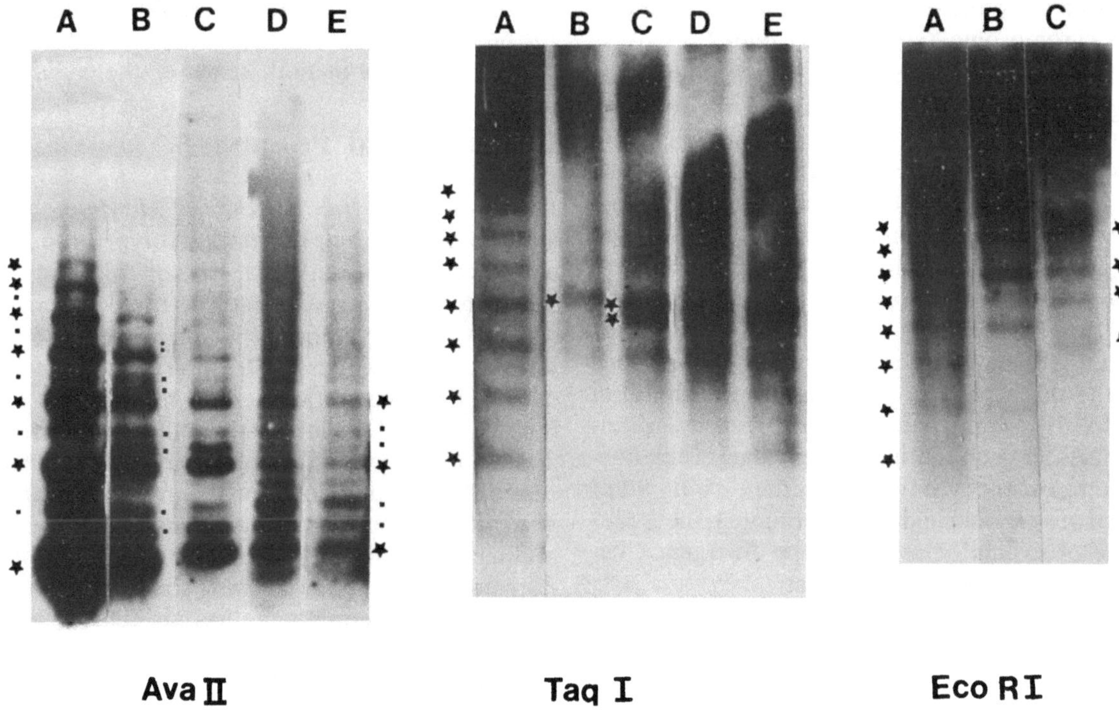

Figure 5—Southern hydridization with *M. musculus* cloned satellite DNA as probe after digesting the genomic DNAs with restriction enzymes. Different lanes were exposed for different times to facilitate comparison of array of fragments produced. Asterisks indicate the monomer and multiples and dots indicte the sub-repeats. Lane A: *M. musculus*. Lane B: *M. booduga*. Lanes C, D and E: *M. dunni* chromosome types I, II and III, respectively.

dunni, they are closer to each other in the basic organization of the satellite DNA. The slight variation observed in the subrepeat patterns between *M. booduga* and *M. dunni*, and even between the three chromosome types of *M. dunni*, reflects that the segmental sequence variants arose after diversification of these taxa. The striking contrast observed earlier in the absence of type A segments in *M. dunni* on treating with Eco RII can be explained by assuming that the amplification of the sequence variant to whatever extent (which was not very great) occurred in the genomes of *M. dunni* after the change in the sequence within the target site of Eco RII had taken place. Since this has been observed in north as well as south Indian populations of *M. dunni*, it will be reasonable to assume that this segmental sequence variation took place before evolutionary differentiation of the chromosome types of *M. dunni*.

The difference observed in the subrepeat organization in *M. musculus* of the type A pattern on digestion with Ava II and Sau 96I in comparision with almost identical patterns in *M. booduga* and the three chromosome types of *M. dunni* on one hand, and the absence of the type A pattern in *M. dunni* taxa on digestion with Eco RII in contrast to both *M. musculus* and *M. booduga* on the other, may be useful in finding the sequence of evolutionary divergence of the three species. Analysis of mitochondrial DNA digestion pattern with restriction endonucleases and immunological analysis by microcomplement fixation, together with other evidence, should aid in determining the sequence of events of evolutionary divergence besides their time of divergence.

The observed behaviour of *M. dunni* females in estrus exhibiting a striking preference to males of their own chromosomal type, strongly suggests that some degree of ethological isolation is already in operation between the different chromosome types. A study on mate preference behaviour in individuals of *M. dunni* chromosomes types I, II and III with females as the choosers was conducted using especially designed cages where males and females were not allowed to come into physical contact (Sharma *et al.*, 1986; Cheong, 1986). Individuals of both sexes belonging to different chromosomal types when allowed to mix were invariably involved in fighting and often went to the extent of killing. Our earlier attempts to obtain hybrids between the chromosome types had failed but lately it has been possible (Bahadur and Sharma, preliminary observation). It has now become experimentally feasible to seek answers to questions which have a bearing on the problem related to heterochromatin's role in the speciation process. In particular, one can try to examine critically the effects of qualitative and/or quantitative variations of centromeric heterochromatin or highly repeated DNA between the homologous chromosomes on segregation by studying meiosis in both sexes of the hybrids.

Inferences and Prospects

In the different populations of *M. dunni* the stability seen in their diverse karyotypes and the existence of some degree of ethological isolation between them suggest that the three chromosomal types are potentially incipient species. It is difficult to say whether the ethological isolation developed concomitantly with the fixation of homozygosity for variations of the heterochromatin, or pre-mating preceded in them. In the case of the 24- and 26-chromosomal races of *M. musculus* post-mating isolation can occur due to reduction in fecundity of the hybrids, but Capanna (1985), after studying social aggression behaviour between them, has speculated that pre-mating isolation may get rapidly selected in sympatric or parapatric contact since the price to be paid in terms of a decrease in the reproductive potential of the populations is too high if the reproductive isolation is entrusted only to post-mating mechanisms. In *S. ehrenbergi* mate selection efficiency has been noted to vary among the chromosomal species in accordance with their time of origin, suggesting that ethological isolation accumulates slowly and gradually (Nevo, 1986). In case of *M. dunni,* whether it is pre or post-mating isolation will be clear on further investigation.

Unlike structural rearrangements, variations in heterochromatin have been generally assumed not to cause reproductive isolation because it is taken that such variations do not lead to reduction in fecundity of the heterozygotes. White (1982), however, remarked that changes in satellite DNA do not lead directly to cladogenesis or splitting of pre-existing ones, but that does not mean that they play no part in speciation. How it plays such role(s) remains to be established. It is clear that pairing between homologous chromosomes is not affected by quantitative and/or qualitative alterations of centromeric heterochromatin (Miklos *et al.*, 1980). However, it might be of interest to study effects of such variations on segregation of chromosomes at meiosis from the viewpoint of the speciation process. Lica *et al.* (1986) have shown, though in a mouse cell line L929, that the presence of satellite DNA in centromeric regions is necessary for maintenance of contact between sister chromatids. Any system in which the influence, if any, of difference in heterochromatin of centromeric and of adjacent regions can be ascertained on segregation of homologues at meiosis by analysing metaphase II in both sexes for occurrence of nondisjunction, might throw light on a functional role of heterochromatin. There is possibility of determining this in the hybrids between the chromosomal types of *M. dunni.* If it is found to show a significant effect, reduction in fecundity and viability due to aneuploidy will result in the hybrids and thus variations in heterochromatin or the tandem array of highly repeated DNA can also be causal to speciation similar to structural rearrangements, which may lead to cladogenesis.

In the actively speciating *Caledia captiva* species complex of grasshoppers, the highly repeated DNA sequences of the heterochromatin present as large interstitial and telcomeric blocks or as small centromeric and telomeric bands which exhibit variations in different taxa indicate various functional roles of evolutionary significance (Arnold and Shaw, 1985). The two tandemly repeated 168 bp and 144 bp sequence families present have a varying number of copies in different populations. An overall negative correlation between the quantity of the 168 bp sequence and the levels of reproductive isolation and genic divergence between the various taxa was observed, but they have presumed by referring to the finding by Shaw *et al.* (1985) that the increase or decrease of a specific sequence family might have the potential for modulating the cell cycle parameters which could result in a differential pattern of development and thereby affect a change in the phenotype of the grasshopper. A significant difference of 20 per cent in embryonic weights between the offspring of different taxa was found which, they speculated, was not due to any difference in the overall time of embryogenesis but might be the consequence of fluctuation in the cell cycle time. Furthermore, they noticed reduction in viability of the hybrid individuals when crossed between different taxa, which are distinctive in the position and amount of the 168 bp sequence within their respective genomes (Arnold and Shaw, 1985). The causal factors could have been genic and structural differences but, according to them, the reduction in viability attributed to the 'genic' component could be partly owing to the differences in the amount or distribution of the 168 bp sequences though it is not possible to separate the effects of these two components on hybrid viability at present. To examine this, the *M. dunni* system appears promising, since polymorphism of highly repeated AT-rich satellite DNA exists within and between the chromosome types.

Peacock *et al.* (1982) have speculated that an increase of particular repeated sequences could result in alterations of chromosomal positioning in the nucleus, and if spatial orientation influences gene interactions, there could be major effects on gene expression due to alterations in chromosome positioning which may have potential for involvement in speciation. For examining such nature of influence, the feasibility of obtaining hybrids between the chromosome types of *M. dunni* with variation in substantial amount of heterochromatin as short arms in particular chromosomes of the complements will be suitable, for which the knowledge of linkage map of the closely allied species *M. musculus* will be of tremendous advantage since the euchromatic portions of the genomes of these species had undergone no major structural changes.

The presence of composite submetacentric X and large acrocentric Y in all the chromosome types of *M. dunni* known so far strongly indicates that acquisition of the heterochromatin in the sex chromosomes might have occurred concomitantly with the evolutionary differentiation of the sympatric *M. dunni* and *M. booduga*. The difference observed in the burrow patterns and preference for sites within the same cultivated fields between *M. dunni* and *M. booduga* (Cheong, 1986) does not rule out the possibility that the sympatric distribution throughout the subcontinent could be a subsequent development. It needs to be further investigated at molecular, chromosomal, cellular, organismal and population levels to know the exact mode of speciation involved. Till our knowledge about the genetic relationships and time of divergence improves, it is difficult to make out the direction of evolutionary change and to know the precise phylogenetic relationships between them. The presence of identical sex chromosomes in all the three chromosomes types and the increase or decrease of heterochromatin as short arms in homozygous condition in pairs 1 and 3 in the type II and pairs 1, 3 and 6 in the type III, leaves hardly any doubt that the evolutionary differentiation of these taxa took place in a stepwise manner. When the zones of contact between these taxa are located, information on the prevailing conditions of the chromosome types there will be helpful in understanding the mode of speciation involved. The absence of any obvious geographical barriers between the chromosome types of *M. dunni* makes the classical allopatric mode unlikely. Our multidimensional study in progress may unravel whether stasipatric or invasive mode (White, 1982) or an entirely novel mode of speciation was involved. One should see if the *M. dunni* complex, like *S. ehrenbergi* (Nevo, 1986), also represents a pluralistic example. At any rate, the actively speciating *M. dunni* complex is an extremely interesting model system to establish the functional role of heterochromatin or tandemly arrayed highly repeated DNA in speciation.

Acknowledgement

The work was supported by the financial assistance received from the University Grants Commission and Department of Atomic Energy, India, and the GSF, West Germany.

References

Arnold ML, Shaw DD (1985). The heterochromatin of grasshoppers from the *Caledia captiva* species complex II. Cytological organization of tandemly repeated DNA sequences. Chromosoma 93; 183–90.

Arrighi FE, Hsu TC, Saunders PP, Saunders GF (1970). Localization of repetitive DNA in the chromosomes of *Microtus agrestis* by means of *in situ* hybridization. Chromosoma 32; 224–36.

Barros MA, Patton JL (1985). Genome evolution in pocket gophers (Genus *Thomomys*) III. Fluorochrome-revealed heterochromatin heterogeneity. Chromosoma 92; 337–43.

Botchan M, Kram R, Schmid CW, Hearst JE (1971). Isolation and chromosomal localization of highly repeated

DNA sequences in *Drosophila melanogaster*. Proc Natl Acad Sci USA 68; 1125–9.

Britten RJ, Kohne DE (1968). Repeated sequences in DNA. Science 161; 529–40.

Britton-Davidian J, Bonhome E, Croset H, Capanna E, Thaler L (1980). Variabilite genetique chez les populations de souris (genure *Mus* L.) a nombre chromosomique reduit. CR Acad Sci 290; 195–8.

Brown SDM, Dover GA (1980). Conservation of segmental variants of satellite DNA of *Mus musculus* in a related species: *Mus spretus*. Nature 285; 47–9.

Brutlag DL (1980). Molecular arrangement and evolution of heterochromatic DNA. Ann Rev Genet 14; 121–44.

Capanna E (1976). Genetic aneuploidy in the mouse hybrids. Chromosomes Today 5; 83–9.

Capanna E (1982). Robertsonian numerical variation in animal speciation: *Mus musculus*, an emblematic model. In : Barigozzi CB (ed.) Mechanisms of Speciation. Alan R Liss, Inc., New York; 155–77.

Capanna E (1985). Karyotype variability and chromosome transilience in rodents: The case of the genus *Mus*. In: Luckett WP, Hartenberger JL (eds) Evolutionary relationships among rodents. Plenum, New York ; 643–69.

Capanna E, Corti M (1982). Reproductive isolation between two chromosomal races of *Mus musculus* in the Rhaethian Alps (northern Italy). Mammalia 46; 107–9.

Capanna E, Corti M, Mainardi D, Parmigiani S, Brain PF (1984). Karyotype and intermale aggression in wild house mice: Ecology and speciation. Behavior Genetics 14; 195–208.

Capanna E, Corti M, Nascetti G (1985). Role of the contact areas in the chromosomal speciation of the European long tailed house mouse (*Mus musculus domesticus*). Boll Zool 52; 153–76.

Cheong N (1986). Evolutionary divergence in Indian pygmy mice: Cytogenetics, habitat and behaviour studies. Ph D thesis, Banaras Hindu University.

Committee on Standardized Genetic Nomenclature for Mice (1972). Standard karyotype of the mouse, *Mus musculus*. J Hered 63; 69–72.

Corti M, Capanna E, Estabrook GF (1986). Microevolutionary sequences in house mouse chromosomal speciation. Syst Zool 35; 163–75.

Duffey PA (1972). Chromosome variation in *Peromyscus*: a new mechanism. Science 176; 1333–4.

Ellerman JR (1961). Fauna of India, Mammalia, Vol. III, Rodentia, 2nd edn. In: Roonwal ML (ed.) Zoological Survey of India.

Fedyk S (1980). Chromosome polymorphism in a population of *Sorex araneus* L. at Biolowieza. Folia Biol 28; 83–120.

Ford CE, Evans EP (1973). Robertsonian translocation in mice: segregational irregularities in male heterozygotes and zygotic imbalance. Chromosomes Today 4; 387–97.

Ford CE, Hamerton JL (1970). Chromosome polymorphism in the common shew, *Sorex araneus*. Symp Zool Soc Lond 26; 223–36.

Fry K, Salser W (1977). Nucleotide sequence of HS alpha satellite DNA from Kangaroo rat *Dipodomys ordii* and characterization of similar sequences in other rodents. Cell 12; 1069–84.

Gall JG, Atherton DD (1974). Satellite DNA sequences in *D. virilis*. J Mol Biol 85; 663–4.

Gall JG, Cohen EH, Polan ML (1971). Repetitive DNA sequences in *Drosophila*. Chromosoma 33; 319–44.

Gamperl R, Ehmann C, Bachmann K (1982). Genome size and heterochromatin variation in rodents. Genetica 58; 199–212.

Gropp A, Tettenborn U, von Lehman E (1970). Chromosomenvariation von Robertson'schen typus bei der Tabakmaus *M. poschiavinus* und ihren Hybriden mit der Labortoriumsmans. Cytogenetics 9; 9–23.

Gropp A, Winking H (1981). Segregation pattern and biological consequences of Robertsonian translocations: Cytology, meiosis. In: Berry RJ (ed.) Biology of the House Mouse. Academic Press, New York; 141–81.

Hall WP (1973). Comparative population cytogenetics, speciation, and evolution of the iguanid lizard Genus *Sceloporus*. Ph D Thesis, Harvard University.

Hausser J, Catzeflis F, Meylan A, Vogel P (1985). Speciation in the *Sorex araneus* complex (Mammalia: Insectivora). Acta Zool Fennica 170; 125–30.

Hawes ML (1976). Odor as a possible isolating mechanism in sympatric species of shrews (*Sorex vagrans* and *Sorex obscurus*). J Mammal 57; 404–7.

Henning W, Walker PMB (1970). Variations in the DNA from two rodent families (Cricetidae and Muridae). Nature 225; 915–9.

Heth G, Nevo E (1981). Origin and evolution of ethological isolation in subterranean mole rats. Evolution 35; 259–74.

Hörz W, Altenburger W (1981). Nucleotide sequence of mouse satellite DNA. Nucleic Acids Res 9; 683–96.

Hörz W, Zachau HG (1977). Characterization of distinct segments in mouse satellite DNA by restriction nucleases. Eur J Biochem 73; 383–92.

Jacobs LL (1978). Fossil rodents (Rhizomyidae and Muridae) from Neogene Siwalik deposits, Pakistan. Museum of Northern Arizona Press Bulletin Series 52; 1–103.

John B, Miklos GLG (1979). Functional aspects of satellite DNA and heterochromatin. In: Bourne GH, Danielli JF (eds) International Rev Cytol. Academic Press 58; 1–114.

Jones KW (1970). Chromosomal and nuclear location of mouse satellite DNA in individual cells. Nature 225; 912–5.

Jones KW, Robertson FW (1970). Localization of reiterated nucleotide sequence in *Drosophila* and mouse by *in situ* hybridization of complementary RNA . Chromosoma 31; 331–45.

Lica LM, Narayanswami S, Hamkalo BA (1986). Mouse satellite DNA, centromere structure and sister chromatid pairing. J Cell Biol 103; 1145–51.

Manjunatha KR, Aswathanarayana NV (1979). Studies on the chromosomes of the genus *Mus:* autosomal polymorphism in the Indian pygmy mouse *Mus dunni* (Wroughton). Curr Sci 48; 657–9.

Markvong A, Marshall JT, Pathak S, Hsu TC (1975). Chromosomes and DNA of *Mus*: The karyotype of *M. fulvidiventris* and *M. dunni.* Cytogenet Cell Genet 14; 116–25.

Matthey R, Petter F (1968). Existence de deux especes distinctes, L'une chromosomiquement polymorphe chez der *Mus* Indiens der groups *booduga.* Etude cytogenetique ettaxonomique. Rev Suisse Zool 75; 461–98.

Mayr E (1969). Species, speciation and chromosomes. In: Benirschke K (ed) Comparative Mammalian Cytogenetics. Springer-Verlag, New York; 1–7.

Miklos GLB, Willcocks DA, Baverstock PR (1980). Restriction endonuclease and moclecular analyses of three rat genomes with special reference to chromosome rearrangement and speciation problems. Chromosoma 76; 339–63.

Miller DA, Gosden JR, Hastie ND, Evans HJ (1984). Mechanism of endonuclease banding of chromosomes. Exp Cell Res 155; 294–8.

Miller DA, Miller OJ (1989). Restriction enzyme banding of metaphase chromosomes. In present volume.

Misonne X (1969). African and Indo-Australian muridae, evolutionary trends. Annls Mus Afr Cent Ser Zool 172; 1–219.

Nevo E (1969). Mole rat *Spalax ehrenbergi*: Mating behaviour and its evolutionary significance. Science 163; 484–6.

Nevo E (1979). Adaptive convergence and divergence of subterranean mammals. Ann Rev Ecol Syst 10; 269–308.

Nevo E (1982). Speciation in subterrean mammals. In: Barigozzi C (ed.) Mechanism of Speciation. Alan R Liss, Inc., New York; 191–218.

Nevo E (1986). Mechanisms of adaptive speciation at the molecular and organismal levels. In: Karlin S, Nevo E (eds) Evolutionary Processes and Theory. Academic Press Inc., Orlando; 439–74.

Nevo E, Cleve H (1978). Genetic differentiation during speciation. Nature 275; 125–6.

Nevo E, Heth G, Beiles A (1982). Populations structure and evolution in subterranean rats. Evolution 36; 1283–9.

Nevo E, Naftali G, Guttmann R (1975). Aggression patterns and speciation. Proc Natl Acad Sci USA 72: 3250–4.

Nevo E, Shaw CR (1972). Genetic variation in a subterranean mammal, *Spalax ehrenbergi.* Biochem Genet 7; 235–41.

Pardue ML, Gall JG (1970). Chromosomal localization of mouse satellite DNA. Science 168; 1356–8.

Pathak S, Hsu TC, Arrighi FE (1973). Chromosomes of *Peromyscus* (Rodentia: Cricetidae) IV. The role of heterochromatin in karyotype evolution. Cytogenet Cell Genet 12; 315–26.

Patton JL (1981). Chromosomal and genic divergence, population structure and speciation potential in *Thomomys bottae* pocket gophers. In: Reig OA (ed) Ecology and Genetics of Animal Speciation (Ecologiay Genetica de la Especiacion Animal). Universidad Siman Bolivar, Caracas, Venezuela; 255–95.

Patton JL, Sherwood SW (1982). Genome evolution in pocket gophers (*Genus Thomomys*). I. Heterochromatin variation and speciation potential. Chromosoma 85; 149–62.

Peacock WJ, Apples R, Dunsmuir P, Lohe AR, Gerlach WL (1977). Highly repeated DNA sequences: chromosomal localization and evolutionary conservation. In: Brinkely BR, Porter KR (eds) International Cell Biology. Rockefellar Univ Press; 494–506.

Peacock WJ, Dennis ES, Gerlach WL (1982). DNA sequence changes and speciation. In: Barigozzi CB (ed) Mechanisms of speciation. Alan R Liss, Inc., New York; 123–42.

Peacock WJ, Dennis ES, Hilliker AJ, Pryor AJ (1981). Differentiation of heterochromatin. In: Atchley WR, Woodruff DS (eds) Evolution and Speciation. Cambridge University Press; 78–100.

Rice NR, Straus NA (1973). Relatedness of mouse satellite deoxyribonucleic acid to deoxyribonucleic acid of various *Mus* species. Proc Natl Acad Sci USA 70; 3546–50.

Schildkraut CL, Maio JJ (1969). Fractions of HeLa DNA differing in their contents of guanine + cytosine. J Mol Biol 46; 305–12.

Searle JB (1984). Nondisjunction frequencies in Robertsonian heterozygotes from natural populations of the common shrew, *Sorex araneus* L. Cytogenet Cell Genet 38, 265–76.

Searle JB (1988a). Karyotypic variation and evolution in the common shrew, *Sorex araneus.* Proceedings Kew Chromosome Conference III, HMSO; 97–107.

Searle JB (1988b). Selection and Robertsonian variation in nature: The case of the common shrew. In: The Cytogenetics of Mammalian Autosomal Rearrangements. Alan R Liss, Inc., New York; 507–31.

Sen S, Sharma T (1980). Quantitative variation of 'Mus musculus-like' constitutive heterochromatin and satellite DNA-sequences in the genus *Mus.* Chromosoma 81; 393–402.

Sen S, Sharma T (1983). Role of constitutive heterochromatin in evolutionary divergence: results of chromosome banding and condensation inhibition studies in *Mus musculus, Mus booduga* and *Mus dunni*. Evolution 37; 628–36.

Sharma T, Cheong N, Sen P, Sen S (1986). Constitutive heterochromatin and evolutionary divergence of *Mus dunni, M. booduga* and *M. musculus*. In: Potter M, Nadeau JH, Cancro MP (eds) The Wild Mouse in Immunology. Springer-Verlag, Berlin, Heidelberg; 35–44.

Sharma T, Balajee AS, Cheong N (1989). Differential response to Alu I digestion between the two X chromosomes of female of *Mus dunni* and polymorphism of *Mus musculus*-like heterochromatin. Abstract 10th International Chromosome Conference Uppsala, Sweden; 106.

Sharma T, Garg GS (1975). Constitutive heterochromatin and karyotype variation in Indian pygmy mouse, *Mus dunni*. Genet Res (Camb) 25; 189–91.

Sharma T, Raman R (1973). Variation of constitutive heterochromatin in the sex chromosomes of the rodent *Bandicota b. bengalensis* (Gray). Chromosoma 41; 75–84.

Shaw DD, Coates DJ, Arnold ML, Wilkinson P (1985). Temporal variation in the chromosomal structure of a hybrid zone and its relationship to karyotypic repatterning. Heredity 55; 293–306.

Southern DI (1970). Polymorphism involving heterochromatic segments in *Metrioptera brachyptera*. Chromosoma 30; 154–68.

Sutton WD, McCallum M (1972). Related satellite DNA's in the genus *Mus*. J Mol Biol 71; 633–56.

Wahrman J, Goiten R, Nevo E (1969a). Georgraphic variation of chromosome forms in *Spalax*, a subterranean mammal of restricted mobility. In: Benirschke K (ed.) Comparative Mammalian Cytogenetics. Springer-Verlag, New York; 30–48.

Wahrman J, Goiten R, Nevo E (1969b). Mole rat *Spalax*: Evolutionary significance of chromosome variation. Science 164; 82–4.

White MJD (1968). Models of speciation. Science 159; 1065–70.

White MJD (1969). Chromosomal rearrangements and speciation in animals. Ann Rev Genet 3; 75–98.

White MJD (1973). Animal Cytology and Evolution. 3rd edn, Cambridge University Press.

White MJD (1974). Speciation in the Australian morabine grass-hoppers: The cytogenetic evidence. In: White MJD (ed.) Genetic Mechanisms of Speciation in Insects. Australia and New Zealand Book Co., Sydney; 57–68.

White MJD (1978a). Modes of Speciation. WH Freeman and Company, San Francisco.

White MJD (1978b). Chain processes in chromosomal speciation. Syst Zool 27; 285–98.

White MJD (1982). Rectangularity, speciation, and chromosome architecture. In: Barigozzi CB (ed.) Mechanisms of Speciation. Alan R Liss, Inc., New York; 75–103.

White MJD, Blackith RE, Blackith RM, Cheney J (1967). Cytogenetics of the *Viatica* group of morabine grasshoppers, I: The 'coastal' species. Austral J Zool 15; 263–302.

Winking H (1986). Some aspects of Robertsonian karyotype variation in European wild mice. In: Potter M, Nadeau JH, Cancro MP (eds) The Wild Mouse in Immunology. Springer-Verlag, Berlin, Heidelberg; 68–74.

Winking H, Gropp A (1976). Meiotic non-disjunction of metacentric heterozygotes in oocytes versus spermatocytes. Serono Symp 1975, Academic Press.

Yamamoto M, Miklos GLB (1978). Genetic studies on heterochromatin in *Drosophila melanogaster* and their implications for the function of satellite DNA. Chromosoma 66; 71–98.

Yunis JJ, Yasmineh WG (1971). Heterochromatin, satellite DNA, and cell function. Science 174; 1200–9.

Author Index

Subject Index

MIX
Papier aus verantwortungsvollen Quellen
Paper from responsible sources
FSC® C105338

If you have any concerns about our products,
you can contact us on
ProductSafety@springernature.com

In case Publisher is established outside the EU,
the EU authorized representative is:
Springer Nature Customer Service Center GmbH
Europaplatz 3, 69115 Heidelberg, Germany

Printed by Libri Plureos GmbH
in Hamburg, Germany